T0234572

Lecture Notes in Mathematics　　　2132

More information about this series at http://www.springer.com/series/304

Antonio Avilés • Félix Cabello Sánchez •
Jesús M.F. Castillo • Manuel González •
Yolanda Moreno

Separably Injective Banach Spaces

Springer

Antonio Avilés
Dpto. de Matemáticas
Universidad de Murcia
Murcia, Spain

Félix Cabello Sánchez
Dpto. de Matemáticas
Universidad de Extremadura
Badajoz, Spain

Jesús M.F. Castillo
Dpto. de Matemáticas
Universidad de Extremadura
Badajoz, Spain

Manuel González
Dpto. de Matemáticas
Universidad de Cantabria
Santander, Spain

Yolanda Moreno
Dpto. de Matemáticas
Universidad de Extremadura
Cáceres, Spain

ISSN 0075-8434   ISSN 1617-9692 (electronic)
Lecture Notes in Mathematics
ISBN 978-3-319-14740-6   ISBN 978-3-319-14741-3 (eBook)
DOI 10.1007/978-3-319-14741-3

Library of Congress Control Number: 2016935425

Mathematics Subject Classification (2010): 46A22, 46B03, 46B08, 46M10, 46M18, 46B26, 54B30

Printed on acid-free paper

This Springer imprint is published by Springer Nature
The registered company is Springer International Publishing AG Switzerland

Preface

The Plot

Injective Banach spaces are those spaces that allow the extension of any operator with values in them to any superspace. Finite dimensional and ℓ_∞ are the simplest examples of injective spaces. When a Banach space is injective, there automatically appears a constant that controls the norms of the extensions. At the other end of the line, we encounter the Banach spaces allowing the "controlled" extension of finite-rank operators, which form the well-known class of $\mathscr{L}_\infty$-spaces.

The main topic of this monograph lies between these two extremes: Banach spaces that allow a controlled extension of operators under certain restrictions on the size of their range or the size of the spaces to where they can be extended. The basic case is that of Banach spaces allowing the extension of operators from subspaces of separable spaces, called separably injective, for which c_0 is the simplest example. The second more important case is that of Banach spaces allowing the extension of operators from separable spaces elsewhere, called universally separably injective, for which the space $\ell_\infty^c(\Gamma)$ is the perfect example.

This monograph contains most of what is currently known about (universally) separably injective spaces; and certainly it contains all we know plus a good part of all we don't. Chapters 1–5, whose content we describe below, give a rather detailed account of the current theory of separable injectivity, with its many connections and applications. At the same time, the "Notes and remarks" sections at the end of each chapter and the entire "Open problems" section in Chap. 6 provide a large-scale map of the land beyond the sea. We have decided not to reproduce here several items already covered in books. The best example of this would be Zippin's theorem asserting that c_0 is the only separable separably injective Banach space, for which an exceptionally limpid exposition can be found in [253]; see also [33]. Other seemingly relevant pieces of information can be found in the literature, but as we were unable to establish precise connections with our topic, we chose to omit them.

Injectivity vs. (Universal) Separable Injectivity

The reader is referred to either the Preliminaries or the Appendix for all unexplained notation.

A Banach space E is said to be injective if for every Banach space X and every subspace Y of X, each operator $t: Y \to E$ admits an extension $T: X \to E$. The space is said to be λ-injective if, besides, T can be chosen so that $\|T\| \leq \lambda \|t\|$.

The space ℓ_∞ is the perfect example of 1-injective space, and a key point is to determine to what extent other injective spaces share the properties of ℓ_∞. On the positive side, one should score the results of Nachbin and Kelley who characterized the 1-injective spaces as those Banach spaces linearly isometric to a $C(K)$ space, with K an extremely disconnected compact space. On the other hand, the research of Argyros, Haydon, Rosenthal, and other authors provided several general structure theorems, good examples of exotic injective spaces, and the feeling that the complete classification of injective Banach spaces is an unmanageable problem.

Most of the research on injective spaces has revolved around the following problems:

(1) Is every injective space isomorphic to a 1-injective space?
(2) Is every injective space isomorphic to some $C(K)$-space?
(3) What is the structure of an injective space?

Admittedly, (3) is rather vague and (2) is a particular case of the more general problem of finding out whether or not a complemented subspace of a $C(K)$-space is again isomorphic to a $C(K)$-space (the compact space may vary). These problems have remained open for 50 years, (1) and (3) even in the case in which the injective space is a $C(K)$-space. We refer the reader to Chap. 1 for a summary of what is known about injective spaces.

In this monograph, we deal with several weak forms of injectivity, mainly *separable injectivity* and *universal separable injectivity*. A Banach space E is said to be separably injective if it satisfies the extension property in the definition of injective spaces under the restriction that X is separable; it is said to be a universally separably injective if it satisfies the extension property when Y is separable. Obviously, injective spaces are universally separably injective, and these, in turn, are separably injectives; the converse implications fail. The corresponding definitions of λ-separably injective and universally λ-separably injective should be clear.

The study of separably injective spaces was initiated by Phillips [216] and Sobczyk [235], who showed that c (resp. c_0) is 2-separably injective. Later Zippin [252] proved that c_0 is the only (infinite dimensional) separable space that is separably injective. Moreover, Ostrovskii [208] proved that a λ-separably injective space with $\lambda < 2$ cannot be separable, Baker [25] and Seever [228] studied separably injective $C(K)$ spaces, and several results in the literature on the extension of operators can also be formulated in terms of separably injective spaces. Separably injective spaces have been studied more recently by several authors such as Rosenthal [225], Zippin [253], Johnson and Oikhberg [146], and also by the present authors in [20], where the notion of universal separable injectivity was formally introduced.

The theory of separably injective and universally separably injective spaces is quite different from that of injective spaces, is much richer in examples, and contains interesting structure results and homological characterizations. For instance, problems (1) and (2) above have a negative answer for separably injective spaces, and quite interesting information about the structure of (universally) separably injective spaces can be offered. Moreover this theory is far from being complete: many open problems can be formulated (see Chap. 6) that we expect can be attractive for Banach spacers and could foster the interest in studying injectivity-like properties of Banach spaces.

A Brief Description of the Contents of This Monograph

After this Preface, "Preliminaries" section contains all due preliminaries about notation and basic definitions. An Appendix at the end of the book describes the basic facts about $\mathscr{L}_\infty$ and $\mathscr{L}_1$-spaces, homological techniques, and transfinite chains that will appear and be used throughout the monograph. Other definitions, properties, or required constructions will be given when needed.

In Chap. 1, we gather together properties of injective Banach spaces, basic examples and counterexamples, and criteria that are useful to prove that certain spaces are not injective. These facts will be used later and will allow the reader to compare the stability properties and the variety of examples of injective spaces with those of (universally) separably injective spaces.

In Chap. 2, we introduce the separably injective and the universally separably injective Banach spaces, as well as their quantitative versions, and obtain their basic properties and characterizations. We establish that infinite-dimensional separably injective spaces are $\mathscr{L}_\infty$-spaces, contain c_0, and have Pełczyński's property (V). Universally separably injective spaces, moreover, are Grothendieck spaces, contain ℓ_∞, and enjoy Rosenthal's property (V). We also prove a number of stability results that allow us to present many natural examples of separably injective spaces, such as $C(K)$-space when K is either an F-space or has finite height, twisted sums and c_0-vector sums of separably injective spaces, etc., including an example of a separably injective space that is not isomorphic to any complemented subspace of a $C(K)$-space, which solves problem (2) above for separable injectivity. In passing, these facts provide a major structural difference between λ-separably injective spaces for various values of λ, something that currently does not exist for injective spaces: 1-separably injective spaces are Grothendieck and Lindenstrauss spaces; hence they must be nonseparable when they are infinite dimensional. However, 2-separably injective spaces can be even separable.

The fundamental structure theorem for universally separably injective spaces is that a Banach space E is universally separably injective if and only if every separable subspace is contained in a copy of ℓ_∞ inside E. This establishes a bridge toward the study of the partially automorphic character of (universally) separably injective spaces, namely, toward determining in which cases an isomorphism between two subspaces extends to an automorphism of the whole space.

Homological characterizations are possible: recall that $\mathrm{Ext}(Z, Y) = 0$ means that whenever a Banach space isomorphic to Y is contained in a Banach space X in such a way that X/Y is isomorphic to Z, it must be complemented. In this language, a space E is separably injective if and only if $\mathrm{Ext}(S, E) = 0$ for every separable space S. For universally separably injective spaces, we have less: if $\mathrm{Ext}(\ell_\infty/S, U) = 0$ for every separable space S, then U is universally separably injective. A problem that is crying out to be solved is whether also the converse holds: Does the identity $\mathrm{Ext}(\ell_\infty, U) = 0$ characterize universally separably injective spaces U? A rather detailed discussion can be found in Sect. 6.2.

Section 2.4 is specifically devoted to 1-separably injective spaces. At this point, set theory axioms enter the game. Indeed, Lindenstrauss obtained in the mid-1960s what can be understood as a proof that, under the continuum hypothesis CH, 1-separably injective spaces are 1-universally separably injective; he left open the question in general. We show how to construct (in a way consistent with ZFC) an example of a Banach space of type $C(K)$ that is 1-separably injective but not 1-universally separably injective. The chapter closes with a detailed study of the separable injectivity and related homological properties of $C(\mathbb{N}^*)$.

Chapter 3 focuses on the study of spaces of universal disposition because they will provide new (and, in the case of p-Banach spaces, the only currently known) examples of (universally) separably injective spaces. We present a basic device to generate such spaces. The device is rather flexible and thus, when performed with the appropriate input data, is able to produce a great variety of examples: the Gurariy space $\mathscr{G}$ [118], the p-Gurariy spaces [58], the Kubiś space [169], new spaces such as $\mathscr{F}^{\omega_1}$ which is of universal disposition for all finite-dimensional spaces but not for separable spaces, or the $\mathscr{L}_\infty$-envelopes obtained in [69]. Remarkable outputs are the examples of a 1-separably injective spaces $\mathscr{S}^{\omega_1}$ and a 1-universally separably injective space $\mathscr{U}^{\omega_1}$ that are not isomorphic to complemented subspaces of any C-space (or an M-space), which solves problem (2) for the classes of (universally) 1-separably injective space in a somewhat surprising way. These spaces turn out to be of universal disposition for separable spaces, which confirms a conjecture made by Gurariy in the 1960s. Moreover, under CH, the space $\mathscr{S}^{\omega_1}$ coincides with the Fraïssé limit in the category of separable Banach spaces considered by Kubiś [169] and with the countable ultrapowers of the Gurariy space.

In Chap. 4, we study injectivity properties of ultraproducts. We show that ultraproducts built over countably incomplete ultrafilters are universally separably injective as long as they are $\mathscr{L}_\infty$-spaces, in spite of the fact that they are never injective. Then, we focus our attention on ultraproducts of Lindenstrauss spaces, with special emphasis in ultrapowers of $C(K)$-spaces and the Gurariy space. With the aid of M-ideal theory, the results can be applied to the lifting of operators and to the study of the behavior of the functor Ext regarding duality. One section is devoted to the Henson-Moore classification problem of isomorphic types of ultrapowers of $\mathscr{L}_\infty$-spaces. A particular concern of the authors has been to make clear which results can be proved without invoking "model theory" and which ones still belong to that domain.

In Chap. 5, we consider a natural generalization of separable injectivity to higher cardinals. Namely, given an infinite cardinal $\aleph$, we say that E is $\aleph$-injective (respectively, universally $\aleph$-injective) if it satisfies the extension property stated at the beginning for spaces X (respectively, for subspaces Y) having density character strictly smaller than $\aleph$. Stopping at the first uncountable cardinal $\aleph_1$, one reencounters the classes of (universally) separably injective spaces. Some facts generalize to the higher cardinal context (say, there exist $(1, \aleph)$-injective spaces that are complemented in no M-space), but some others offer difficulties. For example, we need to restrict to what we call $c_0(\aleph)$-supplemented subspaces in order to extend the results on universally separably injective spaces; or, concerning ultraproducts, we have to deal with $\aleph$-good ultrafilters. The chapter includes a study of spaces $C(K)$ which are $(1, \aleph)$-injective and a study of the interplay with topological properties of the compact space K in the form of a rather satisfactory duality between extension properties of operators into $C(K)$ and lifting properties of continuous maps from K.

At the end of each chapter, we have included a "Notes and Remarks" section containing additional information, mainly considered from the positive side. A final Chap. 6 describes problems related to the topics of this monograph that remain open. We have chosen to give the available information about them—including partial solutions, connections with other results, etc.—in order that it could be helpful for further research.

Murcia, Spain   Antonio Avilés
Badajoz, Spain   Félix Cabello Sánchez
Badajoz, Spain   Jesús M.F. Castillo
Santander, Spain   Manuel González
Cáceres, Spain   Yolanda Moreno

Acknowledgements

The research of the first author has been supported in part by projects MTM2014-54182 funded by MINECO and FEDER 19275/PI/14 (PCTIRM 2011–2014) funded by Fundación Séneca - Agencia de Ciencia y Tecnología de la Región de Murcia.

The research of the other four authors has been supported in part by projects MTM2010-20190 and MTM2013-45643, D.G.I. Spain.

Preliminaries

Banach Spaces and Operators

Throughout the book, the ground field is $\mathbb{R}$ and all Banach spaces are assumed to be real unless explicitly stated otherwise. Most of the results presented here can be adapted to the complex setting, but, in general, we will refrain from doing it. See, however, Sect. 2.7.2.

In a typical situation, we will have an operator $t : Y \to E$ acting between Banach spaces, a further space X containing Y, and we deal with the possibility (or impossibility) of extending t to an operator $T : X \to E$. Operators are always assumed to be linear and continuous. The space of operators from A to B is denoted by $L(A, B)$.

The *Banach-Mazur distance* between the Banach spaces X and Y is

$$d(X, Y) = \inf\{\|u\| \cdot \|u^{-1}\| : u \text{ is an isomorphism between } X \text{ and } Y\}.$$

The spaces X and Y are said to be λ-isomorphic if there is an isomorphism $u : X \to Y$ such that $\|u\|\|u^{-1}\| \leq \lambda$. By writing $X \sim Y$, we indicate that X and Y are isomorphic. By writing $X \approx Y$, we mean that they are isometric.

The density character of a Banach space X is the least cardinal that a subset spanning a dense subspace of X can have, and it is denoted by $\mathrm{dens}(X)$. Note that $\mathrm{dens}(X) = \dim(X)$ for X finite dimensional, and for infinite-dimensional X, our definition of $\mathrm{dens}(X)$ coincides with the usual one in topology: the least cardinal of a dense subset of X.

The twisting process is a rather natural method to obtain new spaces from old. A twisted sum of two Banach spaces Y and Z is a Banach space X containing a subspace Y' isomorphic to Y so that X/Y' is isomorphic to Z. In the homological language, this can be represented with an exact sequence

$$0 \longrightarrow Y \stackrel{j}{\longrightarrow} X \stackrel{q}{\longrightarrow} Z \longrightarrow 0.$$

The twisted sum above is said to be trivial when $j[Y]$ is complemented in X, which implies that $X \sim Y \oplus Z$. Short exact sequences are often used in this book; the reader is referred to Appendix A.4.4 for more information. Here we only recall that, given Banach spaces Z and Y, we write $\mathrm{Ext}(Z, Y) = 0$ to mean that all twisted sums of Y and Z are trivial.

A property P is said to be a three-space property if whenever one has an exact sequence

$$0 \longrightarrow Y \longrightarrow X \longrightarrow Z \longrightarrow 0$$

in which both Y and Z have P, then also X has P. Thus, given a three-space property P, a space X has P whenever a subspace Y and the corresponding quotient X/Y have P. For instance, separability and reflexivity are three-space properties, while "to be a subspace of $c_0(I)$" and Pełczyński's property (V) are not. To determine whether a given property is a three-space property is called a three-space problem. See [61] for a general background on three-space problems.

Classes of Banach Spaces

Given a set Γ, we denote by $\ell_\infty(\Gamma)$ the space of bounded functions $f : \Gamma \to \mathbb{R}$ endowed with the supremum norm, and $c_0(\Gamma)$ is the closed subspace spanned by the characteristic functions of the singletons of Γ. Clearly, the isometric type of $\ell_\infty(\Gamma)$ depends only on $\aleph = |\Gamma|$, and sometimes we write $\ell_\infty(\aleph)$ with the obvious meaning. Similar conventions apply to $c_0(\aleph)$ and $\ell_p(\aleph)$, the usual space of p-summable families. When $\Gamma = \mathbb{N}$ or $\aleph = \aleph_0$, we just write ℓ_p and c_0.

Given a measure μ and $1 \le p < \infty$, we denote by $L_p(\mu)$ the space of functions whose pth power is integrable with respect to μ. As usual we identify functions which agree almost μ-everywhere. The space of essentially bounded functions is denoted by $L_\infty(\mu)$ and carries the essential supremum norm. The space of all continuous functions on a compact space K is denoted by $C(K)$, and, given a locally compact space L, we denote by $C_0(L)$ the space of continuous functions $f : L \to \mathbb{R}$ "vanishing at infinity," that is, such that for every $\varepsilon > 0$, the set $\{x \in L : |f(x)| \ge \varepsilon\}$ is compact. These spaces are equipped with the sup norm.

When dealing with "vector-valued spaces," we will take care of using the following notation: If $(E_i)_{i \in I}$ is a family of Banach spaces, then $\ell_\infty(I, E_i)$ denotes the space of families $(x_i)_{i \in I}$ such that $x_i \in E_i$ for every $i \in I$ and $\sup_{i \in I} \|x_i\| < \infty$, with the obvious norm. The same convention applies to the spaces $c_0(I, E_i)$ and $\ell_p(I, E_i)$. We omit the index set only if I is countable and all spaces E_i coincide with a single space E, in which case we simply write $\ell_\infty(E)$, $c_0(E)$ and $\ell_p(E)$.

By a C-space, we mean a Banach space isometrically isomorphic to $C(K)$ for some (often unspecified) compact Hausdorff space K. This unusual notation is borrowed from lattice theory, where an M-space is a Banach lattice where

$\|x + y\| = \max(\|x\|, \|y\|)$ provided x and y are disjoint, that is, $|x| \wedge |y| = 0$. Each M-space is representable as a (concrete) sublattice in some $C(K)$.

A Banach space X is said to be an $\mathscr{L}_{p,\lambda}$-space (with $1 \leq p \leq \infty$ and $\lambda \geq 1$) if every finite-dimensional subspace of X is contained in another finite-dimensional subspace of X whose Banach-Mazur distance to the corresponding ℓ_p^n is at most λ. A space X is said to be an $\mathscr{L}_p$-space if it is an $\mathscr{L}_{p,\lambda}$-space for some $\lambda \geq 1$; we will say that it is an $\mathscr{L}_{p,\lambda+}$-space when it is an $\mathscr{L}_{p,\lambda'}$-space for all $\lambda' > \lambda$. Similar conventions apply to properties and classes of Banach spaces depending on a real parameter. A Lindenstrauss space is one whose dual is linearly isometric to $L_1(\mu)$ for some measure μ. Lindenstrauss spaces and $\mathscr{L}_{\infty,1+}$-spaces are identical classes.

We are especially interested in $\mathscr{L}_\infty$-spaces and, to some extent, in $\mathscr{L}_1$-spaces (see Appendix). These classes are dual one of the other (see [181]): a Banach space is an $\mathscr{L}_\infty$-space (resp. an $\mathscr{L}_1$-space) if and only if its dual space is an $\mathscr{L}_1$-space (resp. an $\mathscr{L}_\infty$-space).

Approximation Properties

A Banach space X is said to have the λ-approximation property (λ-AP) if for each finite-dimensional subspace $F \subset X$ and every $\lambda' > \lambda$, there is a finite-rank operator $T : X \to X$ such that $\|T\| \leq \lambda'$ and $T(f) = f$ for each $f \in F$. This is not the standard definition, but it is equivalent (see [60, Theorem 3.3]). The space is said to have the bounded approximation property (BAP) if it enjoys the λ-AP for some λ. It is clear that if X has the λ-AP and $Y \subset X$ is complemented by a projection of norm μ, then Y has the $\lambda\mu$-AP. Another useful fact is that the λ-AP passes from X^* to X; see [60, Proposition 3.5]. When X is separable, the λ-AP is equivalent to the existence of a sequence $T_n : X \to X$ of finite-rank operators with $\|T_n\| \leq \lambda$ that is pointwise convergent to the identity.

A Banach space X is said to have the uniform approximation property (UAP) when every ultrapower of X has the BAP. It is clear that the UAP exactly means that X has the λ-AP and there exists a "control function" $f : \mathbb{N} \to \mathbb{N}$ so that, given F and $\lambda' > \lambda$, one can choose T such that $\text{rank}(T) \leq f(\dim F)$ and $Tf = f$ for all $f \in F$, with $\|T\| \leq \lambda'$. Since X^{**} is complemented in some ultrapower of X (Proposition A.8), when X has the UAP, then all even duals have the UAP. And since approximation properties pass from the dual to the space, when X has the UAP, all its duals have the UAP. See either [145, p. 60] or [60, Sect. 7] for details.

Topological Spaces

Any of the books [164, 172, 246] contains the material on general topology needed here and much more. Topological spaces are assumed to be Hausdorff (every two different points are separated by disjoint open sets). The weight of a topological space X, denoted by $w(X)$, is the least cardinality of a base of open sets. We are particularly interested in compact spaces. Every compact space K is homeomorphic to a closed subspace of a product $[0, 1]^X$, where $|X| = w(K)$. For every compact space K, we consider the Banach space $C(K)$ of all real-valued continuous functions on K, endowed with the norm given by the maximum absolute value. This correspondence is a contravariant functor, since each continuous $\varphi : K \longrightarrow L$ gives rise to an operator $\varphi^\circ : C(L) \longrightarrow C(K)$, namely, $\varphi^\circ(f) = f \circ \varphi$. Some especial types of subsets of compact spaces are often used:

1. A G_δ-set is a countable intersection of open sets.
2. An F_σ-set is a countable union of closed sets.
3. A zero set of K is one of the form $\{x \in K : f(x) = 0\}$ for some $f \in C(K)$. Equivalently, a closed G_δ-set.
4. A cozero set is the complement of a zero set. Equivalently, an open F_σ-set.
5. A clopen set is a set which is open and closed.

The family of clopen subsets of K is denoted by $CO(K)$.

If L is a locally compact space, then αL denotes its one-point compactification. The Stone-Čech compactification of a (completely regular) topological space X is denoted by βX. The Stone-Čech compactification is characterized by two simple properties: βX contains a dense subset homeomorphic to X and every bounded continuous function $f : X \longrightarrow \mathbb{R}$ extends to a continuous function $\beta X \longrightarrow \mathbb{R}$.

A point $x \in K$ is said to be a P-point if the intersection of a countable family of neighborhoods of x is again a neighborhood of x.

Zero-Dimensional and Scattered Compacta

A compact space is said to be zero dimensional if its clopen subsets form a base of the topology. Equivalently, if for every two different points $x, y \in K$, there exists a clopen set A such that $x \in A$, $y \notin A$. The topological structure of compact zero-dimensional spaces is completely described by the algebraic structure of its family of clopen sets through Stone duality. Let us explain this. An algebra of sets is a nonempty subfamily of the family $\mathbb{P}(X)$ of all subsets of a set X which contains the sets $\varnothing$ and X and is closed under finite unions, finite intersections, and taking complements. A Boolean algebra is a set $\mathcal{B}$ endowed with two abstract binary operations of "union" (or "join," denoted by $\vee$) and "intersection" (or "meet," denoted by $\wedge$), an operator of "complementation" $A \mapsto A'$, and two distinguished elements 0 and 1, which satisfy the same laws of the union, intersection, and complements of algebras of sets, with 0 in place of the empty set $\varnothing$ and 1 in place of the ambient set X. More precisely, it is required that union and intersection are commutative and associative, and each distributive with respect to the other, the absorption laws $A \vee (A \wedge B) = A$ and $A \wedge (A \vee B) = A$ and the complementation laws $A \wedge A' = 0$ and $A \vee A' = 1$ for every $A, B \in \mathcal{B}$. The simplest Boolean algebras are algebras of sets and, actually, every Boolean algebra is isomorphic to some algebra of sets.

A filter on a Boolean algebra $\mathcal{B}$ is a proper subset $\mathcal{F} \subset \mathcal{B}$ that is closed under finite intersections and does not contain the empty set and one has $A \in \mathcal{F}$ provided $B \wedge A = B$ (i.e., A "contains" B) and $B \in \mathcal{F}$. A filter $\mathcal{F}$ is an ultrafilter if it is maximal among all filters on $\mathcal{B}$ or, equivalently, if for every $A \in \mathcal{B}$, either A or its complement belongs to $\mathcal{F}$. By Zorn's lemma, every filter is contained in an ultrafilter. When one talks about a filter or an ultrafilter on a set X, it means a filter on the algebra of all subsets of X.

If K is a compact space, then $CO(K)$ is a Boolean algebra. An inverse procedure allows to recover a compact space from a Boolean algebra $\mathcal{B}$: The family of all subsets of $\mathcal{B}$ is identified with $2^{\mathcal{B}}$ and can be viewed as a compact space endowed with the product topology. Thus, the Stone space $S(\mathcal{B})$ is defined to be the closed subset of $2^{\mathcal{B}}$ formed by the ultrafilters of $\mathcal{B}$. When K is zero dimensional, the two procedures are inverse of each other in the sense that we have natural identifications $\mathcal{B} = CO(S(\mathcal{B}))$ and $K = S(CO(K))$. These two correspondences are contravariant functors in a natural way.

A topological space is called scattered if every nonempty subset has an isolated point. The derived set of a topological space X is the subset X' of all non-isolated points of X. Given an ordinal α, the αth derived set of X is inductively defined as $X^{(0)} = X$ and

$$X^{(\alpha)} = \bigcap_{\beta < \alpha} \left(X^{(\beta)} \right)'.$$

When K is a scattered compact space, there is a smallest α such that $K^{(\alpha)} = \varnothing$ and this ordinal is the Cantor-Bendixson index of K or the height of K, which is always a successor ordinal, that is, $\alpha = \beta + 1$. Some authors define the height of a compact space as this ordinal β instead of α.

$\beta\mathbb{N}$ and $\mathbb{N}^*$

The Stone space of the algebra $\mathbb{P}(\mathbb{N})$ of all subsets of $\mathbb{N}$ is naturally homeomorphic with the Stone-Čech compactification $\beta\mathbb{N}$ of the set $\mathbb{N}$ equipped with the discrete topology. Indeed, for each natural number n, we have a so-called *principal* ultrafilter $\mathcal{F}_n = \{A \subset \mathbb{N} : n \in A\}$, and by identifying n with $\mathcal{F}_n$, we may consider $\mathbb{N}$ as a subset of $S(\mathbb{P}(\mathbb{N}))$, which is easily seen to be dense. On the other hand, every bounded $f : \mathbb{N} \to \mathbb{R}$ extends to a continuous function on the Stone space of $\mathbb{P}(\mathbb{N})$ by the formula $\mathcal{U} \mapsto \lim_{\mathcal{U}} f(n)$: recall that the elements of $S(\mathbb{P}(\mathbb{N}))$ are the ultrafilters on $\mathbb{N}$. Hence the obvious map $\beta\mathbb{N} \to S(\mathbb{P}(\mathbb{N}))$, being continuous and one-to-one, is a homeomorphism. Since every bounded function $f : \mathbb{N} \longrightarrow \mathbb{R}$ can be seen a continuous function on $\beta\mathbb{N}$, we have a natural isometry that identifies the Banach space ℓ_∞ with the space of continuous functions $C(\beta\mathbb{N})$.

The clopen subsets of $\beta\mathbb{N}$ are the closures $\overline{A}$ of subsets $A \subset \mathbb{N}$. In particular, the points of $\mathbb{N}$ are the isolated points of $\beta\mathbb{N}$. Removing the isolated points of $\beta\mathbb{N}$, one obtains the space $\mathbb{N}^* = \beta\mathbb{N} \setminus \mathbb{N}$. The clopen subsets of $\mathbb{N}^*$ are the sets of the form $[A] = \overline{A} \setminus A$ with $A \subset \mathbb{N}$. We have that $[A] = [B]$ if and only if $(A \setminus B) \cup (B \setminus A)$ is finite, and in this way $CO(\mathbb{N}^*)$ is identified with the quotient Boolean algebra $\mathbb{P}(\mathbb{N})/$ fin of subsets of $\mathbb{N}$ modulo the equivalence relation described above. Two continuous functions on $\beta\mathbb{N}$ coincide on $\mathbb{N}^*$ if and only if their difference converges to 0 on $\mathbb{N}$, and thus we have a natural isometric identification of $C(\mathbb{N}^*)$ and ℓ_∞/c_0.

Other General Conventions and Notations

As a general convention, special spaces are denoted by symbols displayed in "mathcal" fonts. Thus, for instance, the Gurariy space is denoted by $\mathscr{G}$ and Kunen compact is $\mathscr{K}$. Families of sets (topologies, filters, and the like) are displayed in "mathscr" fonts: in this way a typical ultrafilter will be $\mathscr{U}$. As a rule, classes of Banach spaces are displayed in "mathfrak" fonts: for instance, the class of all separable Banach spaces is denoted by $\mathfrak{S}$, while the set of finite-dimensional subspaces of a given Banach space X is $\mathfrak{F}(X)$.

If S is a set, possibly a subset of a larger set, 1_S denotes the identity on S, while $\mathbb{1}_S$ stands for the characteristic function of S. If $S = \{s\}$ is a singleton, we write 1_s instead of $1_{\{s\}}$. We write $|S|$ for the cardinality of S. The power set of S is denoted by $\mathbb{P}(S)$ or 2^S, and if $|S| = \aleph$, then $|2^S| = 2^\aleph$. The subclass of all finite subsets of S will be denoted by fin(S).

The axiomatic system in which we work is ZFC, the usual Zermelo-Fraenkel axioms for set theory, including the axiom of choice. CH is the continuum hypothesis ($2^{\aleph_0} = \aleph_1$) and GCH the generalized continuum hypothesis ($2^\aleph = \aleph^+$ for all infinite cardinals $\aleph$). As long as additional axioms are assumed, they appear in square brackets before the corresponding statement.

Contents

Chapter 1
A Primer on Injective Banach Spaces

To put in a proper context the results in this monograph it will be useful to keep in mind the theory of injective spaces and the general theory of $\mathscr{L}_\infty$-spaces. In this way one can compare the stability properties and the variety of examples of separably injective spaces that will be presented later with those of injective spaces. For the convenience of the reader, in this preparatory chapter we have summarized the basic properties and examples of injective Banach spaces, as well as a few remarkable examples of non-injective spaces, and some criteria that allow one to check whether a space is or is not injective. The results in this chapter have been known for many years and thus proofs will be sketched or omitted. For alternative expositions we refer to [1, Sect. 4.3], [83, Appendix D], [194, Sects. 7–9] and [253, Sect. 2].

1.1 Injective and Locally Injective Spaces

Injective spaces are those enjoying the extension property for any operator; precisely

Definition 1.1 A Banach space E is called *injective* if for every Banach space X and each subspace $Y \subset X$, every operator $t : Y \to E$ extends to an operator $T : X \to E$. If there is some constant $\lambda > 0$ such that an extension with $\|T\| \leq \lambda \|t\|$ always exists, the space E will be called λ-*injective*.

The simplest examples of injective spaces are the finite-dimensional Banach spaces for which one gets the required extension taking a basis and invoking the Hahn-Banach theorem. The same idea works for $\ell_\infty(\Gamma)$, as we see in the following classical result of Phillips [216].

Proposition 1.2 *The space $\ell_\infty(\Gamma)$ is 1-injective.*

© Springer International Publishing Switzerland 2016
A. Avilés et al., *Separably Injective Banach Spaces*, Lecture Notes
in Mathematics 2132, DOI 10.1007/978-3-319-14741-3_1

Proof Each operator $t : Y \to \ell_\infty(\Gamma)$ is defined by a bounded family $(t_\gamma)_\gamma$ of elements of Y^* in the form $t(y) = (t_\gamma(y))_\gamma$, with $\|t\| = \sup\{\|t_\gamma\| : \gamma \in \Gamma\}$. Pick for each γ an extension $T_\gamma \in X^*$ of t_γ with $\|T_\gamma\| = \|t_\gamma\|$ and define $T : X \to \ell_\infty(\Gamma)$ by $T(x) = (T_\gamma(x))_{\gamma \in \Gamma}$. Then T is an extension of t with the same norm.   □

At the other end of the line one encounters the spaces satisfying the λ-extension property for finite-rank operators; precisely:

Definition 1.3 A Banach space E is *locally λ-injective* if for every finite dimensional Banach space G and every subspace F of G, every operator $t : F \to E$ has an extension $T : G \to E$ with $\|T\| \leq \lambda\|t\|$. We say that E is *locally injective* if it is locally λ-injective for some λ.

Locally injective spaces were formally introduced in [58, Definition 5.1], where it is observed that they coincide with the $\mathscr{L}_\infty$-spaces.

Proposition 1.4 *A Banach space is locally injective if and only if it is an $\mathscr{L}_\infty$-space.*

Proof Since ℓ_∞^n is 1-injective, an $\mathscr{L}_{\infty,\lambda}$-space must be locally λ-injective. Conversely, suppose that E is locally λ-injective and embeds as a subspace of X. Pick a finite dimensional subspace $F \subset X$ and consider the inclusion map $F \cap E \to E$. There exists by hypothesis an extension $T : F \to E$ having norm at most λ. Thus E is locally complemented (Definition A.9) in every superspace and, by Lemma A.12, it must be an $\mathscr{L}_\infty$-space.   □

A useful fact connecting local injectivity to injectivity is the following. Recall that "locally λ^+-injective" means "locally λ'-injective for every $\lambda' > \lambda$".

Proposition 1.5 *A Banach space is locally λ^+-injective if and only if its bidual is λ-injective.*

Proof First suppose that E is locally λ^+-injective. Let X be a Banach space, Y a subspace of X, and $t : Y \to E^{**}$ an operator. Let $\mathfrak{F}(X)$ and $\mathfrak{F}(E^*)$ denote the set of all finite dimensional subspaces of X and E^*, respectively. We order $\mathfrak{F}(X) \times \mathfrak{F}(E^*) \times (0, \infty)$ by declaring $(F, G, \varepsilon) \leq (F', G', \varepsilon')$ when $F \subset F', G \subset G'$ and $\varepsilon' \leq \varepsilon$. For each $(F, G, \varepsilon) \in \mathfrak{F}(X) \times \mathfrak{F}(E^*) \times (0, \infty)$ choose an operator $T_{(F,G,\varepsilon)} : t[F] \to E$ as in the principle of local reflexivity (Theorem A.7), that is,

- $\|T_{(F,G,\varepsilon)}\| \|T_{(F,G,\varepsilon)}^{-1}\| \leq 1 + \varepsilon$,
- $T_{(F,G,\varepsilon)} f = f$ for every $f \in t[F] \cap E$, and
- $f^{**}(g^*) = g^*(T_{(F,G,\varepsilon)} f^{**})$ for every $g^* \in G$ and every $f^{**} \in t[F]$.

For each (F, G, ε) we consider the composition $T_{(F,G,\varepsilon)} \circ t : F \cap Y \to E^{**} \to E$, and we extend it to an operator $\tau_{(F,G,\varepsilon)} : F \to E$ with

$$\|\tau_{(F,G,\varepsilon)}\| \leq (1 + \varepsilon)\lambda\|T_{(F,G,\varepsilon)} \circ T\| \leq (1 + \varepsilon)^2 \lambda\|T\|.$$

Now, let $\mathcal{U}$ be an ultrafilter refining the order filter on $\mathfrak{F}(X) \times \mathfrak{F}(E^*) \times (0, \infty)$ and define $T : X \to E^{**}$ by letting

$$T(x) = \text{weak*-} \lim_{\mathcal{U}} \tau_{(F,G,\varepsilon)}(x).$$

This definition makes sense because each x belongs to F "eventually". Besides, it is clear that T is a linear operator and $\|T\| \leq \lambda \|t\|$. To see that T extends t just note that for every $y \in Y$ one has $\tau_{(F,G,\varepsilon)}(y) = T_{(F,G,\varepsilon)}(t(y))$ and that, for each $f^{**} \in t[Y]$

$$f^{**} = \text{weak*-} \lim_{\mathcal{U}} T_{(F,G,\varepsilon)}(f^{**})$$

since each $g^* \in E^*$ falls eventually in G and $f^{**}(g^*) = g^*(T_{(F,G,\varepsilon)} f^{**})$ provided $g^* \in G$. This shows that E^{**} is λ-injective.

The converse is easier. Suppose E^{**} is merely locally λ^+-injective and let us prove that so is E. Let G be a finite dimensional space, F a subspace of G and $t : F \to E$ a norm-one operator. Fix $\varepsilon > 0$ and consider t as an E^{**}-valued operator. The hypothesis gives an extension $\tau : G \to E^{**}$ with $\|\tau\| \leq \lambda + \varepsilon$. As $\tau[G]$ is finite dimensional the principle of local reflexivity yields an operator $T : \tau[G] \to E$ such that $T(f) = f$ for $f \in E \cap \tau[G]$, and $\|T\| \leq 1 + \varepsilon$. Clearly $T \circ \tau$ is an extension of t, and $\|T \circ \tau\| \leq (1 + \varepsilon)(\lambda + \varepsilon)$ and since ε is arbitrary the proof is complete. □

The parameter λ is clearly necessary in Definition 1.3 since all spaces enjoy the extension property for finite-rank operators. In fact, when the class of operators one extends is "closed", a uniform bound appears automatically (cf. Proposition 1.32 in Sect. 1.6.3). In particular:

Proposition 1.6 *Every injective Banach space is λ-injective for some $\lambda \geq 1$.*

Proof Suppose that, for every $n \in \mathbb{N}$, the space E is not n-injective. Then we can find Banach spaces X_n containing subspaces Y_n and norm-one operators $t_n : Y_n \to E$ that do not admit extension to X_n with norm less than or equal to n. Since $\ell_1(\mathbb{N}, Y_n)$ is a closed subspace of $\ell_1(\mathbb{N}, X_n)$, and clearly $t\big((y_n)\big) = \sum_{n=1}^{\infty} t_n(y_n)$ defines an operator $t : \ell_1(\mathbb{N}, Y_n) \to E$ that does not admit extension to $\ell_1(\mathbb{N}, X_n)$, the space E cannot be injective. □

1.2 Basic Properties and Examples

Let us present now a few more examples of injective Banach spaces and describe their basic properties. It is not difficult to show that the finite products and the complemented subspaces of injective spaces are injective. As for infinite products, one has the following generalization of Proposition 1.2. Recall that if $(E_i)_{i \in I}$ is a family of Banach spaces, $\ell_\infty(I, E_i)$ denotes the space of bounded families (x_i), with $x_i \in E_i$ for every $i \in I$, and $\|(x_i)\| = \sup\{\|x_i\| : i \in I\}$.

Proposition 1.7 *If $(E_i)_{i\in I}$ is a family of λ-injective Banach spaces, then $\ell_\infty(I, E_i)$ is λ-injective.*

Proof It is enough to observe that every operator $T : X \to \ell_\infty(I, E_i)$ is given by $T(x) = (T_i(x))_{i\in I}$ with $T_i : X \to E_i$ and $\|T\| = \sup_{i\in I} \|T_i\|$. □

Other simple examples of 1-injective spaces are presented in the following result of Kantorovič [160]. To avoid any measure-theoretic pathology we consider σ-finite measures only.

Proposition 1.8 *If μ is a σ-finite measure, then $L_\infty(\mu)$ is 1-injective.*

Proof Suppose that μ is a measure on a σ-algebra of subsets of Ω. Let $\mathscr{P}(\Omega)$ denote the family of all partitions of Ω into countably many measurable subsets of finite positive measure. Note that $\mathscr{P}(\Omega)$ is a directed set for the order given by $\mathcal{Q} \le \mathcal{P}$: each element of $\mathcal{P}$ is contained in an element of $\mathcal{Q}$. For every $\mathcal{P} \in \mathscr{P}(\Omega)$ we consider the "conditional expectation" operator $\mathbb{E}_\mathcal{P}$ defined by

$$\mathbb{E}_\mathcal{P}(f) = \sum_{A\in\mathcal{P}} \left(\frac{1}{\mu(A)} \int_A f d\mu \right) 1_A.$$

Clearly, $\mathbb{E}_\mathcal{P}$ is a contractive projection on $L_\infty(\mu)$ whose range consists of those functions that are constant on every set of $\mathcal{P}$.

Now, let Y be a subspace of X and let $t : Y \to L_\infty(\mu)$ be a norm one operator. Given $\mathcal{P} \in \mathscr{P}(\Omega)$ we consider the composition $t_\mathcal{P} = \mathbb{E}_\mathcal{P} \circ t$ so that

$$t_\mathcal{P}(y) = \mathbb{E}_\mathcal{P}(t(y)) = \sum_{A\in\mathcal{P}} \left(\frac{1}{\mu(A)} \int_A t(y) d\mu \right) 1_A.$$

Note that $y \mapsto \mu(A)^{-1} \int_A t(y) d\mu$ is a linear functional on Y with norm at most 1. We take extensions $x_A^* \in X^*$ (depending on t and A) with the same norm and define $T_\mathcal{P} : X \to L_\infty(\mu)$ by

$$T_\mathcal{P}(x) = \sum_{A\in\mathcal{P}} x_A^*(x) 1_A.$$

Clearly, $\|T_\mathcal{P}\| \le \|t_\mathcal{P}\| \le 1$.

For each $r \ge 0$ we denote by B_r the closed ball of $L_\infty(\mu)$ with center 0 and radius r, endowed with the weak* topology induced by $L_1(\mu)$. By Tychonoff's theorem, $\Pi = \prod_{x\in X} B_{\|x\|}$ is compact in the product topology. Moreover, we can identify each contractive operator $T : X \to L_\infty(\mu)$ with a point of Π just considering the family $(T(x))_{x\in X}$. Let T be any limit point of the net $(T_\mathcal{P})_{\mathcal{P}\in\mathscr{P}(\Omega)}$ in Π. It is not difficult to check that T is an operator from X to $L_\infty(\mu)$ with $\|T\| \le 1$. Moreover, since for every $f \in L_\infty(\mu)$ the net $(\mathbb{E}_\mathcal{P}(f))_{\mathcal{P}\in\mathscr{P}(\Omega)}$ is weak* convergent to f, the operator T is an extension of t. □

More general conditions on μ work: when $L_\infty(\mu)$ is the dual of $L_1(\mu)$ then it is injective. In fact,

Proposition 1.9 *An $\mathscr{L}_\infty$-space isomorphic to a dual space is injective.*

Proof We shall prove a more precise result, namely that every dual $\mathscr{L}_{\infty,\lambda+}$-space is λ-injective.

Let E^* be an $\mathscr{L}_{\infty,\lambda}$-space, let Y be a subspace of a Banach space X and let $t : Y \to E^*$ be a norm one operator. Let $\mathfrak{F}(Y)$ be the set of all finite dimensional subspaces of Y. We consider the order given in $\mathfrak{F}(Y) \times (0, \infty)$ by $(F, \varepsilon) \le (G, \delta)$ if $F \subset G$ and $\delta \le \varepsilon$. Let $\mathcal{U}$ be a ultrafilter on $\mathfrak{F}(Y) \times (0, \infty)$ refining the order filter. For each $F \in \mathfrak{F}(Y)$ and each $\varepsilon > 0$, let $T_{(F,\varepsilon)} : X \to E^*$ be an extension of $t|_F$ with norm at most $\lambda + \varepsilon$, which exists since $t[F]$ falls inside a subspace of E^* which is $(\lambda + \varepsilon)$-isomorphic to ℓ_∞^n for some $n \in \mathbb{N}$. The operator

$$T(x) = \text{weak}^* - \lim_{\mathcal{U}} T_{(F,\varepsilon)}(x).$$

is an extension of t with $\|T\| \le \lambda$. $\square$

Every infinite dimensional Banach space X is isometric to a subspace of $\ell_\infty(\Gamma)$, with $|\Gamma| = \text{dens}(X)$. Indeed, let $\{x_i : i \in \Gamma\}$ be a dense subset of X and choose, for each $i \in \Gamma$, a norm-one $f_i \in X^*$ such that $f_i(x_i) = \|x_i\|$. Then the operator $f : X \longrightarrow \ell_\infty(\Gamma)$ defined by $f(x) = (f_i(x))_{i \in \Gamma}$ is an isometric embedding. It immediately follows that:

Proposition 1.10 *The injective Banach spaces coincide with the complemented subspaces of the spaces $\ell_\infty(\Gamma)$.*

This result should however be compared with Corollary 1.17, in which it is shown that no injective subspace of ℓ_∞ exists, apart from itself.

The next result gathers several equivalent formulations of λ-injectivity for Banach spaces.

Proposition 1.11 *Let E be a Banach space and let $\lambda \ge 1$. The following assertions are equivalent:*

1. *E is λ-injective.*
2. *For every space X containing E isometrically, there exists a projection from X onto E of norm at most λ.*
3. *For every space X containing E isometrically, each operator $t : E \to Y$ admits an extension $T : X \to Y$ with $\|T\| \le \lambda \|t\|$.*

Proof

$(1) \Rightarrow (2)$ and $(3) \Rightarrow (2)$:   If E is contained in X isometrically, using (1) or (3) we can extend the identity on E to an operator $P : X \to E$ with $\|P\| \le \lambda$, and P is the required projection.

$(2) \Rightarrow (1)$:   Suppose Y is a subspace of X and $t : Y \to E$ is an operator. We saw before that there exists an isometric operator $u : E \to \ell_\infty(\Gamma)$ for some set Γ, and (2) implies the existence of a projection P on $\ell_\infty(\Gamma)$ onto E with $\|P\| \le \lambda$. Since $\ell_\infty(\Gamma)$ is 1-injective, the operator $u \circ t : Y \to \ell_\infty(\Gamma)$ admits an extension $\tau : X \to \ell_\infty(\Gamma)$ with $\|\tau\| = \|t\|$, and it is clear that $T = P \circ \tau : X \to E$ is the required extension of t.

(2) $\Rightarrow$ (3):   If X contains E isometrically, (2) implies the existence of a projection P on X onto E with $\|P\| \le \lambda$, and given an operator $t : E \to Y$, $T = t \circ P : X \to Y$ is an extension of t with $\|T\| \le \lambda \|t\|$. $\square$

Goodner [112] introduced the $\mathscr{P}_\lambda$-spaces ($\lambda \ge 1$) as those Banach spaces satisfying (2) in Proposition 1.11. Thus, $\mathscr{P}_\lambda$ and λ-injective spaces coincide.

Regarding universality issues, the density character is important. Observe that $\mathrm{dens}(\ell_\infty(\aleph)) = 2^\aleph$ because if $|\Gamma| = \aleph$ the characteristic functions $\{1_A : A \subset \Gamma\}$ form a 1-separated set of cardinality $2^\aleph$ whose linear span is dense in $\ell_\infty(\aleph)$. Rosenthal [222, Theorem 5.1.f] showed that the space $\ell_\infty(\aleph)$ is "quotient universal" for injective spaces of density character $2^\aleph$ or less:

Proposition 1.12 *Every injective Banach space having density character at most* $2^\aleph$ *is isomorphic to a quotient of* $\ell_\infty(\aleph)$.

It follows from the embedding $X \to \ell_\infty(\mathrm{dens}\,X)$ above that every injective Banach space having density character at most $\aleph$ is a subspace of $\ell_\infty(\aleph)$. But it is not however true that an injective Banach space with density character at most $2^\aleph$ must be a subspace of $\ell_\infty(\aleph)$: In Proposition 1.21 we will show an injective space which is a quotient of ℓ_∞ but it is not isomorphic to a dual space; in particular it is not isomorphic to ℓ_∞ and thus it cannot be a subspace of ℓ_∞ by Corollary 1.17.

We need now a technical result of Rosenthal [223, Lemma 1.1], which we present in a proof due to Kupka [171]. Recall that a *finitely additive measure* on the set $\mathbb{P}(\Gamma)$ of all subsets of Γ is a map $\mu : \mathbb{P}(\Gamma) \to \mathbb{R}$ which satisfies $\mu(\varnothing) = 0$ and $\mu(A \cup B) = \mu(A) + \mu(B)$ if A and B are disjoint subsets of Γ. The variation $|\mu|$ of μ is defined by $|\mu|(A) = \sup \sum_{k=1}^n |\mu(A_k)|$, where the supremum is taken over all finite partitions $A_1, \ldots, A_n$ of A. It is easy to show that every $x^* \in \ell_\infty(\Gamma)^*$ induces a finitely additive measure μ_{x^*} on $\mathbb{P}(\Gamma)$, given by

$$\mu_{x^*}(A) = x^*(1_A) \qquad A \in \mathbb{P}(\Gamma),$$

and one has $|\mu_{x^*}|(\Gamma) = \|x^*\|$.

Lemma 1.13 *Let* Γ *be an infinite set, let* $\{\mu_\gamma : \gamma \in \Gamma\}$ *be a set of positive finitely additive measures defined on* $\mathbb{P}(\Gamma)$ *with* $\sup_{\gamma \in \Gamma} \mu_\gamma(\Gamma) < \infty$, *and let* $\varepsilon > 0$. *Then there exists a subset* Γ' *of* Γ *with* $|\Gamma'| = |\Gamma|$ *such that*

$$\mu_\gamma(\Gamma' \setminus \{\gamma\}) < \varepsilon \quad \text{for each } \gamma \in \Gamma'.$$

Proof Assume that for some $\varepsilon > 0$ we cannot find such a set Γ'. The axiom of choice implies the existence of a partition $\{I_\gamma : \gamma \in \Gamma\}$ of Γ with $|I_\gamma| = |\Gamma|$ for all $\gamma \in \Gamma$, and we can select $\delta \in \Gamma$ such that $\mu_i(\Gamma \setminus I_\delta) \ge \varepsilon$ for all $i \in I_\delta$. Otherwise, for all $\gamma \in \Gamma$ we can find $i_\gamma \in I_\Gamma$ such that $\mu_{i_\gamma}(\Gamma \setminus I_\gamma) < \varepsilon$, and the set

$I = \{i_\gamma : \gamma \in \Gamma\}$ would contradict the assumption. Repeating this procedure with I_δ in place of Γ, and iterating the process a finite number of times, we contradict the hypothesis $\sup_{\gamma \in \Gamma} \mu_\gamma(\Gamma) < \infty$.   □

With this result hand we get the following fundamental result due to Rosenthal [223, Theorem 1.3]:

Theorem 1.14 *Let E be a Banach space complemented in E^{**} and let Γ be an infinite set. Let $T : E \longrightarrow Y$ be an operator for which there exists a subspace M of E isomorphic to $c_0(\Gamma)$ such that $T|_M$ is an isomorphism. Then there exists a subspace N of E isomorphic to $\ell_\infty(\Gamma)$ such that $T|_N$ is an isomorphism.*

Proof of Theorem 1.14 First we consider the case in which $E = \ell_\infty(\Gamma)$ and M is the natural copy of $c_0(\Gamma)$.

The hypothesis implies that there exists a constant $C > 0$ such that $\|Tx\| \geq C\|x\|$ for each $x \in c_0(\Gamma)$. Let $\{e_\gamma : \gamma \in \Gamma\}$ denote the unit vector basis in $c_0(\Gamma)$. For each $\gamma \in \Gamma$ we select $y_\gamma^* \in Y^*$ with $\|y_\gamma^*\| = \|Te_\gamma\|^{-1}$ and $y_\gamma^*(Te_\gamma) = 1$, and denote by μ_γ the finitely additive measure associated to $T^* y_\gamma^*$. Then

$$|\mu_\gamma|(\Gamma) = \|T^* y_\gamma^*\| \leq \|T\| C^{-1} \quad \text{for all } \gamma.$$

Now we apply Lemma 1.13 with $\varepsilon = 1/2$, and get a subset Γ' of Γ with $|\Gamma'| = |\Gamma|$ so that, for each $x \in \ell_\infty(\Gamma')$ and $\gamma \in \Gamma'$,

$$T^* y_\gamma^*(x) = \int_{\Gamma'} x \, d\mu_\gamma = x(\gamma) + \int_{\Gamma' \setminus \gamma} x \, d\mu_\gamma$$

because $\mu_\gamma(\gamma) = T^* y_\gamma^*(e_\gamma) = 1$. Hence

$$C^{-1}\|Tx\| \geq \|y_\gamma^*\| \cdot \|Tx\| \geq |x(\gamma)| - \|x\|/2,$$

and taking the supremum over $\gamma \in \Gamma'$ we conclude $\|Tx\| \geq (C/2)\|x\|$, hence T is an isomorphism on $\ell_\infty(\Gamma')$, and the proof is done in this case.

In the general case we denote by P a projection from E^{**} onto E, and consider an isomorphism U from the natural copy of $c_0(\Gamma)$ in $\ell_\infty(\Gamma)$ onto the subspace M of E. The second conjugate $U^{**} : \ell_\infty(\Gamma) \to E^{**}$ is an extension of the operator U, and the restriction of the operator $TPU^{**} : \ell_\infty(\Gamma) \to E$ to $c_0(\Gamma)$ is an isomorphism. Applying the result for $E = \ell_\infty(\Gamma)$ and $M = c_0(\Gamma)$ to this operator, we get a subset Γ' of Γ with $|\Gamma'| = |\Gamma|$ so that the restriction of TPU^{**} to $\ell_\infty(\Gamma')$ is an isomorphism, and clearly the subspace $N = PU^{**}[\ell_\infty(\Gamma')]$ is isomorphic to $\ell_\infty(\Gamma')$ and $T|_N$ is an isomorphism.   □

Let us see some consequences of Theorem 1.14. Part (3) is a useful tool to detect that a space is not injective:

Proposition 1.15 *Let E be an injective Banach space.*

1. *Given a non-weakly compact operator $T : E \to Y$, there exists a subspace N of E isomorphic to ℓ_∞ such that the restriction $T|_N$ is an isomorphism.*
2. *If E is infinite dimensional then it contains a subspace isomorphic to ℓ_∞.*
3. *If E contains a subspace isomorphic to $c_0(\Gamma)$ for some set Γ then it contains a subspace isomorphic to $\ell_\infty(\Gamma)$.*

Proof

1. Since E is injective, it is a complemented subspace of some $\ell_\infty(\Gamma)$ via some projection P. The projection P is not weakly compact (Proposition A.14) and thus the operator TP cannot be weakly compact. Since $\ell_\infty(\Gamma)$ has property (V) (Proposition A.4), the operator TP must be an isomorphism on some subspace M isomorphic to c_0. Thus $P[M]$ is isomorphic to c_0 and $T|_{P[M]}$ is an isomorphism, and we can apply Theorem 1.14.
2. Apply (1) to the identity of E, which is a non-weakly compact operator.
3. It is enough to apply Theorem 1.14 to the identity of E. □

Proposition 1.16 *Every complemented subspace of $\ell_\infty(\Gamma)$ containing a subspace isomorphic to $c_0(\Gamma)$ is itself isomorphic to $\ell_\infty(\Gamma)$.*

Proof If E is a complemented subspace of $\ell_\infty(\Gamma)$ containing a subspace isomorphic to $c_0(\Gamma)$ then, by Proposition 1.15, it contains a (necessarily complemented) subspace isomorphic to $\ell_\infty(\Gamma)$. Since $\ell_\infty(\Gamma)$ is isomorphic to $\ell_\infty(\ell_\infty(\Gamma))$ applying Pełczyński's decomposition technique we conclude that X and $\ell_\infty(\Gamma)$ are isomorphic. □

Therefore, since every complemented subspace of ℓ_∞ contains c_0 (Proposition A.5):

Corollary 1.17 *Every infinite dimensional complemented subspace of ℓ_∞ is isomorphic to ℓ_∞.*

1.3   Isometric Theory: 1-Injective Spaces

In this section we give some examples and characterizations of 1-injective Banach spaces.

Definition 1.18 A compact space K is said to be *extremely disconnected (or Stonian)* if the closure of every open subset of K is open.

If K is extremely disconnected, then its clopen sets form a complete Boolean algebra. Conversely, every extremely disconnected compact space is the Stone compact of some complete Boolean algebra. Analogously to what is mentioned in the Preliminaries for $\beta\mathbb{N}$, the *Stone-Čech compactification $\beta\Gamma$* of a discrete space Γ is the unique compact Hausdorff space containing Γ as a dense subset so that every bounded function on Γ extends to a continuous function on $\beta\Gamma$. The compact $\beta\Gamma$ is extremely disconnected, and the space of continuous functions $C(\beta\Gamma)$ is linearly isometric in the natural way with $\ell_\infty(\Gamma)$.

Proposition 1.19 (Zippin [253, Theorem 2.1]) *For a Banach space E, the following assertions are equivalent:*

1. *E is 1-injective;*
2. *E is linearly isometric to a C(K) space with K extremely disconnected.*
3. *Every family of mutually intersecting closed balls in E has a common point.*

Thus, there is an exact correspondence between 1-injective Banach spaces and complete Boolean algebras. Recall that from an arbitrary Boolean algebra $\mathcal{B}$ we can always obtain a complete Boolean algebra by the procedure of completion, adding all the suprema of families of subsets of $\mathcal{B}$ which did not possess one. It is worth to mention some prominent examples of such complete Boolean algebras:

- First of all, we have $\mathbb{P}(\Gamma)$, the family of all subsets of a set Γ, whose corresponding Banach space is $\ell_\infty(\Gamma)$. Note that each n dimensional 1-injective Banach space is linearly isometric to ℓ_∞^n, which corresponds to $\ell_\infty(\Gamma)$ for $|\Gamma| = n$.
- If (Ω, Σ, μ) is a finite measure space then the quotient of Σ by the ideal of μ-null sets is a complete Boolean algebra, corresponding to $L_\infty(\mu)$.
- If K is a compact space, then the family of all regular open sets of K (open sets that equal the interior of their closures) constitute a complete Boolean algebra, whose Stone space is called the *Gleason space* $G(K)$ of K. The space $G(K)$ maps continuously onto K in a natural way, providing a canonical embedding of the space $C(K)$ into a 1-injective space $C(G(K))$. See [83, Theorem D.2.6] and [245] for additional information.

To give an idea of how different the 1-injective Banach spaces associated to complete Boolean algebras can be let us consider the following property: A Boolean algebra is called *ccc* (a short-cut for "countable chain condition") if every disjoint family contained in it is countable. There are ccc complete Boolean algebras of arbitrarily large size: measure algebras and the completion of ccc Boolean algebras are ccc. On the other hand, it was proved by Rosenthal [222, Theorem 4.5] that a Boolean algebra $\mathcal{B}$ is ccc if and only if every weakly compact subset in $C(S(\mathcal{B}))$ is separable. Equivalently, $C(S(\mathcal{B}))$ contains no subspace isomorphic to $c_0(\Gamma)$ for uncountable Γ. Thus, if we consider a ccc complete Boolean algebra $\mathcal{B}$, then $C(S(\mathcal{B}))$ is an injective Banach space for which all weakly compact subsets are separable. This contrasts with the spaces $\ell_\infty(\Gamma)$ for Γ uncountable.

Several characterizations of 1-injective $C(K)$-spaces can be found in [233, 22.4] and [197, Proposition 2.1.4]:

Proposition 1.20 *Let K be a compact space. The following assertions are equivalent:*

1. *K is extremely disconnected.*
2. *The clopen subsets of K form a complete Boolean algebra.*
3. *Every bounded family of C(K) has a supremum in C(K).*

1.4   Examples of Injective Spaces and Duality

We consider now the isomorphic relations between injective and dual spaces. We will show that among injective spaces we can find non-dual spaces, dual but not bidual spaces, and bidual spaces, which must be isomorphic to some $\ell_\infty(\Gamma)$.

Proposition 1.21 (Rosenthal [222, Corollary 4.4]) *There exists a 1-injective Banach space that is not isomorphic to any dual space.*

The example is the space of continuous functions on certain extremely disconnected compact space G constructed by Gaifman [99]. Moreover the topological weight of G is the continuum and, by Rosenthal [222, Theorem 5.1.e], the space $C(G)$ is isometric to a quotient algebra of ℓ_∞; equivalently, G is homeomorphic to a closed subset of $\beta\mathbb{N}$. However, since $C(G)$ is not isomorphic to ℓ_∞ it cannot be isomorphic to a subspace of ℓ_∞ (cf. Corollary 1.17).

We have already shown that all dual $\mathscr{L}_\infty$ spaces are injective (Proposition 1.9). And there are plenty of them since, in particular, all duals of $\mathscr{L}_1$-spaces are $\mathscr{L}_\infty$. We present now a characterization due to Haydon [124, 2.6 Corollary] of the injective spaces that are isomorphic to a bidual space:

Proposition 1.22 *An injective Banach space is isomorphic to a second dual space if and only if it is isomorphic to $\ell_\infty(\Gamma)$ for some set Γ.*

Rosenthal [222, Theorem 4.8] provides the following equivalent conditions on an injective dual space:

Lemma 1.23 *Let E be an injective Banach space that is isomorphic to a dual space. The following conditions are equivalent:*

1. *E is isomorphic to a subspace of $L_\infty(\mu)$ for some finite measure μ.*
2. *$\ell_\infty(\aleph_1)$ is not isomorphic to a subspace of E.*
3. *Every weakly compact subset of E is separable.*

A combination of Haydon and Rosenthal criteria yields:

Proposition 1.24 *Let μ be a finite measure. Then $L_\infty(\mu)$ is isomorphic to a second dual space if and only if $L_1(\mu)$ is separable.*

Proof If $L_1(\mu)$ is separable then $L_\infty(\mu)$ embeds into ℓ_∞. By Proposition 1.9, $L_\infty(\mu)$ is injective hence a complemented subspace of ℓ_∞ and the "if part" follows from Corollary 1.17.

The converse is as follows. Every weakly compact subset of $L_\infty(\mu)$ is separable by Lemma 1.23 and therefore $L_\infty(\mu)$ cannot be isomorphic to $\ell_\infty(\Gamma)$ for uncountable Γ. On the other hand $L_\infty(\mu)$ is not isomorphic to ℓ_∞ by Rosenthal [222, Theorem 3.5]. Thus, Proposition 1.22 implies that $L_\infty(\mu)$ is not isomorphic to a second dual space.   □

We exhibit now examples of each kind:

- There are measures μ for which the spaces $L_\infty(\mu)$ is not injective (in particular, not isomorphic to a dual space by Proposition 1.8): If μ is be the counting measure on an uncountable set Γ restricted to the σ-algebra of sets that are countable or have countable complement. In this case, $L_\infty(\mu)$ is the unitization of the space $\ell_\infty^c(\Gamma)$, studied in the next Section, which is not injective (Proposition 1.28).
- Spaces $L_\infty(\mu)$ isomorphic to a dual space (hence injective) but not to a bidual space: let $\mathbf{2}$ be a two point space with the usual probability. Let Γ be any set and let μ be the product probability on $\mathbf{2}^\Gamma$. Then dens $L_1(\mu) = |\Gamma|$. So, if Γ is uncountable, then $L_\infty(\mu)$ is a 1-injective dual space, but not isomorphic to a second dual space.
- The space $L_\infty(0, 1)$ is 1-injective and isomorphic to ℓ_∞.

1.5 Examples of Non-Injective Spaces

Here we give several examples of Banach spaces that are not injective but are "close" to be; indeed, they will appear later as examples of (universally) separably injective spaces. In the following result we consider $c_0(\Gamma)$ as a subspace of $\ell_\infty(\Gamma)$ in the obvious way.

Theorem 1.25 *The spaces $c_0(\Gamma)$ and $\ell_\infty(\Gamma)/c_0(\Gamma)$ are not injective unless Γ is finite.*

Proof First of all we observe that one can assume Γ countable in both cases. Indeed, fixing a countable $\Gamma_0 \subset \Gamma$ and working with the elements supported in Γ_0 we see that $c_0 = c_0(\Gamma_0)$ is a complemented subspace of $c_0(\Gamma)$ and ℓ_∞/c_0 is a complemented subspace of $\ell_\infty(\Gamma)/c_0(\Gamma)$.

Now, let us consider the exact sequence

$$0 \longrightarrow c_0 \longrightarrow \ell_\infty \longrightarrow \ell_\infty/c_0 \longrightarrow 0. \tag{1.1}$$

Let $\mathcal{M}$ be a family of infinite subsets of the integers with $|\mathcal{M}| = \mathfrak{c}$, the cardinality of the continuum, and such that $N \cap M$ is finite for $N \neq M$. A classical way to show the existence of this family: we identify $\mathbb{N}$ with the set of rational numbers $\mathbb{Q}$. Then we denote by I the irrational numbers, and for every $i \in I$ we take as N_i (the set underlying) a sequence of rational numbers converging to i and we set $\mathcal{M} = \{N_i : i \in I\}$. The characteristic function 1_N of each N can be seen as an element of ℓ_∞, and it is easy to see that the corresponding images in ℓ_∞/c_0 generate a subspace isometric to $c_0(\mathcal{M})$. In fact $\{1_N : N \in \mathcal{M}\}$ corresponds to the unit vector basis of $c_0(\mathcal{M})$.

The sequence (1.1) cannot split since ℓ_∞/c_0 is not a subspace of ℓ_∞ because $c_0(\Gamma)$ is not a subspace of ℓ_∞ for uncountable Γ: actually, no operator $c_0(\Gamma) \to \ell_\infty$ can be injective since ℓ_∞ can be separated by a countable set of functionals and

$c_0(\Gamma)$ cannot. This already shows that c_0 is not complemented in ℓ_∞, hence it is not injective.

The space ℓ_∞/c_0 is not injective either because it contains $c_0(\mathcal{M}) = c_0(\mathfrak{c})$ but it does not contain a copy of $\ell_\infty(\mathfrak{c})$ because

$$\mathrm{dens}(\ell_\infty/c_0) = \mathfrak{c} < 2^{\mathfrak{c}} = \mathrm{dens}\big(\ell_\infty(\mathcal{M})\big).$$

According to Proposition 1.15 (3), the space ℓ_∞/c_0 cannot be injective.   □

The part of Proposition 1.25 concerning c_0 appeared in Phillips' paper [216] and about the same time in Sobczyk's [235]; see also [56] for an account. The second part of Theorem 1.25 is due to Amir [5], but the simple proof we give here is taken from [223].

Further examples of non-injective Banach spaces are provided by Pełczyński and Sudakov [214], as we describe next. Given an uncountable set Γ, we denote by $\ell_\infty^<(\Gamma)$ the set of all bounded functions $x : \Gamma \to \mathbb{R}$ such that, for every $\varepsilon > 0$, one has $|\{\gamma \in \Gamma : |x(\gamma)| > \varepsilon\}| < |\Gamma|$. Clearly $\ell_\infty^<(\Gamma)$ is a closed subspace of $\ell_\infty(\Gamma)$, and if $|\Gamma|$ has uncountable cofinality one has

$$\ell_\infty^<(\Gamma) = \big\{x \in \ell_\infty(\Gamma) : |\{\gamma \in \Gamma : x(\gamma) \neq 0\}| < |\Gamma|\big\}.$$

The following result was proved in [214].

Theorem 1.26 *If Γ is uncountable, then $\ell_\infty^<(\Gamma)$ is not injective.*

Proof We prove that $\ell_\infty^<(\Gamma)$ is not complemented in $\ell_\infty(\Gamma)$. Recall that no injective operator $c_0(\aleph) \to \ell_\infty(\Gamma)$ exists when $\aleph > |\Gamma|$ since the evaluation functionals $\{\delta_\gamma^* : \gamma \in \Gamma\}$ form a total subset in the dual space of $\ell_\infty(\Gamma)$. Therefore the canonical exact sequence

$$0 \longrightarrow \ell_\infty^<(\Gamma) \longrightarrow \ell_\infty(\Gamma) \longrightarrow \ell_\infty(\Gamma)/\ell_\infty^<(\Gamma) \longrightarrow 0$$

cannot split if we prove that the quotient $\ell_\infty(\Gamma)/\ell_\infty^<(\Gamma)$ contains a copy of $c_0(\aleph)$ for some $\aleph > |\Gamma|$.

We argue as in the proof of Theorem 1.25, but this time we need a result of Sierpiński (see [232, Sect. XVII.3, Theorem 1]) asserting that given any infinite set Γ there exist a family $\mathcal{M}$ of subsets of Γ so that:

- $|\mathcal{M}| > |\Gamma|$;
- for each $N \in \mathcal{M}$ one has $|N| = |\Gamma|$;
- $|N \cap M| < |\Gamma|$ for $N, M \in \mathcal{M}, N \neq M$.

It is clear that the images of the characteristic functions 1_N in $\ell_\infty(\Gamma)/\ell_\infty^<(\Gamma)$ generate an isometric copy of $c_0(\mathcal{M})$ inside $\ell_\infty(\Gamma)/\ell_\infty^<(\Gamma)$.   □

Let Γ be an uncountable set, and let $\aleph$ be a cardinal with $\aleph_0 < \aleph \leq |\Gamma|$. We denote by $\ell_\infty^{<\aleph}(\Gamma)$ the set of all $(x_\gamma)_{\gamma \in \Gamma} \in \ell_\infty(\Gamma)$ such that $|\{\gamma \in \Gamma : |x_\gamma| > \varepsilon\}| < \aleph$ for every $\varepsilon > 0$.

Corollary 1.27 *If $\aleph_0 < \aleph \le |\Gamma|$, then $\ell_\infty^{<\aleph}(\Gamma)$ is not injective.*

Proof Clearly $\ell_\infty^{<\aleph}(\Gamma)$ is a closed subspace of $\ell_\infty(\Gamma)$. Suppose that it is complemented, and take a subset G of Γ with $|G| = \aleph$. Then $\ell_\infty^{<\aleph}(\Gamma) \cap \ell_\infty(G) = \ell_\infty^{<}(G)$ is complemented in $\ell_\infty(G)$, in contradiction with Theorem 1.26.   □

Given an uncountable set Γ, the space $\ell_\infty^{<\aleph_1}(\Gamma)$ is usually denoted by $\ell_\infty^c(\Gamma)$, the space of all bounded scalar functions on Γ with countable support.

Proposition 1.28 *Let Γ be an uncountable set.*

1. *The space $\ell_\infty^c(\Gamma)$ is not injective.*
2. *Every separable subspace of $\ell_\infty^c(\Gamma)$ is contained in a copy of ℓ_∞ inside $\ell_\infty^c(\Gamma)$.*
3. *Every operator $T : Y \to \ell_\infty^c(\Gamma)$ with separable range admits an extension to any superspace.*

Proof (1) follows from Corollary 1.27. To prove (2) observe that, given a separable subspace S of $\ell_\infty^c(\Gamma)$, there exist a countable subset Γ_0 of Γ such that the support of each $f \in S$ is contained in Γ_0. (3) is an obvious consequence of (2) and the injectivity of ℓ_∞.   □

In the case $|\Gamma| = \mathfrak{c}$, part (1) of Proposition 1.28 is a consequence of Proposition 1.15, since $\ell_\infty^c(\Gamma)$ has density character equal to $\mathfrak{c}^{\aleph_0} = \mathfrak{c}$ and contains $c_0(\Gamma)$ as a subspace, while $\ell_\infty(\Gamma)$ has density character equal to $2^{\mathfrak{c}}$.

The basic structure of the space $\ell_\infty^c(\Gamma)$ is described in Example 2.4. The spaces $\ell_\infty^c(\Gamma)$ were studied by Johnson et al. in [149]. They show that every infinite dimensional complemented subspace of $\ell_\infty^c(\Gamma)$ is isomorphic to $\ell_\infty^c(\kappa)$ for some cardinal $\kappa \le |\Gamma|$. They also sow that every copy of $\ell_\infty^c(\Gamma)$ inside $\ell_\infty^c(\Gamma)$ contains a further complemented copy of $\ell_\infty^c(\Gamma)$.

We close this section with a few remarks on spaces of measurable functions. Answering a question posed by Rosenthal, Argyros [11] showed:

Proposition 1.29 *The space of all bounded Borel (respectively, Lebesgue) measurable functions on the line is not injective.*

Let us add that the traditional identification between functions which agree almost everywhere does not apply here. The Borel case of the preceding Proposition easily follows from part (3) of Proposition 1.15: the characteristic functions of the singletons generate a copy of $c_0(\mathbb{R})$ in the space of bounded Borel functions. The density character of the latter space is the continuum, as there are $\mathfrak{c}$ Borel subsets. Therefore it cannot contain a copy of $\ell_\infty(\mathbb{R})$, whose density character is $2^{\mathfrak{c}}$. The full force of Argyros' argument is necessary only for Lebesgue measurable functions, where the preceding proof does not work: there are $2^{\mathfrak{c}}$ Lebesgue measurable sets in the line. We cannot resist to mention the key ingredient of the proof (cf. [11, Lemma]):

Lemma 1.30 *For every operator $T : \ell_\infty(\mathbb{R}) \to \ell_\infty(\mathbb{R})$ satisfying $T1_t = 1_t$ for each $t \in \mathbb{R}$, there exists $f \in \ell_\infty(\mathbb{R})$ such that Tf fails to be Lebesgue measurable.*

Proposition 1.29 was extended in [38] as follows. Recall that a topological space M is *completely metrizable* if there exists a distance d on M such that (M, d) is a complete metric space and d induces the topology of M.

Proposition 1.31 (Blasco and Ivorra [38, Corollary 6]) *Let M be a separable completely metrizable space. Then the space of bounded Borel measurable functions on M is injective if and only if M is countable.*

1.6   Notes and Remarks

1.6.1   The Basic Problem for Injective Spaces

The characterization of 1-injective spaces as those Banach spaces linearly isometric to a $C(K)$ space with K extremely disconnected is due to Nachbin [201] and Kelley [163]. However, despite the deep investigations of Argyros [8–11], Haydon [124], Rosenthal [222, 223] and other authors, the isomorphic theory of injective spaces is far from being satisfactory. Probably the main open question in the theory of injective Banach spaces is whether each injective space isomorphic to a 1-injective space. The answer to this problem is not known even for $C(K)$ spaces. Some partial answers are collected now:

1. Amir [4] If the space $C(K)$ is λ-injective for some $\lambda < 2$, then the compact K is extremely disconnected.
2. Amir [5] and Isbell and Semadeni[138] There exists a 2-injective space $C(K)$ space with K non-extremely disconnected.

We refer to [250, Sect. 6] for plenty of questions about the structure of injective spaces and $C(K)$ spaces. A few additional partial answers can be found in [39].

1.6.2   Injective Spaces that Are Not Dual Spaces

Proposition 1.21 asserts the existence of an example of a 1-injective Banach space that is not isomorphic to a dual space. There are older examples of 1-injective spaces that are not isometric to dual spaces, like the one given by Dixmier in [85]. A detailed construction following the ideas of [85] can be found in [151, Theorem 3.5], and a succinct description is given in [1, Problems 4.8 and 4.9]. We give below just the first steps of a special case of the construction:

Consider the space B of all bounded Borel functions on the unit interval with the sup norm. Let N denote the subspace of functions having nowhere dense support:

$$N = \{f \in B : \overline{\{t : f(t) \neq 0\}} \text{ has empty interior}\}$$

The quotient space B/N (with the quotient norm) is 1-injective and not isometric to a dual space.

1.6.3 The (Weakly) Compact Extension Property

A Banach space E has the *compact extension property* if for every Banach space X and each subspace $Y \subset X$, every compact operator $t : Y \to E$ has a compact extension $T : X \to E$. The following result already appeared in Lindenstrauss' memoir [175, Theorem 2.1].

Proposition 1.32 *A Banach space is an $\mathscr{L}_\infty$-space if and only if it has the compact extension property.*

Proof It is an easy consequence of the definition that $\mathscr{L}_\infty$-spaces enjoy the bounded approximation property. Thus, compact operators into $\mathscr{L}_\infty$-spaces can be approximated by finite-rank operators, and consequently can also be extended.

To show that the compact extension property of implies local injectivity it is enough to show that there exists some λ so that the compact extension T can always be obtained with $\|T\| \leq \lambda\|t\|$. Which is simple because compact operators form a closed subspace of the space of all operators, so the open mapping theorem applies. Indeed, given any set of Banach spaces $(X_i)_{i \in I}$ with subspaces Y_i, we form the space $c_0(I, Y_i)$ and consider it as a closed subspace of $c_0(I, X_i)$. Each compact operator $Y_i \to E$ yields a compact operator $c_0(I, Y_i) \to E$ by composition with the canonical projection $c_0(I, Y_i) \to Y_i$ onto the i th coordinate, and every (compact) operator $c_0(I, X_i) \to E$ induces a (compact) operator $X_i \to E$ by "restriction". Since there is a uniform constant λ so that every norm one compact operator $c_0(I, Y_i) \to E$ admits a compact extension $c_0(I, X_i) \to E$ with norm at most λ, the same occurs to each couple $Y_i \subset X_i$. $\square$

Replacing "compact" by "weakly compact" everywhere in the definition of the compact extension property one defines the *weakly compact extension property*. It can be shown [45, Proposition 1.34] that a space has the weakly compact extension property if and only if it is an $\mathscr{L}_\infty$-space in which weakly convergent sequences are norm convergent. The existence of infinite dimensional spaces of this kind was proved by Bourgain and Delbaen [46] (see also [45]).

1.6.4 The Quotient $\ell_\infty(\Gamma)/\ell_\infty^{\leqslant}(\Gamma)$

Pełczyński and Sudakov identify in [214] the quotient space $\ell_\infty(\Gamma)/\ell_\infty^{\leqslant}(\Gamma)$ with the space of continuous functions on a certain closed subset $\upsilon\Gamma$ of the Stone-Čech compactification $\beta\Gamma$ that we now describe.

Recall that $\beta\Gamma$ is a topological space whose points are the ultrafilters on Γ. The family of all sets

$$F_A = \{p \in \beta\Gamma : A \in p\} \quad (A \subset \Gamma)$$

is a base of the topology of $\beta\Gamma$. If $\gamma \in \Gamma$, let then $p_\gamma = \{A \subset \Gamma : \gamma \in A\}$ is a (trivial) ultrafilter, the map $h : \Gamma \to \beta\Gamma$ defined by $h(\gamma) = p_\gamma$ is the embedding of Γ in $\beta\Gamma$, and the isometric isomorphism

$$x = (x_\gamma) \in \ell_\infty(\Gamma) \to f_x \in C(\beta\Gamma)$$

is determined by $f_p(\gamma) = x_\gamma$. Let us denote $\nu\Gamma = \{p \in \beta\Gamma : A \in p \Rightarrow |A| = |\Gamma|\}$, the set of the so-called *uniform ultrafilters*. Note that $\mathcal{F} = \{A \subset \Gamma : |\Gamma \setminus A| < |\Gamma|\}$ is a filter on Γ, and each ultrafilter refining $\mathcal{F}$ belongs to $\nu\Gamma$. Moreover, given an ultrafilter $p \in \beta\Gamma \setminus \nu\Gamma$, there exists $A \in p$ with $|A| < |\Gamma|$, and F_A is a neighborhood of p which does not meet $\nu\Gamma$. Therefore $\nu\Gamma$ is a non-empty closed subset of $\beta\Gamma$.

Since $\nu\Gamma$ is closed in $\beta\Gamma$, Tietze's extension theorem implies that the restriction operator $r : C(\beta\Gamma) \to C(\nu\Gamma)$ is surjective. Let $x = (x_\gamma) \in \ell_\infty(\Gamma)$ and let $f_x \in C(\beta\Gamma)$ be the corresponding function. It is not difficult to check that $x \in \ell_\infty^<(\Gamma)$ if and only if $f_p) = 0$ for each $p \in \nu\Gamma$; i.e., $rf_x = 0$. Thus $C(\nu\Gamma)$ is isometrically isomorphic to $\ell_\infty(\Gamma)/\ell_\infty^<(\Gamma)$.

1.6.5 Finite Dimensional Injective Spaces

As Zippin says [253, p. 1716]: "The nature of a finite-dimensional $\mathcal{P}_\lambda$-space is a fascinating mystery". Recall from Proposition 1.11 that $\mathcal{P}_\lambda$-spaces ($\lambda \geq 1$) coincide with the λ-injective spaces.

Probably the core problem is to know something about the function $f(n, m, \lambda)$ that makes an n-dimensional λ-complemented subspace of ℓ_∞^m to be $f(n, m, \lambda)$-isomorphic to ℓ_∞^n. Asymptotically speaking, to know if a sequence E_n of λ-complemented subspaces of ℓ_∞^n must be uniformly isomorphic to the corresponding ℓ_∞^k. Or else, assume that for each $n \in \mathbb{N}$ one picks an n-dimensional λ-injective space F_n. Is it true that

$$\sup_n d(F_n, \ell_\infty^n) < +\infty?$$

We refer to [253, Problem 2.13 and 2.14] for a more detailed exposition of this problem and some indications suggesting that it could have a positive solution. An infinite dimensional reformulation could be this: Assume that (F_n) is a sequence of n-dimensional Banach spaces so that $c_0(\mathbb{N}, F_n) \sim c_0$. Must the spaces F_n be uniformly isomorphic to ℓ_∞^n?

Of course, the problem for $\lambda = 1$ is trivial: It follows from Proposition 1.19 that a finite-dimensional space is 1-injective if and only if it is linearly isometric to ℓ_∞^n.

Chapter 2
Separably Injective Banach Spaces

It is no exaggeration to say that the theory of separably injective spaces is quite different from that of injective spaces. In this chapter we will explain why. Indeed, we will enter now in the main topic of the monograph, namely, separably injective spaces and their "universal" version. After giving the main definitions and taking a look at the first natural examples one encounters, we present the basic characterizations and a number of structural properties of (universally) separable injective Banach spaces. We will show, among other things, that 1-separably injective spaces are not necessarily isometric to C-spaces, that (universally) separably injective spaces are not necessarily complemented in any C-space—the separably injective part of the assertion will be shown here while the "universal" part can be found in the next chapter—and that there exist essential differences between 1-separably injective and 2-separably injective spaces.

Moreover, in contrast with the scarcity of examples and general results concerning the class of injective Banach spaces, there exist many different types of separably injective spaces and a rich theory around them. In fact, most of the chapter is devoted to examples: Some of them are rather natural, while others are Banach spaces introduced elsewhere for different purposes and that, at the end of the day, turn out to be separable injective.

Definition 2.1 A Banach space E is *separably injective* if for every separable Banach space X and each subspace $Y \subset X$, every operator $t : Y \to E$ extends to an operator $T : X \to E$. If some extension T exists with $\|T\| \leq \lambda\|t\|$ we say that E is λ-*separably injective*.

We are especially interested in the following subclass of separably injective spaces.

Definition 2.2 A Banach space E is said to be *universally separably injective* if for every Banach space X and each separable subspace $Y \subset X$, every operator $t : Y \to E$ extends to an operator $T : X \to E$. If some extension T exists with $\|T\| \leq \lambda\|t\|$ we say that E is *universally λ-separably injective*.

© Springer International Publishing Switzerland 2016

A. Avilés et al., *Separably Injective Banach Spaces*, Lecture Notes in Mathematics 2132, DOI 10.1007/978-3-319-14741-3_2

Before going any further we will present a couple of examples to give the flavor of the chapter. Recall that $c_0(I)$ denotes the space of all functions $f : I \to \mathbb{R}$ such that, for every $\varepsilon > 0$, the set $\{i \in I : |x(i)| > \varepsilon\}$ is finite. We present first Sobczyk's theorem [236], with Veech's proof [244]. See also [56] for an account of different proofs for this result.

Theorem 2.3 (Sobczyk's Theorem) *The space $c_0(I)$ is 2-separably injective in the sup norm for every index set I.*

Proof Since the elements of $c_0(I)$ have countable support, every $c_0(I)$-valued operator from a separable space has its range contained in a copy of c_0. So, it suffices to prove the result when I is countable; i.e., when $c_0(I)$ is c_0, the space of null sequences. So, let X be a separable Banach space and $t : Y \to c_0$ a norm one operator, where Y is a subspace of X. Write t as a sequence of functionals $t_n \in Y^*$, so that $t(y) = (t_n(y))$ for every $y \in Y^*$ and $\|t_n\| \le 1$ for every $n \in \mathbb{N}$. The sequence (t_n) is weakly* null in Y^* and one has to find a sequence of extensions (T_n) which is again weakly* null in X^*, with $\|T_n\| \le 2$. For each n, let $\tau_n : X \to \mathbb{R}$ be a Hahn-Banach extension of $t_n : Y \to \mathbb{R}$. Recall that the weak* topology is metrizable on every bounded subset of X^* by a translation-invariant metric d.

If Λ is the set of weak* accumulation points of the sequence (τ_n), then $d(\tau_n, \Lambda) \to 0$ as $n \to \infty$ (a sequence such that every subsequence contains a further subsequence converging to zero is itself convergent to zero). Choose $\lambda_n \in \Lambda$ such that $d(\tau_n, \lambda_n) \le d(\tau_n, \Lambda) + 1/n$. Then $\tau_n - \lambda_n$ is an extension of t_n (since any functional in Λ vanishes on Y) and $\|\tau_n - \lambda_n\| \le \|\tau_n\| + \|\lambda_n\| \le 2$. Clearly, the sequence $(\tau_n - \lambda_n)_n$ is weakly*-null in X^*. The operator $T : X \to c_0$ defined by $T(x) = ((\tau_n - \lambda_n)(x))$ is an extension of t and $\|T\| \le 2$.   □

The space $c_0(I)$ is not universally separably injective (unless I is finite) since c_0 is not complemented in ℓ_∞ (Example 1.25). By the same token, and Corollary 1.17, no separable space can be universally separably injective. A deep result of Zippin [252] puts an end to the story for separable spaces: every infinite dimensional separable separably injective space is isomorphic to c_0. Zippin's theorem has a long and delicate proof; we refer to [253] for what is perhaps the simplest proof due to Benyamini [33].

Thus, the results in this monograph belong naturally to the theory of non-separable Banach spaces. The "basic case" of Pełczyński-Sudakov spaces (see Proposition 1.28) provides a typical universally separably injective space. Although simple, this natural example shows that the theory of universally separably injective spaces does not run parallel with that of injective spaces: contrary to what happens in the injective case, 1-universally separably injective spaces need not be isometric to any $C(K)$ space.

Example 2.4 Let Γ be an uncountable set and let $\ell_\infty^c(\Gamma)$ denote the space of countably supported bounded functions $f : \Gamma \to \mathbb{R}$. Then $\ell_\infty^c(\Gamma)$ is:

1. 1-universally separably injective,
2. not isometric to any C-space,

3. isomorphic to a C-space,
4. not injective.

Proof

1. Every separable subspace of $\ell_\infty^c(\Gamma)$ is contained in another subspace isometric to ℓ_∞.
2. The unit ball of every $C(K)$ has extreme points. In fact f is an extreme point if and only if $|f(x)| = 1$ for every $x \in K$. Quite clearly, the ball of $\ell_\infty^c(\Gamma)$ has no extreme points.
3. Consider the unitization of $\ell_\infty^c(\Gamma)$ inside $\ell_\infty(\Gamma)$, that is,

$$\ell_\infty^c(\Gamma)_+ = \{f \in \ell_\infty(\Gamma) : f = \lambda 1_\Gamma + g : \lambda \in \mathbb{R}, g \in \ell_\infty^c(\Gamma)\}.$$

It is clear that $\ell_\infty^c(\Gamma)_+$ is 2-isomorphic to $\ell_\infty^c(\Gamma) \oplus \mathbb{R}$, endowed with the sup-norm, and this is in turn isomorphic to $\ell_\infty^c(\Gamma)$; and, as every unital subalgebra of $\ell_\infty(\Gamma)$, it is isometrically isomorphic to the algebra of all continuous real-valued functions on certain compact space K (much more general results are available, see Sect. 2.2.1). In fact, if A is a unital subalgebra of $\ell_\infty(\Gamma) = C(\beta\Gamma)$, we can identify A with $C(K)$, where K is the quotient space of $\beta\Gamma$ by the equivalence $x \sim y$ if $f(x) = f(y)$ for every $f \in A$.
4. The space $\ell_\infty^c(\Gamma)$ contains a complemented subspace isometric to $\ell_\infty^c(\aleph_1) = \ell_\infty^<(\aleph_1)$, which is not injective by the result of Pełczyński and Sudakov quoted in Theorem 1.26. $\qquad\square$

2.1 Basic Properties

2.1.1 Characterizations

Separably injective spaces can be characterized as follows.

Proposition 2.5 *For a Banach space E the following properties are equivalent.*

1. *E is separably injective.*
2. *Every operator from a subspace of ℓ_1 into E extends to ℓ_1.*
3. *For every Banach space X and each subspace Y such that X/Y is separable, every operator $t : Y \to E$ extends to X.*
4. *If Z is a Banach space containing E and Z/E is separable, then E is complemented in Z.*
5. *For every separable space S one has $\mathrm{Ext}(S, E) = 0$.*

Proof It is clear that (3) $\Rightarrow$ (1) $\Rightarrow$ (2) and (3) $\Rightarrow$ (4) $\Leftrightarrow$ (5). We prove now that (2) $\Rightarrow$ (1) and (2) $\Rightarrow$ (3). Since every separable space X/Y can be set as a quotient $q : \ell_1 \to X/Y$ of ℓ_1, the lifting property of ℓ_1 provides an operator $Q : \ell_1 \to X$

yielding a commutative diagram

$$
\begin{array}{ccccccccc}
0 & \longrightarrow & \ker q & \overset{j}{\longrightarrow} & \ell_1 & \overset{q}{\longrightarrow} & X/Y & \longrightarrow & 0 \\
 & & \phi\downarrow & & \varrho\downarrow & & \| & & \\
0 & \longrightarrow & Y & \longrightarrow & X & \underset{p}{\longrightarrow} & X/Y & \longrightarrow & 0
\end{array}
\tag{2.1}
$$

Let $t : Y \to E$ be an operator, for which there must be an extension $\tau : \ell_1 \to E$ of $t\phi$ provided by (2). When X is separable (case (2) $\Rightarrow$ (1)) then Q can be chosen surjective and then an extension $T : X \to E$ of t can be defined as follows: if $x = Q(\xi)$ then

$$
T(x) = \tau\xi.
$$

The map is well defined because if $0 = Q\xi$ then $q\xi = 0$ and thus also $\phi\xi = 0$ from where it follows $\tau\xi = t\phi\xi = 0$. The map T is continuous since the open mapping theorem yields the existence of some μ so that norm one elements $x \in X$ are images of $x = Q\xi$ of some ξ with $\|\xi\| \le \mu$. Thus $\|Tx\| = \|Q\xi\| \le \|Q\|\mu$. It extends t because $T(y) = t(y)$ choosing the representation $y = \phi(\xi)$.

But even if X is not separable (case (2) $\Rightarrow$ (3)), diagram (2.1) implies that X is a quotient of $Y \oplus_1 \ell_1$ via the operator $\overline{Q}(y, \xi) = y + Q\xi$. Thus, yields an extension $T : X \to E$ defined as

$$
T(x) = ty + \tau\xi.
$$

Indeed, the map is well defined because if $0 = y + Q\xi$ then $Q\xi = -y$ and thus $q\xi = pQ\xi = p(-y) = 0$ from where $\xi \in \ker q$ and moreover $\phi\xi = Q\xi = -y$; therefore $ty + \tau\xi = ty + t\phi\xi = ty - ty = 0$. The map T is continuous since the open mapping theorem yields the existence of some μ so that norm one elements $x \in X$ admit a representation as $x = y + Q\xi$ with $\|y\| + \|\xi\| \le \mu$. Thus $\|Tx\| = \|ty + Q\xi\| \le \max(\|t\|, \|Q\|\mu)$. It extends t because $T(y) = ty$ choosing the representation $y + Q(0)$.

That (4) $\Rightarrow$ (3) follows from the existence of the push-out diagram: given an operator $t : Y \to E$ one gets

$$
\begin{array}{ccccccccc}
0 & \longrightarrow & Y & \overset{\iota}{\longrightarrow} & X & \overset{\pi}{\longrightarrow} & X/Y & \longrightarrow & 0 \\
 & & t\downarrow & & \downarrow t' & & \| & & \\
0 & \longrightarrow & E & \overset{\iota'}{\longrightarrow} & PO & \longrightarrow & PO/E & \longrightarrow & 0
\end{array}
$$

and thus the existence of a projection p' through ι' yields the existence of an extension $p't' : X \to E$ of t.   □

Analogous characterizations can be given for universal separable injectivity.

Proposition 2.6 *For a Banach space E the following properties are equivalent.*

1. *E is universally separably injective.*
2. *Every operator $t : S \to E$ from a separable Banach space S can be extended to an operator $T : \ell_\infty \to E$ through any embedding $S \to \ell_\infty$.*
3. *For every Banach space X and each subspace Y, every operator $t : Y \to E$ with separable range extends to X.*

Proof The equivalence of (1) and (2) is clear: since ℓ_∞ is injective, once an operator can be extended from S to ℓ_∞ it can be extended anywhere. That (1) implies (3) only requires to draw a push-out diagram:

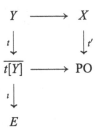

where $\imath$ denotes the canonical inclusion. Since $\imath$ can be extended to an operator $I : \mathrm{PO} \to E$, the composition It' yields an extension of t. $\qquad\square$

Proposition 2.7 *Every (universally) separably injective Banach space is (universally) λ-separably injective for some $\lambda \geq 1$.*

Proof One only has to modify the proof of Proposition 1.6 assuming X_n separable. For the part concerning universally separably injective spaces just shift the separability assumption from X_n to Y_n. $\qquad\square$

2.1.2 First Structural Properties

Recall that a Banach space X has *Pełczyński's property* (V) if each operator defined on X is either weakly compact or it is an isomorphism on a subspace isomorphic to c_0. The indulgent reader (and Rosenthal, we hope) will forgive us for saying that X has *Rosenthal's property* (V) if it satisfies the preceding condition with ℓ_∞ replacing c_0.

All C-spaces as well as their complemented subspaces have Pełczyński's property (V) [212]. Lindenstrauss spaces (i.e., $\mathcal{L}_{\infty,1+}$-spaces) also have this property [147], although there are $\mathcal{L}_\infty$-spaces that do not have it. For example, the ones constructed by Bourgain and Delbaen [46] that contain no copies of c_0, or the space Ω constructed in [57] as a twisted sum

$$0 \longrightarrow C[0,1] \longrightarrow \Omega \longrightarrow c_0 \longrightarrow 0$$

with strictly singular quotient map. Of course Argyros-Haydon's hereditarily indecomposable $\mathscr{L}_\infty$ space is also a counter-example, although this is a clear case of using a sledgehammer to crack an almond.

We say that X is a *Grothendieck space* if every operator from X to a separable Banach space (equivalently, to c_0) is weakly compact; equivalently, weak* and weak convergent sequences in X^* coincide. Clearly, a Banach space with property (V) is a Grothendieck space if and only if it has no complemented subspace isomorphic to c_0. It is well-known that ℓ_∞ is a Grothendieck space. In fact, it has Rosenthal's property (V) (see Proposition 1.15), which is clearly stronger.

Proposition 2.8

1. *A separably injective space is of type $\mathscr{L}_\infty$, has Pełczyński's property (V) and, when it is infinite dimensional, contains copies of c_0.*
2. *A universally separably injective space is a Grothendieck space of type $\mathscr{L}_\infty$, it has Rosenthal's property (V) and, when it is infinite dimensional, contains ℓ_∞.*

Proof

1. A separably injective space is obviously locally injective and thus (see Proposition 1.4) an $\mathscr{L}_\infty$-space.

   To show that E contains c_0 and has property (V), let $T : E \to X$ be a non-weakly compact operator (E being an infinite dimensional $\mathscr{L}_\infty$ space cannot be reflexive). Choose a bounded sequence (x_n) in E such that (Tx_n) has no weakly convergent subsequences and let Y be the subspace spanned by (x_n) in E. As Y is separable we can regard it as a subspace of $C[0,1]$. Let $J : C[0,1] \to E$ be any operator extending the inclusion of Y into E. We already mentioned that C-spaces have property (V), so since $TJ : C[0,1] \to E$ is not weakly compact, TJ is an isomorphism on some subspace isomorphic to c_0 ; and the same occurs to T.
2. To show that an universally separably injective space E has Rosenthal's property (V) we may take $T : E \to Z$ and $Y \subset E$ as in the previous argument, but this time we consider Y as a subspace of ℓ_∞. If $J : \ell_\infty \to E$ is any extension of the inclusion of Y into E, then $TJ : \ell_\infty \to Z$ is not weakly compact. Hence it is an isomorphism on some subspace isomorphic to ℓ_∞ and so is T. $\square$

The list of spaces with Pełczyński's property (V) includes Lindenstrauss spaces (see [147]) and, by Proposition 2.8(1), separably injective spaces. Consequently:

Corollary 2.9 *A separably injective space is a Grothendieck space if and only if it does not contain complemented copies of c_0.*

Let us mention another similarity between separably injective spaces and complemented subspaces of $C[0,1]$.

Proposition 2.10 *If a separably injective space contains a subspace with nonseparable dual then it also contains $C[0, 1]$.*

Proof Assume that a separably injective space X contains a subspace Z with nonseparable dual through some embedding j. Consider an embedding $i : Z \to C[0, 1]$ and get an extension $J : C[0, 1] \to X$ of j through i, that is $Ji = j$. The operator J^* must have nonseparable range; hence, a result of Rosenthal [224] yields that J fixes a copy of $C[0, 1]$. $\square$

2.1.3 Stability Properties

In this section we study the stability properties of (universally) separably injective spaces under some natural "operations" such as taking subspaces and quotients, forming direct products and twisted sums. This will allow us to present many natural examples of (universally) separably injective spaces as soon as we have the basic ingredients to start.

Proposition 2.11 *Let* $0 \longrightarrow A \overset{i}{\longrightarrow} B \overset{q}{\longrightarrow} C \longrightarrow 0$ *be an exact sequence of Banach spaces.*

1. *If A and C are separably injective, then so is B.*
2. *If A and B are separably injective, then so is C.*
3. *If A is separably injective and B is universally separably injective then C is universally separably injective.*

In particular, products and complemented subspaces of (universally) separably injective spaces are (universally) separably injective. Moreover, 1-complemented subspaces of (universally) λ-separably injective spaces are (universally) λ-separably injective.

Proof The simplest proof for (1) follows from characterization (2) in Proposition 2.5. Let $j : K \to \ell_1$ be an isomorphic embedding and let $\phi : K \to B$ be an operator. Then $q\phi$ can be extended to an operator $\Phi : \ell_1 \to C$, which can in turn be lifted to an operator $\Psi : \ell_1 \to B$. The difference $\phi - \Psi j$ takes values in A and can thus be extended to an operator $\upsilon : \ell_1 \to A$. The desired operator is $\Psi + i\upsilon$.

To prove (2) and (3) suppose A is separably injective and B is (resp. universally) separably injective. Let Y be a subspace of a separable (resp. arbitrary) space X and let $\phi : Y \to C$ be an operator. Consider the pull-back diagram

$$
\begin{array}{ccccccccc}
0 & \longrightarrow & A & \longrightarrow & B & \overset{q}{\longrightarrow} & C & \longrightarrow & 0 \\
& & \| & & \uparrow{\scriptstyle '\phi} & & \uparrow{\scriptstyle \phi} & & \\
0 & \longrightarrow & C & \longrightarrow & \text{PB} & \overset{'q}{\longrightarrow} & Y & \longrightarrow & 0
\end{array}
$$

Since C is separably injective, the lower exact sequence splits, so $'q$ admits a linear continuous selection $s : Y \to PB$. By the assumption about B, the operator $'\phi s$ can be extended to an operator $T : X \to E$. Thus, $qT : X \to C$ is the desired extension of ϕ. $\square$

Thus, if $0 \to A \to B \to C \to 0$ is an exact sequence of Banach spaces, we know that B is separably injective if the other two relevant spaces are; and the same happens with C. What about A? Bourgain showed in [44] that ℓ_1 contains an uncomplemented subspace isomorphic to ℓ_1 which yields an exact sequence $0 \to \ell_1 \to \ell_1 \to B \to 0$ that does not split (see Sect. 6.3). By Lindenstrauss' lifting (Proposition A.18) B is not an $\mathscr{L}_1$ space. Its dual sequence $0 \to B^* \to \ell_\infty \to \ell_\infty \to 0$ shows that the kernel of a quotient mapping between two injective spaces may fail to be even an $\mathscr{L}_\infty$-space.

In [20, Proposition 5.3] it was claimed that universal separable injectivity is a 3-space property; but the proof contains a gap we have been unable to fill. Consequently, other claims also remain without proper justification, namely Propositions 5.4 and 5.6 and Theorem 5.5 in [20] and Example 4.5(a) and the second part of Proposition 5.1 in [21]. See Sect. 6.2 for a more detailed account of the situation.

Several variations of these results can be seen in [70]. Regarding infinite products, it is obvious that if $(E_i)_{i \in I}$ is a family of λ-separably injective Banach spaces, then $\ell_\infty(I, E_i)$ is λ-separably injective. The non-obvious fact that also $c_0(I, E_i)$ is separably injective can be considered as a vector valued version of Sobczyk's theorem. Proofs for this result were obtained by Johnson-Oikhberg [146], Rosenthal [225], Cabello Sánchez [52] and Castillo-Moreno [65], each with its own estimate for the constant: respectively, $2\lambda^2$ (implicitly), $\lambda(1+\lambda)^+$, $(3\lambda^2)^+$ and $6\lambda^+$. Here we present a proof like that of Castillo and Moreno [65] based on an idea of Sánchez et al. [57] and giving the same bound as [225].

Proposition 2.12 *If $(E_i)_{i \in I}$ is a family of λ-separably injective spaces, then $c_0(I, E_i)$ is $\lambda(1 + \lambda)^+$-separably injective.*

Proof Since the elements of $c_0(I, E_i)$ have countable "supports" it suffices to prove the result for countable families. So, let (E_n) be a sequence of λ-separably injective spaces, X a separable Banach space, Y a subspace of X, and $t : Y \to c_0(\mathbb{N}, E_n)$ a norm one operator that we can write as $t = (t_n)$, where each $t_n : Y \to E_n$ has norm at most 1.

Fix $\varepsilon \in (0, 1)$. Set $Z = X/Y$ and let $\pi : X \to Z$ denote the natural quotient map. Let (Z_k) be an increasing sequence of finite dimensional subspaces of Z whose union is dense in Z. For each k, let X_k be a finite dimensional subspace of X so that $\pi[X_k] = Z_k$. We may assume that (X_k) is an increasing sequence whose union is dense in X. We require, moreover, that for every $z \in Z_k$ there is $x \in X_k$ such that $\pi(x) = z$, with $\|x\| \leq (1 + \varepsilon)\|z\|$. This implies that Z_k is $(1 + \varepsilon)$-isomorphic to the quotient of X_k by $Y_k = Y \cap X_k$ through the obvious map. It is clear that, for every $k \in \mathbb{N}$, the diagram

$$
\begin{array}{ccccccccc}
0 & \longrightarrow & Y_k & \longrightarrow & X_k & \xrightarrow{\ \pi_k\ } & Z_k & \longrightarrow & 0 \\
 & & \downarrow & & \downarrow & & \downarrow & & \\
0 & \longrightarrow & Y & \longrightarrow & X & \xrightarrow{\ \pi\ } & Z & \longrightarrow & 0
\end{array}
$$

in which π_k is the restriction of π to X_k and the vertical arrows are the canonical embeddings is commutative.

For each n, let $\tau_n : X \to E_n$ be an extension of t_n with $\|\tau_n\| \le \lambda \|t_n\|$, which exists by hypothesis. Let $t_{n,k}$ denote the restriction of t_n to Y_k. Let $\tau_{n,k} : X \to E_n$ be an extension of $t_{n,k}$ such that $\|\tau_{n,k}\| \le \lambda \|t_{n,k}\|$ which, once again, exists by hypothesis. Since $\tau_n - \tau_{n,k}$ vanishes on Y_k there is an operator $\phi_{n,k} : Z_k \to E_n$ such that $\tau_n - \tau_{n,k} = \phi_{n,k} \circ \pi_k$. Besides, the norm of $\phi_{n,k}$ on X_k/Y_k is $\|\tau_n - \tau_{n,k}\|$ and we have

$$\|\phi_{n,k} : Z_k \longrightarrow E_n\| \le (1+\varepsilon)\|\tau_n - \tau_{n,k}\|.$$

Let $\Phi_{n,k} : Z \to E_n$ be an extension of $\phi_{n,k}$ with $\|\Phi_{n,k}\| \le \lambda \|\phi_{n,k}\|$.

Since for every $y \in Y$ one has $\lim \|t_n(y)\| = 0$ and Y_k is finite dimensional, for fixed k, one has $\lim_n \|t_{n,k}\| = 0$. Put $N(k) = \max\{n : \|t_{n,k}\| > \varepsilon^k\}$. Then $N(k)$ is increasing, and $N(k) \to \infty$ as $k \to \infty$.

We define a sequence of operators $T_n : X \to E_n$ as follows:

$$
T_n(x) = \begin{cases} \tau_n(x) - \Phi_{n,k}(\pi(x)) & \text{for } N(k) < n \le N(k+1), \\ \tau_n(x) & \text{if } n \le N(1). \end{cases}
$$

These T_n are uniformly bounded and thus define an operator $T : X \longrightarrow \ell_\infty(\mathbb{N}, E_n)$ given by $T(x) = (T_n(x))$. Let us see that T is the desired extension of t:

1. To check that T takes values in $c_0(\mathbb{N}, E_n)$ it is sufficient to work on $\bigcup_k X_k$. So, take $x \in X_k$, with $\|x\| = 1$. Then for $n > N(k)$ one has

$$T_n(x) = \tau_n(x) - \Phi_{n,i}(\pi(x)) = \tau_n(x) - \Phi_{n,k}(\pi_k(x)) = \tau_n(x) - \tau_n(x) + \tau_{n,k}(x) = \tau_{n,k}(x).$$

Thus, for $n > N(k)$, one has

$$\|T_n(x)\| \le \|\tau_{n,k}\| \le \lambda \|t_{n,k}\| \le \lambda \varepsilon^k.$$

2. The operator T is an extension of t. Indeed, if $y \in Y$, then for every n one has $T_n(y) = \tau_n(y) = t_n(y)$, by the very definitions.

3. To estimate $\|T\|$ it is enough to bound each coordinate. If $n \le N(1)$, then $T_n = \tau_n$, so $\|T_n\| \le \lambda \|t_n\| \le \lambda$. Otherwise $N(k) < n \le N(k+1)$ for some $k \ge 1$ and we have $T_n = \tau_n - \Phi_{n,k} \circ \pi$ and so $\|T_n\| \le \|\tau_n\| + \|\Phi_{n,k}\|$. But $\|\tau_n\| \le \lambda \|t_n\| \le \lambda$;

as for the other chunk, we have

$$\|\Phi_{n,k}\| \le \lambda \|\phi_{n,k}\| \le \lambda(1 + \varepsilon)\|\tau_n - \tau_{n,k}\|$$

$$\le \lambda(1 + \varepsilon)(\|\tau_n\| + \|\tau_{n,k}\|) \le \lambda^2(1 + \varepsilon)(\|t_n\| + \|t_{n,k}\|)$$

$$\le \lambda^2(1 + \varepsilon)(\|t_n\| + \varepsilon^k) \le \lambda^2(1 + \varepsilon)^2,$$

from where it follows that $c_0(\mathbb{N}, E_n)$ is $(\lambda + \lambda^2)^+$-separably injective. □

2.2 Examples of Separably Injective Spaces

In this section we will present a number of separably injective spaces appearing in nature. The first obvious example, since ℓ_∞ is injective and c_0 is separably injective, follows from Proposition 2.11: ℓ_∞/c_0 is universally separably injective. In fact, it will be shown later that ℓ_∞/c_0 is 1-universally separably injective (Theorem 2.40 and Corollary 2.41) and non injective (Theorem 1.25 and also Proposition 2.43).

The non isomorphic [74] spaces $c_0(\ell_\infty)$ and $\ell_\infty(c_0)$ are also separably injective and not universally separably injective; the quotients $\ell_\infty/c_0(\ell_\infty)$ and $\ell_\infty/\ell_\infty(c_0)$ are universally separably injective as well. It also follows from Proposition 2.11 that for Γ an uncountable set, $\ell_\infty^c(\Gamma)/c_0(\Gamma)$ is universally separably injective non-injective. It is worth noticing that it is possible to identify such spaces with $C(K)$ spaces (perhaps after unitization, see Sect. 2.2.1 below).

$C(K)$-spaces (and their ideals) in which K is either of finite height or an F-space, twisted sums of separably injective spaces and quotients of separably injective spaces will be our next examples. We will also show the first examples of separably injective spaces that are not isomorphic to a complemented subspace of any C-space (which is clearly impossible for an injective space). Further examples will be exhibited in Chaps. 4 and 5, when other important classes of separably injective spaces will be presented.

2.2.1 C(K)-Spaces When K Is an F-Space

There are close connections between the 1-separable injectivity of $C(K)$, the topological properties of K and the lattice structure of $C(K)$. Let us recall some separation conditions that compacta may or may not have.

Definition 2.13 A compact Hausdorff space is said to be:

- An F-space if disjoint open F_σ sets (equivalently, cozeroes) have disjoint closures.
- Basically disconnected (or σ-Stonian) if the closure of every open F_σ set is open.

- Extremely disconnected (or Stonian) if the closure of every open set is open.
- Zero-dimensional if the topology has a base of clopen sets.

Recall that a cozero set of K is one of the form $\{x \in K : f(x) \neq 0\}$, for some continuous function f. Cozeroes and open F_σ sets agree on a normal space. Indeed, for any $f \in C(K)$ one has

$$\{x \in K : f(x) \neq 0\} = \bigcup_{n=1}^{\infty} |f|^{-1}\big([1/n, \infty)\big).$$

Thus each cozero is F_σ. Conversely, if $V = \bigcup_n V_n$ is open with all V_n closed, according to Tietze, we may take for each n a continuous $0 \leq f_n \leq 2^{-n}$ vanishing off V and such that $f_n = 2^{-n}$ on V_n. Clearly V is the cozero set of $\sum_n f_n$.

Of course Stonian implies σ-Stonian and this implies F-space. $\beta\mathbb{N}$ is perhaps the most natural example of extremely disconnected compactum. It is obvious that closed sets of F-spaces are F-spaces, so $\mathbb{N}^* = \beta\mathbb{N}\backslash\mathbb{N}$ is an F-space.

Theorem 2.14 *Let K be a compact space. The following conditions are equivalent:*

1. *$C(K)$ is 1-separably injective.*
2. *Given sequences (f_i) and (g_j) in $C(K)$ such that $f_i \leq g_j$ for each $i, j \in \mathbb{N}$, there exists $h \in C(K)$ such that $f_i \leq h \leq g_j$ for each $i, j \in \mathbb{N}$.*
3. *Every sequence of mutually intersecting balls in $C(K)$ has nonempty intersection.*
4. *K is a F-space.*
5. *For every $f \in C(K)$ there is $u \in C(K)$ such that $f = u|f|$.*
6. *Every operator from a two-dimensional space into $C(K)$ has a norm preserving extension to any three-dimensional space.*

Proof The equivalence of (1), (2), (3) and (4) is a special case of the equivalence of the corresponding conditions in Theorem 5.16, where details and accurate references are provided. The proof that (4) and (5) are equivalent is based on the fact that open F_σ sets and cozeroes agree on a normal space:

That (5) holds when K is an F-space is clear: take $f \in C(K)$ and consider the sets $P = f^{-1}(0, \infty)$ and $N = f^{-1}(-\infty, 0)$. These are disjoint cozeroes and so they have disjoint closures. Therefore there is $u \in C(K)$ such that $u = 1$ on P and $u = -1$ on N. Clearly, $f = u|f|$. The converse is also easy: let P and N be disjoint cozero sets and take $f, g \in C(K)$ such that $P = \{x \in K : f(x) \neq 0\}$ and $N = \{x \in K : g(x) \neq 0\}$. Define $h(x) = |f(x)| - |g(x)|$. Now, if $h = u|h|$ for some continuous u, then since $u = 1$ on P and $u = -1$ on N we see that P and N have disjoint closures.

That (6) implies the separable injectivity of $C(K)$ is proved in [177], and the converse implication is trivial.   □

The correspondence between "K is an F-space" and "$C(K)$ is 1-separably injective" does not extend to $C_0(L)$, the space of continuous functions vanishing

at infinity on a locally compact space L: indeed, $\mathbb{N}$ is an F-space while c_0 is not
1-separably injective. However, one has the following:

Proposition 2.15 *Let L be a locally compact space. Then $C_0(L)$ is 1-separably
injective if and only if every compact subset of L is an F-space and the infinity
point is a P-point in αL.*

Proof Assume first that $C_0(L)$ is 1-separably injective. If K is a compact subset of
L we have an exact sequence

$$0 \longrightarrow \ker r \longrightarrow C_0(L) \overset{r}{\longrightarrow} C(K) \longrightarrow 0, \tag{2.2}$$

where r is the restriction map and since $\ker r = \{f \in C_0(L) : f|_K = 0\}$ is an M-ideal
in $C_0(L)$ we have that $C(K)$ is 1-separably injective (Theorem 2.21) and so K is an
F-space (Theorem 2.14).

To prove that the infinity point is a P-point in αL let us assume on the contrary
that there is a sequence (x_n) in L such that $x_n \to \infty$ in αL. We may and do assume
that $x_n \neq x_m$ for $n \neq m$. Then the evaluation map $\pi : C_0(L) \to c_0$ given by
$\pi f = (f(x_n))_n$ is an "isometric quotient" whose kernel is an M-ideal in $C_0(L)$ and
reasoning as before the space c_0 would be 1-separably injective, a contradiction.

As for the other implication, let $t : Y \to C_0(L)$ be an operator, where Y is a closed
subspace of a separable space X. As the infinity point is a P-point in αL and Y is
separable it is clear that there is a compact $K \subset L$ such that $\operatorname{supp} t(y) \subset K$ for every
$y \in Y$. Let us define $\tau : Y \to C(K)$ by $\tau(y) = t(y)|_K$. Since K is an F-space τ has an
extension $\hat{\tau} : X \to C(K)$ with $\|\hat{\tau}\| = \|\tau\| = \|t\|$. But $\hat{\tau}[X]$ is a separable subspace
of $C(K)$ and Proposition 2.20 applied to the sequence (2.2) provides a "lifting" of $\hat{\tau}$
to $C_0(L)$ which is the required extension of t.   □

It is not true that K is an F-space when $C(K)$ is only *isomorphic to* a 1-separably
injective space. To see this we proceed as follows: identify two points $u, v \in \mathbb{N}^*$ that
we may consider as two free ultrafilters $\mathcal{U}$ and $\mathcal{V}$ on $\mathbb{N}$ and let us call $\beta(u, v)$ to the
corresponding quotient space of $\beta\mathbb{N}$. The space $C(\beta(u, v)) = \{f \in C(\beta\mathbb{N}) : f(u) =
f(v)\}$ is a closed hyperplane of $C(\beta\mathbb{N})$ and thus it is 2-isomorphic to ℓ_∞. However,
$\beta(u, v)$ is not an F-space: pick $U \in \mathcal{U}\backslash\mathcal{V}$, so that $V = \mathbb{N}\backslash U$ belongs to $\mathcal{V}$. Set the
function $f : \mathbb{N} \to \mathbb{R}$ given by

$$f(n) = \frac{1_U(n) - 1_V(n)}{n}$$

and extend it to a continuous function on $\beta\mathbb{N}$ denoted again f. As $f(u) = f(v) = 0$
we have $f \in C(\beta(u, v))$. However there is no factorization $f = g|f|$ with $g \in
C(\beta(u, v))$ since it this case we would have $g(u) = 1$ and $g(v) = -1$.

It is important to realize that many Banach algebras are $C(K)$ spaces though given in a disguised form. The most convenient characterization of the algebras of all continuous functions on compact spaces in our real-valued setting is the one due to Albiac and Kalton [1, 2]: if A is a (real, unital) Banach algebra whose norm satisfies the inequality

$$2\|fg\| \le \|f^2 + g^2\| \qquad (f, g \in A),$$

then, as a Banach algebra, A is isometrically isomorphic to $C(K)$, for some compact space K. See [1, 2] for the remarkably simple proof. The next example is just one application.

Proposition 2.16 *The space of all bounded Borel (respectively, Lebesgue) measurable functions on the line is 1-separably injective in the sup norm.*

Proof Clearly, the given spaces are in fact Banach algebras satisfying the inequality required by Albiac-Kalton characterization. Thus they can be represented as $C(K)$ spaces. On the other hand, each measurable function can be decomposed as $f = u|f|$, with u (and $|f|$, of course) measurable. This clearly implies that the corresponding compacta satisfy the fifth condition in Theorem 2.14. □

2.2.2 M-ideals of Separably Injective Spaces

Let M be a closed subset of the compact space K. By Tietze's extension theorem each continuous function on M is the restriction of some continuous function on K having the same norm. The space $L = K \setminus M$ is locally compact and one has the exact sequence

$$0 \longrightarrow C_0(L) \longrightarrow C(K) \overset{r}{\longrightarrow} C(M) \longrightarrow 0, \qquad (2.3)$$

where the map r is plain restriction. Even if this sequence does not split (as a rule), one has the following result, which can be regarded as a linear version of Tietze's extension theorem.

Proposition 2.17 (Borsuk-Dugundji Theorem) *Let M be a closed set in the compact space K. For every separable subspace $S \subset C(M)$ there is a norm-one operator $s : S \longrightarrow C(K)$ such that $rs = 1_S$.*

Borsuk proved this result in [42] for K a metrizable and separable space (not necessarily compact), setting as $C(K)$ the space of continuous bounded functions and $S = C(M)$. Separability was removed by Dugundji in [91, Theorem 5.1], see [230, Sect. 21]. The version of the theorem as it is stated in Proposition 2.17 is a corollary of the more general Proposition 2.20 that we shall discuss later. We can rephrase Borsuk-Dugundji Theorem by saying that, with the same notations as before, if $t : X \rightarrow C(M)$ has separable range, then the lower sequence in the pull back diagram

$$
\begin{array}{ccccccccc}
0 & \longrightarrow & C_0(L) & \longrightarrow & C(K) & \overset{r}{\longrightarrow} & C(M) & \longrightarrow & 0 \\
 & & \| & & \uparrow & & \uparrow{\scriptstyle t} & & \\
0 & \longrightarrow & C_0(L) & \longrightarrow & \mathrm{PB} & \longrightarrow & X & \longrightarrow & 0
\end{array}
$$

splits.

Theorem 2.18 *Let K be a compact space, M a closed subset of K and $L = K \backslash M$.*

1. *If $C(K)$ is (universally) λ-separably injective, then so is $C(M)$.*
2. *If $C(K)$ is λ-separably injective, then $C_0(L)$ is 2λ-separably injective.*

Proof

1. Let Y be a separable subspace of X and $t : Y \longrightarrow C(M)$ an operator. Let $S \subset C(M)$ any separable subspace containing the image of t and $s : S \longrightarrow C(K)$ the lifting provided by the Borsuk-Dugundji theorem. If $T : X \longrightarrow C(K)$ is an extension of st, then $rT : X \longrightarrow C(M)$ is an extension of t, and $\|rT\| = \|T\|$.
2. Let us remark that if S is a subspace of $C(K)$ containing $C_0(L)$ and $S/C_0(L)$ is separable, then there is a projection $p : S \longrightarrow C_0(L)$ of norm at most 2. Indeed, $S/C_0(L)$ is a separable subspace of $C(M)$ and there is a lifting $s : S/C_0(L) \longrightarrow C(K)$, with $\|s\| = 1$, and $p = \mathbf{1}_S - sr$ is the required projection. Now, let $t : Y \longrightarrow C_0(L)$ be an operator, where Y is a subspace of a separable Banach space X. Considering t as taking values in $C(K)$, there is an extension $T : X \longrightarrow C(K)$ with $\|T\| \leq \lambda\|t\|$. Let S denote the least closed subspace of $C(K)$ containing the range of T and $C_0(L)$ and $p : S \longrightarrow C_0(L)$ a projection with $\|p\| \leq 2$. The composition $pT : X \longrightarrow C_0(L)$ is an extension of t and thus $\|pT\| \leq 2\lambda\|t\|$. □

It is easy to see that every closed ideal of $C(K)$ has the form $\{f \in C(K) : f|_S = 0\}$ for some closed subset $S \subset K$ (see [249, III.D.1]). Thus, part (1) of the theorem above can be reformulated as:

Corollary 2.19 *Let K be a compact space and let J be an ideal of $C(K)$. If $C(K)$ is (universally) λ-separably injective, then so is $C(K)/J$.*

Let us consider the following construction introduced by Dashiell and Lindenstrauss [80] with the declared purpose of exhibiting spaces admitting a strictly convex renorming but no injective operator into any $c_0(\Gamma)$. Take $\mathbb{I} = [0, 1]$ in its natural topology. For every $A \subset \mathbb{I}$ and every countable ordinal α, let $A^{(\alpha)}$ be the αth-derived set of A. Given $\varepsilon > 0$ and $f \in \ell^c_\infty(\mathbb{I})$, let $\sigma_\varepsilon(f) = \{t \in \mathbb{I} : |f(t)| \geq \varepsilon\}$. For each countable ordinal α we set

$$
X_\alpha = \{f \in \ell_\infty(\mathbb{I}) : \sigma_\varepsilon(f)^{(\alpha)} = \varnothing \ \forall \, \varepsilon > 0\}.
$$

If $X = \bigcup_{\alpha < \omega_1} X_\alpha$ one has the chain

$$c_0(\mathbb{I}) = X_1 \subset X_2 \subset \cdots \subset X_\alpha \subset X_{\alpha+1} \subset \cdots \subset X \subset \ell_\infty^c(\mathbb{I}) \subset \ell_\infty(\mathbb{I}).$$

The function spaces in the preceding chain are all ideals in $\ell_\infty(\mathbb{I})$. Let Y denote any of them. After representing $\ell_\infty(\mathbb{I})$ as a suitable $C(K)$ space (notice that K is just the Stone-Čech compactification of $\mathbb{I}$ viewed as a discrete set) we have $Y = C_0(L)$, where $L = \{k \in K : f(k) \neq 0 \text{ for some } f \in Y\}$. As $\ell_\infty(\mathbb{I})$ is 1-injective, we get from Theorem 2.18 that Y is 2-separably injective.

These spaces are all different—in fact, none is complemented in the next—since [80, Theorem 2]: for $\alpha < \beta$ there is no linear continuous operator $T : X_\beta \to X_\alpha$ whose restriction to $c_0(\mathbb{I})$ is injective; the same is true for any operator $\ell_\infty^c(\mathbb{I}) \to X$. Moreover, Dashiell and Lindenstrauss show that X is the space of Baire 1 class functions having countable support, namely

$$X = B_1 \cap \ell_\infty^c(\mathbb{I}).$$

This should be compared with Proposition 6.10 where we show that B_1 is not 1-separably injective.

A remarkable generalization of Borsuk-Dugundji theorem for M-ideals was provided by Ando [7] and, independently, Choi and Effros [76]. In order to state it let us recall that a closed subspace $J \subset X$ is called an M-ideal (see [121, Definition 1.1]) if its annihilator $J^\perp = \{x^* \in X^* : \langle x^*, x \rangle = 0 \text{ for every } x \in J\}$ is an L-summand in X^*. This just means that there is a linear projection P on X^* whose range is $J^\perp$ and such that $\|x^*\| = \|P(x^*)\| + \|x^* - P(x^*)\|$ for all $x^* \in X^*$. The easier examples of M-ideals are just ideals in $C(K)$-spaces, which arise as in (2.3). The fact that such a $C_0(L)$ is an M-ideal in $C(K)$ is straightforward from the Riesz representation of $C(K)^*$.

Proposition 2.20 *Let J be an M-ideal in the Banach space E and $\pi : E \to E/J$ the natural quotient map. Let Y be a separable Banach space and $t : Y \to E/J$ be an operator. Assume further that one of the following conditions is satisfied:*

1. *Y has the λ-AP.*
2. *J is a Lindenstrauss space.*

Then t can be lifted to E, that is, there is an operator $L : Y \to E$ such that $\pi L = t$. Moreover one can get $\|L\| \leq \lambda \|t\|$ under the assumption (1) and $\|L\| = \|t\|$ under (2).

We refer the reader to [121, Theorem 2.1] for a proof. In a similar way as Theorem 2.18 was deduced from the Borsuk-Dugundji Theorem (Proposition 2.17 above), one gets from Proposition 2.20:

Theorem 2.21 *Let J be an M-ideal in a Banach space E.*

1. *If E is (universally) λ-separably injective, then E/J is (universally) λ^2-separably injective.*
2. *If E is λ-separably injective, then J is $2\lambda^2$-separably injective.*

Proof

1. By Proposition 1.5 E^{**} is λ-injective and so it has the λ-AP. As $E^{**} = J^{**} \oplus_\infty$ $(E/J)^{**}$ we see that also J^{**} and $(E/J)^{**}$ have the λ-AP. Hence both J and (E/J) have the λ-AP. Let Y be a separable subspace of X and $t : Y \longrightarrow E/J$ an operator. Let S be a separable subspace of E/J containing the image of t. By [60, Theorem 9.7] we may assume S has the λ-AP. Let $s : S \longrightarrow E$ be the lifting provided by Proposition 2.20, so that $\|s\| \leq \lambda$. Now, if $T : X \longrightarrow E$ is an extension of st, then $\pi T : X \longrightarrow E/J$ is an extension of t, and this can be achieved with $\|\pi T\| = \|T\| \leq \lambda^2 \|t\|$.
2. The proof is similar to that of Theorem 2.18(2) and is left to the reader.   □

Observe that when E is a Lindenstrauss space then J is also a Lindenstrauss space and then the exponent 2 can be eliminated everywhere in Theorem 2.21. This result also provides a different proof for Proposition 2.12. Indeed, suppose E_i are λ-separably injective for every $i \in I$. Then so is $E = \ell_\infty(I, E_i)$ and therefore its M-ideal $J = c_0(I, E_i)$ is $2\lambda^2$-separably injective. This argument, taken from [146], gives the best constant when each E_i is 1-separably injective; otherwise the value $\lambda(1 + \lambda)^+$ we got in the proof of Proposition 2.12 is smaller than $2\lambda^2$.

As we mentioned in Proposition 1.21, Rosenthal constructed in [222] the first injective Banach space not isomorphic to a dual space. The example appears as a space $C(G)$ where G is a closed part of $\beta\mathbb{N}$. One therefore has an exact sequence

$$0 \longrightarrow J_G \longrightarrow \ell_\infty \longrightarrow C(G) \longrightarrow 0$$

in which J_G is an M-ideal, hence separably injective. In the remarks after the proof of Proposition 2.11 it was already noticed that the kernel of a quotient map $\ell_\infty \to \ell_\infty$ need not to be an $\mathscr{L}_\infty$ space.

2.2.3 Compact Spaces of Finite Height

Given a compact space K, recall that we write K' for its derived set, that is, the set of non-isolated points of K. This process can be iterated to define $K^{(n+1)}$ as $(K^{(n)})'$. We say that K has finite height if $K^{(n)} = \varnothing$ for some $n \in \mathbb{N}$, the least of which is called the (Cantor-Bendixson) height of K.

Proposition 2.22 *Let K be an infinite compact space of finite height. Then $C(K)$ is separably injective but not universally separably injective.*

Proof Let us show that $C(K)$ is separably injective if and only if $C(K')$ is separably injective; which yields the result since after finite number of derivations one necessarily arrives to a finite compact set. Let I be the set of isolated points of K. The restriction operator $C(K) \longrightarrow C(K')$ induces a short exact sequence

$$0 \longrightarrow c_0(I) \longrightarrow C(K) \longrightarrow C(K') \longrightarrow 0.$$

Since separable injectivity is a 3-space property (Proposition 2.11(1)) and $c_0(I)$ is separably injective, if $C(K')$ is separably injective then also $C(K)$ is separably injective. When K is scattered (in particular, of finite height) then the dual of every separable subspace is separable [92], hence $C(K)$ does not contain ℓ_∞ and thus it follows from Proposition 2.8(2) that it cannot be universally separably injective. The only if follows from Proposition 2.11(2). □

Of course that spaces of continuous functions on countable height compacta, such as $C(\omega^\omega)$, need not be separably injective. An alternative proof for the result above provides more information about the constants involved:

Proposition 2.23 *If K is a compact space of height n, then $C(K)$ is $(2n-1)$-separably injective.*

Proof Let $Y \subset X$ with X separable and let $t : Y \to C(K)$ be a norm one operator. The range of t is separable and every separable subspace of a $C(K)$ is contained in an isometric copy of $C(L)$, where L is the quotient of K after identifying k and k' when $y(k) = y(k')$ for all $y \in Y$. This L is metrizable because Y is separable. Moreover, if K has height n, then L has height at most n and so it is homeomorphic to $[0, \omega^r \cdot k]$ with $r < n$, $k < \omega$ (see [36]; or else [120, Theorem 2.56]). Since $C[0, \omega^r \cdot k]$ is $(2r+1)$-separably injective [25], our operator can be extended to an operator $T : X \to C(K)$ with norm

$$\|T\| \le (2r+1)\|t\| \le (2n-1)\|t\|,$$

concluding the proof. □

When K is a metrizable compact of finite height n, Baker [25] showed that $2n-1$ is the best constant for separable injectivity, using arguments from Amir [5]. There are some difficulties in generalizing those arguments for nonmetrizable compact spaces, so we do not know if it could exist a nonmetrizable compact space K of height n such that $C(K)$ is λ-separably injective for some $\lambda < 2n-1$.

2.2.4 Twisted Sums of $c_0(I)$

By Proposition 2.11, twisted sums of separably injective spaces are separably injective, so making twisted sums is an effective method to obtain new separably injective spaces. The simplest examples will be provided by twisted sums of two $c_0(\aleph)$. There exist in the literature several examples of nontrivial twisted sums of the type

$$0 \longrightarrow c_0(I) \longrightarrow E \longrightarrow c_0(J) \longrightarrow 0 \qquad (2.4)$$

with different properties. The twisted sum space E is separably injective but not universally separably injective, just because E cannot be a Grothendieck space unless both I and J are finite [see Proposition 2.8(2)]. All the examples of such twisted sums E that exist in the literature are of the form $C(K)$ with K a compact space of finite height as in Sect. 2.2.3. It is an open problem whether a twisted sum E of $c_0(I)$ and $c_0(J)$ exists that is not a $C(K)$-space. It is shown in [68] that every twisted sum of $c_0(\Gamma)$ and a space with property (V) has property (V).

When J is countable the sequence splits since $c_0(I)$ is separably injective. For $I = \mathbb{N}$ and $\aleph_0 < |J| \leq \mathfrak{c}$ a nontrivial extension can be obtained (see [144, Example 2]; and also [61]) from an almost-disjoint family $\mathcal{M}$ of size $|J|$ of infinite subsets of $\mathbb{N}$; which means that $M \cap N$ is finite for different $M, N \in \mathcal{M}$. The existence of such a family was first observed by Sierpiński; see the proof of Theorem 1.25. Let $E(\mathcal{M})$ be the closure of the linear span in ℓ_∞ of the characteristic functions $\{1_n : n \in \mathbb{N}\}$ and $\{1_M : M \in \mathcal{M}\}$. Since the images of $\{1_M : M \in \mathcal{M}\}$ in ℓ_∞/c_0 generate a copy of $c_0(J)$ we have the pull-back diagram

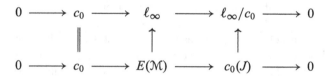

Recall that weakly compactly generated (in short, WCG) subspaces of ℓ_∞ are separable: if K is weakly compact in ℓ_∞, then the coordinates of ℓ_∞ provide countably many real-valued continuous functions on the compact K that separate the points, hence K is metrizable and separable. From this, we get that $E(\mathcal{M})$ is not WCG. Hence the lower sequence in the preceding diagram does not split because c_0 and $c_0(J)$ are WCG and the product of two WCG spaces is WCG.

It is easily seen that $E(\mathcal{M})$ is a subring of ℓ_∞ and so it can be represented as (that is, it is isometric through a ring isomorphism to) certain $C_0(L)$, where L is a locally compact space. It is actually simpler to consider the unitization of $E(\mathcal{M})$ in ℓ_∞, that is,

$$E(\mathcal{M})_+ = \{\lambda 1_{\mathbb{N}} + f : \lambda \in \mathbb{R}, f \in E(\mathcal{M})\}.$$

In this way $E(\mathcal{M})_+$ is a (closed, unital) subalgebra of ℓ_∞ that can be identified with a $C(K)$ for a compact space $K = K_{\mathcal{M}}$ (the one-point compactification of the just mentioned L). The description of $K_{\mathcal{M}}$ is an amusing exercise. It has three levels: isolated points, that correspond to natural numbers; points in the second level correspond to elements of the family $\mathcal{M}$, and a neighborhood of M contains M together with almost all elements of M. The point in the third level is the infinity point in the one-point compactification of the first two levels. One has a diagram

$$
\begin{array}{ccccccccc}
0 & \longrightarrow & c_0 & \longrightarrow & C(\beta\mathbb{N}) & \longrightarrow & C(\mathbb{N}^*) & \longrightarrow & 0 \\
& & \| & & \uparrow & & \uparrow & & \\
0 & \longrightarrow & c_0 & \longrightarrow & C(K_{\mathcal{M}}) & \longrightarrow & C(K'_{\mathcal{M}}) & \longrightarrow & 0
\end{array}
$$

where, moreover, $K'_{\mathcal{M}}$ is the one-point compactification of J.

Other twisted sums of $c_0(I)$ and $c_0(J)$ spaces were obtained by Ciesielski and Pol (see [81, Definition 8.8.2]). They are C-spaces $C(\mathcal{CP})$, where the Ciesielski-Pol compacta $\mathcal{CP}$ have both the derived set $\mathcal{CP}'$ and its complement $\mathcal{CP} \setminus \mathcal{CP}'$ uncountable, and the second derived set $\mathcal{CP}''$ is a singleton. Moreover, $C(\mathcal{CP})$ has a subspace Y isometric to $c_0(I)$ with $C(\mathcal{CP})/Y$ isomorphic to $c_0(J)$, for some uncountable sets I and J. They have the property that there is no injective operator from $C(\mathcal{CP})$ into $c_0(\Gamma)$, for any Γ, so they are not WCG.

Nontrival WCG twisted sums of $c_0(\Gamma)$ also exist. In [13] it is obtained an exact sequence

$$
0 \longrightarrow c_0(\aleph) \longrightarrow C(K) \longrightarrow c_0(\aleph) \longrightarrow 0
$$

in which K is an Eberlein compact. Under GCH one can choose $\aleph = \aleph_\omega$ (and this is the smallest cardinal allowing a WCG nontrivial twisted sum of $c_0(\Gamma)$). Bell and Marciszewski construct in [29] an Eberlein compact $\mathcal{BM}$ of weight $\mathfrak{c}$ and height 3 that cannot be embedded into the space of all characteristic functions of subsets of cardinality lesser than or equal to n of a given set; Marciszewski shows in [192] that $C(\mathcal{BM})$ is actually a nontrivial twisted sum of two $c_0(\Gamma)$. On the other hand, given a compact space K of weight smaller than $\aleph_\omega$, the space $C(K)$ is isomorphic to $c_0(I)$ if and only if K is an Eberlein compact of finite height [109, 192].

2.2.5 Twisted Sums of c_0 and ℓ_∞

The next simplest twisted sum of separably injective spaces are those of c_0 and ℓ_∞. Nontrivial twisted sums $0 \to c_0 \to X \to \ell_\infty \to 0$ exist and explicit examples can be seen in [54], obtained as the lower sequence in a certain pull-back diagram

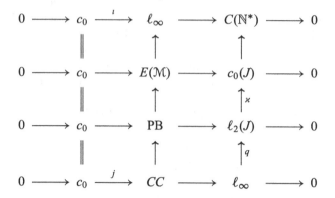

Here $E(\mathcal{M})$ is obtained using an almost disjoint family of size $\mathfrak{c} = |J|$, $\varkappa$ is any operator providing a non WCG pull-back space PB (such as the canonical inclusion, in which case PB is the Johnson-Lindenstrauss space [144, Example 1]) and q is a quotient map. The twisted sum space in the lower sequence was baptized CC in [157].

The lower sequence cannot split since otherwise there would be a quotient map $Q : c_0 \oplus \ell_\infty \to PB$. The restriction of Q to ℓ_∞ cannot be weakly compact, since otherwise PB would be WCG; therefore, Q must be an isomorphism on a copy of ℓ_∞; but PB does not contain ℓ_∞ because "not containing ℓ_∞" is a 3-space property [61, Theorem 3.2.f]. The space CC cannot be universally separably injective: since ι admits the obvious extension through j, if j would also extend through ι then the diagonal principles (Proposition A.22) would yield an isomorphism $\ell_\infty \oplus \ell_\infty = CC \oplus C(\mathbb{N}^*)$, which makes $C(\mathbb{N}^*)$ complemented in ℓ_∞ which is not.

2.2.6 A Separably Injective Space Not Isomorphic to a Complemented Subspace of Any $C(K)$

This counterexample depends on Benyamini's construction appearing in [32] of an M-space not isomorphic to any complemented subspace of a C-space. The basic element in that construction can be described as follows. Let $\tilde{\mathbb{N}}$ denote a copy of the set of the integers. Given $x \in \beta\mathbb{N}$, we denote by $\tilde{x}$ the corresponding element in $\beta\tilde{\mathbb{N}}$. Set $\tilde{\mathbb{N}}^* = \beta\tilde{\mathbb{N}}\backslash\tilde{\mathbb{N}}$ and put $B = \beta\mathbb{N} \oplus \tilde{\mathbb{N}}^*$. Now, for $0 < \tau < 1$, consider

$$\mathscr{B}_\tau = \{f \in C(B) : f(x) = \tau f(\tilde{x}) \text{ for all } x \in \mathbb{N}^*\},$$

equipped with the restriction of the sup norm in $C(B)$. Quite clearly, $\mathscr{B}_\tau$ is a renorming of ℓ_∞. However, and this is the crux, $\mathscr{B}_\tau$ is far away from the complemented subspaces of any $C(K)$ space in the following precise sense: if K is a compact space, $u : \mathscr{B}_\tau \to C(K)$ is an isomorphic embedding and p is a projection of $C(K)$ onto the range of u, then $\|u\|\|u^{-1}\|\|p\| \geq 1/\tau$.

Example 2.24 Suppose $\tau(n) \to 0$. Then the spaces $c_0(\mathbb{N}, \mathscr{B}_{\tau(n)})$ and $\ell_\infty(\mathbb{N}, \mathscr{B}_{\tau(n)})$ are separably injective yet they are isomorphic to no direct factor of a C-space. They are not universally separably injective and $\ell_\infty(\mathbb{N}, \mathscr{B}_{\tau(n)})$ is a Grothendieck space.

Proof It suffices to see that $\mathscr{B}_\tau$ is 5-separably injective for $0 < \tau \leq 1$. Notice that the characteristic functions of the points of $\mathbb{N}$ generate an ideal in $C(B)$ which is fact an isometric copy of c_0 in B_τ that we will denote $c_0(\mathbb{N})$. Clearly, $c_0(\mathbb{N})$ is an M-ideal in B_τ. After a moment's reflection one realizes that the quotient $B_\tau/c_0(\mathbb{N})$ is isometric to $\ell_\infty/c_0 = C(\mathbb{N}^*)$. Thus, even if $\mathscr{B}_\tau$ is badly isomorphic to ℓ_∞ we have an isometric exact sequence

$$0 \longrightarrow c_0(\mathbb{N}) \overset{\iota}{\longrightarrow} \mathscr{B}_\tau \overset{\pi}{\longrightarrow} C(\mathbb{N}^*) \longrightarrow 0$$

whose kernel is an M-ideal.

Let now X be a separable Banach space and $t : Y \to \mathscr{B}_\tau$ be a norm one operator, where Y is a subspace of X. As $C(\mathbb{N}^*)$ is 1-separably injective one can find a norm one $T : X \to C(\mathbb{N}^*)$ extending the composition πt. As T has separable range, by Proposition 2.20, T can be lifted to an operator $L : X \to \mathscr{B}_\tau$, again with $\|L\| = 1$. Clearly, $t - L|_Y$ takes values in $c_0(\mathbb{N})$ and it can be extended to an operator $S : X \to c_0(\mathbb{N})$, with $\|S\| \leq 2\|t - L|_Y\| \leq 4$. Hence $S + L$ is an extension of t to X, and has norm at most 5. Thus we see that $\ell_\infty(\mathbb{N}, \mathscr{B}_{\tau(n)})$ is 5-separably injective and $c_0(\mathbb{N}, \mathscr{B}_{\tau(n)})$ is 10-separably injective.

As for the last statement, $c_0(\mathbb{N}, \mathscr{B}_{\tau(n)})$ cannot be universally separably injective since it contains a complemented copy of c_0, which is not. To see that $\ell_\infty(\mathbb{N}, \mathscr{B}_{\tau(n)})$ is not universally separably injective, observe that $\mathscr{B}_\tau$ is (isometric to) the pull-back space in the diagram

$$
\begin{array}{ccccccccc}
0 & \longrightarrow & c_0(\mathbb{N}) & \longrightarrow & C(\beta\mathbb{N}) & \overset{r}{\longrightarrow} & C(\mathbb{N}^*) & \longrightarrow & 0 \\
 & & \Big\| & & \Big\uparrow & & \Big\uparrow{\scriptstyle\tau} & & \\
0 & \longrightarrow & c_0(\mathbb{N}) & \longrightarrow & \mathrm{PB}(r,\tau) & \longrightarrow & C(\tilde{\mathbb{N}}^*) & \longrightarrow & 0
\end{array}
\qquad (2.5)
$$

where r is plain restriction and τ denotes multiplication (by τ). Indeed, by the very definition we have

$$\mathrm{PB}(r,\tau) = \{(f,g) \in C(\beta\mathbb{N}) \oplus_\infty C(\tilde{\mathbb{N}}^*) : rf = \tau g\}$$
$$= \{(f,g) \in C(\beta\mathbb{N}) \oplus_\infty C(\tilde{\mathbb{N}}^*) : f(x) = \tau g(\tilde{x}) \text{ for every } x \in \mathbb{N}^*\}$$
$$= \mathscr{B}_\tau.$$

Therefore, for each n, we have a commutative diagram

$$
\begin{array}{ccccccccc}
0 & \longrightarrow & c_0 & \longrightarrow & \ell_\infty & \longrightarrow & C(\mathbb{N}^*) & \longrightarrow & 0 \\
& & \big\| & & \big\uparrow & & \big\uparrow & & \\
0 & \longrightarrow & c_0 & \xrightarrow{j_n} & \mathscr{B}_{\tau(n)} & \longrightarrow & C(\mathbb{N}^*) & \longrightarrow & 0
\end{array}
\tag{2.6}
$$

All these can be amalgamated into a unique diagram

$$
\begin{array}{ccccccccc}
0 & \longrightarrow & \ell_\infty(c_0) & \longrightarrow & \ell_\infty(\ell_\infty) & \longrightarrow & \ell_\infty(C(\mathbb{N}^*)) & \longrightarrow & 0 \\
& & \big\| & & \big\uparrow & & \big\uparrow & & \\
0 & \longrightarrow & \ell_\infty(c_0) & \xrightarrow{j} & \ell_\infty(\mathbb{N}, \mathscr{B}_{\tau(n)}) & \longrightarrow & \ell_\infty(C(\mathbb{N}^*)) & \longrightarrow & 0
\end{array}
\tag{2.7}
$$

If $\ell_\infty(\mathbb{N}, \mathscr{B}_{\tau(n)})$ were universally separably injective, then it should be λ-universally separably injective, for some λ. This would imply that every $\mathscr{B}_{\tau(n)}$ is λ-universally separably injective and so the operator j_n in (2.6) admits an extension $J_n : \ell_\infty \to \mathscr{B}_{\tau(n)}$, with $\|J_n\| \le \lambda$. The "diagonal" operator $J : \ell_\infty(\ell_\infty) \to \ell_\infty(\mathbb{N}, \mathscr{B}_{\tau(n)})$ given by $J((f_n)) = (J_n(f_n))$ is then an extension of the operator j in diagram (2.7). Applying Proposition A.22 we would obtain an isomorphism

$$
\ell_\infty(\ell_\infty) \oplus \ell_\infty(C(\mathbb{N}^*)) = E \oplus \ell_\infty(C(\mathbb{N}^*)).
$$

This is impossible, since $\ell_\infty(\mathbb{N}, \mathscr{B}_{\tau(n)})$ is not complemented in any C-space.

It follows from results of Leung and Räbiger in [174] that $\ell_\infty(\mathbb{N}, \mathscr{B}_{\tau(n)})$ is a Grothendieck space: A set I is said to have real-valued measurable cardinal if there exists a countably additive measure $\mu : \mathbb{P}(I) \to [0, 1]$ vanishing on the singletons of I. The existence of real-valued measurable cardinals cannot be proved in ZFC and the fact that $\aleph_0$ is not real-valued measurable is obvious. Leung and Räbiger proved in [174, Theorem] that if (E_i) is a family of Banach spaces indexed by a set I whose cardinal is not real-valued measurable, then the Banach space product $\ell_\infty(I, E_i)$ contains a complemented copy of c_0 (if and) only if some E_i does. As each $\mathscr{B}_{\tau(n)}$ is a renorming of ℓ_∞ we see that $\ell_\infty(\mathbb{N}, \mathscr{B}_{\tau(n)})$ has no complemented subspace isomorphic to c_0. Since it is separably injective, has Pełczyński's property (V) and, consequently, is a Grothendieck space. □

2.3 Universally Separably Injective Spaces

It was proved in Proposition 2.8(2) that universally separably injective spaces contain ℓ_∞. In this section we will show that they are in fact ℓ_∞-upper-saturated, according to the next definition.

Definition 2.25 Let X and Y be Banach spaces. We say that X is Y-*upper-saturated* if every separable subspace of X is contained in some (isomorphic) copy of Y inside X.

It is clear that c_0-upper-saturated spaces are separably injective and ℓ_∞-upper-saturated spaces are universally separably injective. One moreover has:

Theorem 2.26 *An infinite-dimensional Banach space is universally separably injective if and only if it is ℓ_∞-upper-saturated.*

Proof The sufficiency is a clear consequence of the injectivity of ℓ_∞. In order to show the necessity, let Y be a separable subspace of a universally separably injective space X. We consider a subspace Y_0 of ℓ_∞ isomorphic to Y and an isomorphism $t : Y_0 \to Y$. We can find projections p on X and q on ℓ_∞ such that $Y \subset \ker p$, $Y_0 \subset \ker q$, and both p and q have range isomorphic to ℓ_∞. Indeed, let $\pi : X \to X/Y$ be the quotient map. Since X contains ℓ_∞ and Y is separable, π is not weakly compact so, by Proposition 2.8(2), there exists a subspace M of X isomorphic to ℓ_∞ where π is an isomorphism. Now $X/Y = \pi[M] \oplus N$, with N a closed subspace. Hence $X = M \oplus \pi^{-1}[N]$, and it is enough to take p as the projection with range M and kernel $\pi^{-1}[N]$.

Since $\ker p$ and $\ker q$ are universally separably injective spaces, we can take operators $u : X \to \ker q$ and $v : \ell_\infty \to \ker p$ such that $v = t$ on Y_0 and $u = t^{-1}$ on Y. Let $w : \ell_\infty \to \operatorname{ran} p$ be an operator satisfying $\|w(x)\| \geq \|x\|$ for all $x \in \ell_\infty$. We will show that the operator $T = v + w(1_{\ell_\infty} - uv)$ is an into isomorphism $\ell_\infty \to X$. This suffices to end the proof since $\operatorname{ran} T$ is isomorphic to ℓ_∞ and both T and v agree with t on Y_0, so $Y \subset \operatorname{ran} T \subset X$. Since $\operatorname{ran} v \subset \ker p$ and $\operatorname{ran} w \subset \operatorname{ran} p$, there exists $C > 0$ such that for all $x \in \ell_\infty$ one has

$$\|Tx\| \geq C \max\{\|v(x)\|, \|w(1_{\ell_\infty} - uv)x\|\}.$$

Now, if $\|vx\| < (2\|u\|)^{-1}\|x\|$, then $\|uvx\| < \frac{1}{2}\|x\|$; hence

$$\|w(1_{\ell_\infty} - uv)x\| \geq \|(1_{\ell_\infty} - uv)x\| > \frac{1}{2}\|x\|.$$

Thus $\|Tx\| \geq C(2\|u\|)^{-1}\|x\|$ for every $x \in X$.   □

Another similarity between ℓ_∞ and universally separably injective spaces is given in the next Proposition 2.27, which extends [182, Proposition 2.f.12(iii)]. Recall that an operator is Fredholm if its kernel and its cokernel are finite dimensional. Here, the cokernel of an operator $T : X \to Y$ is defined as $\operatorname{coker} T = Y/\operatorname{ran} T$. The index of a Fredholm operator T is defined by

$$\operatorname{ind}(T) = \dim \ker T - \dim \operatorname{coker} T.$$

Note that if $Y/\operatorname{ran}(T)$ is finite dimensional, then T has closed range [242, Theorem IV.5.10].

Proposition 2.27 *Let X be universally separably injective and let $\imath : Y \to X$ and $j : Y \to X$ be two into isomorphisms. Suppose that $X/j[Y]$ and $X/\imath[Y]$ are separable. Then every extension $I : X \to X$ of $\imath$ through j (i.e., $Ij = i$) is a Fredholm operator and all these extensions have the same index.*

Proof Since X is separably injective, we can find $u : X \to X$ and $v : X \to X$ operators such that $uj = \imath$ and $v\imath = j$. Let us denote $w = 1_X - vu$. Since $j(Y)$ is contained in the kernel of w, the operator w factors through $X/j[Y]$. Recall that $\mathscr{L}_\infty$ spaces have the Dunford-Pettis property (every weakly compact operator defined on those spaces takes weakly convergent sequences into convergent ones; see Proposition A.2). Thus, X has the Dunford-Pettis property and its separable quotients must be reflexive by Proposition 2.8(2). Therefore, the operator w is weakly compact and completely continuous; hence w^2 is compact. From this fact it follows that $vu = 1_X - w$ is a Fredholm operator with $\mathrm{ind}(vu) = 0$. Similarly we can show that uv is a Fredholm operator with $\mathrm{ind}(uv) = 0$. Thus u and v are Fredholm operators with $\mathrm{ind}(u) + \mathrm{ind}(v) = 0$, and the proof is done.   □

Proposition 2.27 remains valid for X separably injective provided one asks the quotients to be separable and reflexive (e.g., when X is Grothendieck). Recall that two Banach spaces X and Y are said to be essentially incomparable (see [110]) if for each pair of operators $t : X \to Y$ and $s : Y \to X$, $1_X - st$ is a Fredholm operator. Since it follows from Proposition 2.8(2) that a quotient of a universally separably injective space is either reflexive or it contains copies of ℓ_∞, the proof of Proposition 2.27 shows that universally separable injective spaces and spaces containing no copies of ℓ_∞ are essentially incomparable.

2.4   1-Separably Injective Spaces

While regarding injectivity it is unknown whether the parameter λ in "λ-injective" has real content (after all, it could still be true that every λ-injective space can be renormed to become 1-injective) in this section we shall see that the parameter λ in "λ-separably injective" has *some* meaning (but we do not know which). For instance, 1-separably injective spaces enjoy several properties that, say, 2-separably injective spaces lack; and spaces such as $c_0(\Gamma)$ are 2-separably injective but not λ-separably injective for $\lambda < 2$; at the same time, $C(K)$-spaces λ-separably injective for $\lambda < 2$ are automatically 1-separably injective (Proposition 2.34).

Keeping in mind that separably injective spaces are Grothendieck if and only if they do not contain c_0 complemented, it is possible to establish a major difference between 1-separably injective and general separably injective spaces: 1-separably injective spaces are Grothendieck (hence they cannot be separable or WCG)—see Proposition 2.31 below—while a 2-separably injective space, such as c_0, can be even separable.

To prove that 1-separably injective spaces cannot contain c_0 complemented, the following lemma due to Lindenstrauss [177, p. 221, proof of (i) $\Rightarrow$ (v)] provides a useful technique.

Lemma 2.28 *Let E be a 1-separably injective space, X a Banach space of density $\aleph_1$, and Y a separable subspace of X. Then every operator $t : Y \to E$ can be extended to an operator $T : X \to E$ with the same norm.*

Proof We write X as the union of a continuous ω_1-sequence of separable spaces $(X_\alpha)_{\alpha < \omega_1}$ beginning with $X_0 = Y$. This just means (see Appendix A.6)

- $X_\alpha \subset X_\beta$ if $\alpha \leq \beta$.
- $X = \bigcup_{\alpha < \omega_1} X_\alpha$.
- For every limit ordinal $\beta < \omega_1$ one has $X_\beta = \overline{\bigcup_{\alpha < \beta} X_\alpha}$.

Then we define inductively a coherent family of operators $T_\alpha : X_\alpha \longrightarrow E$, all of them with the same norm as $T_0 = t$. We can do this using the 1-separable-injectivity of E and, in the limit ordinals, using that given $T_{\alpha_n} : X_{\alpha_n} \longrightarrow E$, a coherent sequence of operators of norm $\|t\|$, they determine a unique operator $\bigcup_n X_{\alpha_n} \longrightarrow E$ of norm $\|t\|$. $\square$

Proposition 2.29 (CH) *Every 1-separably injective Banach space is universally 1-separably injective and therefore a Grothendieck space.*

Proof Let E be 1-separably injective, X an arbitrary Banach space and $t : Y \to E$ an operator, where Y is a separable subspace of X. Then $\overline{t[Y]}$, the closure of the image of t, is a separable subspace of E and so there is an isometric embedding $u : \overline{t[Y]} \to \ell_\infty$. As ℓ_∞ is 1-injective there is an operator $T : X \to \ell_\infty$ whose restriction to Y agrees with ut. Thus it suffices to extend the inclusion of $\overline{t[Y]}$ into E to ℓ_∞. But, under CH, the density character of ℓ_∞ is $\aleph_1$ and Lemma 2.28 applies. The "therefore" part is now a consequence of Proposition 2.8(2). $\square$

We will prove later (Theorem 2.39) that CH cannot be dropped in general from Proposition 2.29. However the "therefore" part survives in ZFC. The following characterization of 1-separable injectivity, apart from its intrinsic interest, will help with the proof. Its general version will be stated and proved in Proposition 5.12.

Proposition 2.30 *A Banach space E is 1-separably injective if and only if every countable family of mutually intersecting balls has nonempty intersection.*

Proof SUFFICIENCY. Take an operator $t : Y \to E$, where Y is a closed subspace of a separable space X. We may and do assume $\|t\| = 1$. Let $z \in X \backslash Y$ and let Y_0 be a dense countable subset of Y and, for each $y \in Y_0$, consider the ball $B(ty, \|y - z\|)$ in E. Any two of these balls intersect, since for $y_1, y_2 \in Y_0$ we have

$$\|ty_2 - ty_1\| \leq \|t\| \|y_2 - y_1\| \leq \|y_2 - z\| + \|y_1 - z\|.$$

The hypothesis is that there is

$$f \in \bigcap_{y \in Y_0} B(ty, \|y - z\|) = \bigcap_{y \in Y} B(ty, \|y - z\|).$$

It is clear that the map $T : Y + [z] \rightarrow E$ defined by $T(y + cz) = ty + cf$ is an extension of t with $\|T\| = 1$. The rest is clear: use induction.

NECESSITY. We begin with the observation that if two closed balls of any Banach (or metric) space have a common point, then the distance between the centers is at most the sum of the radii. In ℓ_∞ that necessary condition suffices and every family of mutually intersecting balls has nonempty intersection.

Let E be 1-separably injective and let $B(e_n, r_n)$ be a sequence of mutually intersecting balls in E. Let Y be the closed separable subspace of E spanned by the centers. Let $\kappa : Y \rightarrow \ell_\infty$ be any isometric embedding. Notice that even if $B_Y(e_n, r_n) = B(e_n, r_n) \cap Y$ need not be mutually intersecting in Y, any two balls of the sequence $B(\kappa(e_n), r_n)$ meet in ℓ_∞ because the distance between the centers does not exceed the sum of the radii. Therefore the intersection

$$\bigcap_n B(\kappa(e_n), r_n)$$

contains some point, say $x \in \ell_\infty$. Let X be the subspace spanned by x and $\kappa(Y)$ in ℓ_∞ so that $\dim X/Y \leq 1$. The hypothesis on E allows one to extend the inclusion of Y into E to X through $\kappa : Y \rightarrow X$ without increasing the norms. The image of x in E under any such extension belongs to the intersection of all the $B(e_n, r_n)$.   □

Proposition 2.31 *Every 1-separably injective space is a Grothendieck and a Lindenstrauss space.*

Proof To prove that a 1-separably injective space is Lindenstrauss we recall that a Banach space is a Lindenstrauss space if and only if every finite set of mutually intersecting balls has nonempty intersection [175]. Proposition 2.30 now concludes. A different argument can be derived from Proposition 1.5 that yields the bidual of a 1-separably injective space X is 1-injective, hence X^{**} and so X is a Lindenstrauss space.

It remains to prove that a 1-separably injective space X must be Grothendieck. Since X has property (V) by Proposition 2.8, it suffices to show that c_0 is not complemented in X, so let $j : c_0 \longrightarrow X$ be an embedding. Consider an almost-disjoint family $\mathcal{M}$ of size $\aleph_1$ formed by infinite subsets of $\mathbb{N}$. Proceeding as in Sect. 2.2.4 we get a nontrivial exact sequence

$$0 \longrightarrow c_0 \longrightarrow E(\mathcal{M}) \longrightarrow c_0(\aleph_1) \longrightarrow 0$$

where the space $E(\mathcal{M})$ has density character $\aleph_1$. The embedding j can be extended to all of $E(\mathcal{M})$ by Lemma 2.28, which yields a commutative diagram

$$
\begin{array}{ccccccccc}
0 & \longrightarrow & c_0 & \longrightarrow & E(\mathcal{M}) & \longrightarrow & c_0(\aleph_1) & \longrightarrow & 0 \\
 & & \| & & \downarrow & & \downarrow & & \\
0 & \longrightarrow & c_0 & \overset{j}{\longrightarrow} & X & \longrightarrow & X/j[c_0] & \longrightarrow & 0
\end{array}
$$

Thus, were c_0 complemented in X it would be complemented in $E(\mathcal{M})$ as well, which is not. $\square$

2.4.1 On λ-Separably Injective Spaces When $\lambda < 2$

As we have already mentioned, it is an open problem whether a λ-injective space is isomorphic to a 1-injective space. From Proposition 2.31 it is clear that 2-separably injective spaces cannot be, in general, be renormed to become 1-separably injective. We do not know whether a λ-separably injective space, $\lambda < 2$ must be (isomorphic to a) 1-separably injective or, at least, a Grothendieck space. We have, however, the following result, based on an idea of Ostrovskii [208]:

Proposition 2.32 *A λ-separably injective space with $\lambda < 2$ is either finite dimensional or has density character at least $\mathfrak{c}$.*

Proof Let X be an infinite dimensional λ-separably injective space for $\lambda < 2$. In Proposition 2.8 it is shown that X contains c_0, and thus by a result of James [139] it contains, for each $\varepsilon > 0$, an $(1 + \varepsilon)$-isomorphic copy of c_0. With a standard renorming [211, Proposition 1] we may assume X contains c_0 isometrically and it is λ'-separably injective, still with $\lambda' < 2$. So, let $u : c_0 \to X$ be an isometric embedding and let $u_n = u(e_n)$, where (e_n) is the unit basis of c_0. For each element $f \in \ell_\infty$ with all coordinates ± 1, let $T_f : c_0 + [f] \to X$ be an extension of u with norm at most λ'. For two different f, g pick n so that $f(n) = 1$ and $g(n) = -1$. One has $\|u_n - T_f(f/2)\| = \|u_n + T_g(g/2)\| \leq \lambda'/2$, and thus

$$
\begin{aligned}
\|T_f(f/2) - T_g(g/2)\| &= \|T_f(f/2) - u_n - u_n - T_g(g/2) + 2u_n\| \\
&\geq 2 - \|T_f(f/2) - u_n - u_n - T_g(g/2)\| \\
&\geq 2 - \lambda'/2 - \lambda'/2 \\
&= 2 - \lambda'.
\end{aligned}
$$

So $\operatorname{dens} X \geq \mathfrak{c}$. $\square$

In any case, it seems that some break occurs at $\lambda = 2$. As a preparation for the following result, let us record the following observation:

Lemma 2.33 *If E is λ-separably injective, then given a countable family of mutually intersecting balls $B(e_n, r_n)$ one has $\bigcap_n B(e_n, \lambda r_n) \neq \emptyset$.*

Proof Just read the "necessity part" of the proof of Proposition 2.30.   □

According to Lindenstrausss (see [177, Remarks 3]) the following result "is similar to a result due to Amir [4] and Isbell and Semadeni [138] that if a C-space has projection constant $\lambda < 2$ then it has projection constant 1" (i.e., it is 1-injective).

Proposition 2.34 *If a C-space is λ-separably injective for some $\lambda < 2$, then it is 1-separably injective.*

Proof What one actually proves is that if a $C(K)$-space is λ-separably injective for some $\lambda < 2$ then K actually is an F-space, in the formulation: for every $f \in C(K)$ there is $g \in C(K)$ and $\delta > 0$ such that $f(k) > 0$ implies $g(k) \geq \delta$ and $f(k) < 0$ implies $g(k) \leq -\delta$. Now, if $C(K)$ is λ-separably injective then it has property (c_λ), and therefore any family $B(x_\alpha, r_\alpha)$ of mutually intersecting balls whose centers lie on a separable subspace is such that $\bigcap_\alpha B(x_\alpha, \lambda r_\alpha) \neq \emptyset$.

Pick now $f \in C(K)$ and set $r_n(t) = 1$ for $t \geq 1/n$ and $r_n(t) = -1$ for $t \leq -1/n$ and linear in $[-1/n, 1/n]$. The balls in the sequence $B(r_n \circ f, 1/2)$ are mutually intersecting. By the preceding Lemma there exists $g \in \bigcap_n B(r_n \circ f, \lambda/2)$. Since $\lambda < 2$, set $\delta = 1 - \lambda/2 > 0$ and observe that $g(k) \geq \delta$ when $f(k) > 0$ and $g(k) \leq -\delta$ when $f(k) < 0$.   □

2.4.2   A C-space 1-Separably Injective But Not Universally 1-Separably Injective

We show now that without **CH**, 1-separably injectivity does not longer imply universal 1-separable injectivity. To this end we will produce, assuming that $\mathfrak{c} = \aleph_2$ and also Martin's axiom, a 1-separably injective space $C(K)$ and an operator $c \to C(K)$ that does not admit norm-preserving extensions to ℓ_∞. In order to state Martin's axiom we need a few definitions. Suppose that we have a partially ordered set P. Two elements $p, q \in P$ are compatible if there exists $r \in P$ such that $r < p$ and $r < q$. A filter is a subset $\mathscr{F} \subset P$ of pairwise compatible elements such that if $p \in \mathscr{F}$ and $p < q$, then $q \in \mathscr{F}$. A subset $D \subset P$ is called dense if for every $p \in P$ there exists $q \in D$ such that $q < p$. We say that P has the countable chain condition (or ccc) if every uncountable subset of P contains a pair of compatible elements.

Martin's Axiom [MA] If P is a ccc partially ordered set and $\{D_i : i \in I\}$ is a family of dense subsets of P with $|I| < \mathfrak{c}$, then there exists a filter $\mathscr{F} \subset P$ such that $\mathscr{F} \cap D_i \neq \emptyset$ for every $i \in I$.

This axiom has become a standard tool with a number of applications in analysis. It is compatible with the **ZFC** system of axioms of set theory, and it is also compatible with different values of the continuum, in particular with $\mathfrak{c} = \aleph_2$; that is what we shall use.

Definition 2.35 Let L be a zero-dimensional compact space. An $\aleph_2$-Lusin family on L is a family $\mathcal{F}$ of pairwise disjoint nonempty clopen subsets of L with $|\mathcal{F}| = \aleph_2$, such that whenever $\mathcal{G}$ and $\mathcal{H}$ are subfamilies of $\mathcal{F}$ with $|\mathcal{G}| = |\mathcal{H}| = \aleph_2$, then

$$\overline{\bigcup\{G \in \mathcal{G}\}} \cap \overline{\bigcup\{G \in \mathcal{H}\}} \neq \varnothing.$$

Lemma 2.36 (MA, $\mathfrak{c} = \aleph_2$) *There exists an $\aleph_2$-Lusin family on $\mathbb{N}^*$.*

Proof We are going to construct a family $\{A_\alpha\}_{\alpha<\omega_2}$ of infinite subsets of $\mathbb{N}$ such that

1. $A_\alpha \cap A_\beta$ is finite for $\alpha < \beta < \omega_2$,
2. for every $B \subset \mathbb{N}$ either $\{\alpha : |A_\alpha \setminus B|$ is finite$\}$ or $\{\alpha : |A_\alpha \cap B|$ is finite$\}$ has cardinality strictly lesser than $\aleph_2$.

Once we obtain this family, we can consider the family the clopens $C_\alpha = \{\mathcal{U} \in \mathbb{N}^* : A_\alpha \in \mathcal{U}\}$ of $\mathbb{N}^*$. The family $\mathcal{C} = \{C_\alpha : \alpha < \omega_2\}$ is an $\aleph_2$-Lusin family on $\mathbb{N}^*$, because they are disjoint by (1), and if we have $\mathcal{G}$ and $\mathcal{H}$ subfamilies of $\mathcal{C}$ whose unions have disjoint closures, then these unions can be separated by a clopen set of $\mathbb{N}^*$, which is of the form $\{\mathcal{U} \in \mathbb{N}^* : B \in \mathcal{U}\}$. Property (2) of our family prevents that both $\mathcal{G}$ and $\mathcal{H}$ have cardinality $\aleph_2$.

So let us proceed now to the construction of the sets A_α. Let $\{B_\alpha : \alpha < \omega_2\}$ be an enumeration of all infinite subsets of $\mathbb{N}$. We construct the A_α's inductively on α. Suppose A_γ has been constructed for $\gamma < \alpha$. We define a partially ordered set $\mathbb{P}_\alpha$ whose elements are pairs $p = (f_p, F_p)$ where f_p is a $\{0, 1\}$-valued function on a finite subset $\mathrm{dom}(f_p)$ of $\mathbb{N}$ and F_p is a finite subset of α. The order relation is that $p < q$ if

- $\mathrm{dom}(f_p) \supset \mathrm{dom}(f_q)$ and $f_p|_{\mathrm{dom}(q)} = f_q$,
- $F_p \supset F_q$,
- f_p vanishes in $A_\gamma \cap \mathrm{dom}(f_p) \setminus \mathrm{dom}(f_q)$ for $\gamma \in F_q$.

First, notice that this partially ordered set is ccc. This is simply because if $Q \subset \mathbb{P}_\alpha$ is an uncountable set, we can find $p, q \in Q$ with $f_p = f_q$, and any two such functions are compatible, since $r = (f_p, F_p \cup F_q) = (f_q, F_p \cup F_q)$ satisfies $r < p$ and $r < q$. Thus for any family of $\aleph_1$ many dense subsets we can find a filter $\mathcal{F} \subset \mathbb{P}_\alpha$ that intersects all of them. The family of dense subsets is the following:

- $D_n = \{p \in \mathbb{P}_\alpha : n \in \mathrm{dom}(p)\}$, for $n \in \mathbb{N}$,
- $D'_\beta = \{p \in \mathbb{P}_\alpha : \beta \in F_p\}$, for $\beta < \alpha$,
- $D''_{\gamma,m} = \{p \in \mathbb{P}_\alpha : \text{there is } n > m \text{ such that } n \in B_\gamma \cap \mathrm{dom} f_p \text{ and } f_p(n) = 1\}$, where $m \in \mathbb{N}$ and $\gamma < \alpha$ are such that $B_\gamma \setminus \{0, \ldots, m\}$ is not contained in any finite union of A_β's with $\beta < \alpha$.

It is easily seen that all these sets are dense in $\mathbb{P}_\alpha$. Let $\mathscr{F} \subset \mathbb{P}_\alpha$ be the filter provided by MA and take $A_\alpha = \{n \in \mathbb{N} : \text{there is } p \in \mathscr{F} \text{ such that } f_p(n) = 1\}$. We check:

1. $A_\alpha \cap A_\beta$ is finite for every $\beta < \alpha$. To check this, pick $q \in \mathscr{F} \cap D'_\beta$. We claim that $A_\alpha \cap A_\beta \subset \text{dom}(f_q)$. Suppose on the contrary that we have $n \in A_\alpha \cap A_\beta \setminus \text{dom}(f_q)$. Since $n \in A_\alpha$ we can find $p \in \mathscr{F}$ such that $f_p(n) = 1$. Since $\mathscr{F}$ is a filter, p and q must be compatible, so pick $r \in \mathbb{P}_\alpha$ such that $r < p$ and $r < q$. We can now apply the third condition of the definition of the order relation, because $\beta \in F_q$, and $n \in A_\beta \cap \text{dom}(f_r) \setminus \text{dom}(f_q)$. So we conclude that $f_r(n) = 0$. But we supposed that $f_p(n) = 1$ and $r < p$, a contradiction.

2. For every $\gamma < \alpha$, if B_γ is not contained in any finite union of A_δ's and a finite set then $A_\alpha \cap B_\gamma$ is infinite. To prove this, it is enough to check that $A_\alpha \cap B_\gamma$ contains some $n > m$ for every $m \in \mathbb{N}$. For this, just use $p \in \mathscr{F} \cap D''_{\beta,m}$.

This finishes the inductive construction of the A_α's. They form indeed an almost disjoint family by property 1 above. It remains to check the second property that we claimed about the family $\{A_\alpha\}_{\alpha < \omega_2}$ at the beginning of the proof. So pick $B \subset \mathbb{N}$. If B is contained in a finite union of A_δ's and a finite set F, $B \subset F \cup (\bigcup_{\delta \in \Delta} A_\delta)$, then just using the almost disjointness, we check that $\{\alpha < \omega_2 : |A_\alpha \setminus B| \text{ is finite}\} \subset \Delta$ so we are done. Similarly, if $\mathbb{N} \setminus B$ is contained in a union $F \cup \bigcup_{\delta \in \Delta} A_\delta$, then $\{\alpha < \omega_2 : |A_\alpha \cap B| \text{ is finite}\} \subset \Delta$. So we assume that neither B nor $\mathbb{N} \setminus B$ is contained in a finite union of A_δ's and a finite set. Then pick α_0 such that $B = B_\beta$ and $\mathbb{N} \setminus B = B_\gamma$ with $\beta, \gamma < \alpha_0$. Then, using condition 2 above, we get that for every $\alpha > \alpha_0$, both $A_\alpha \cap B = A_\alpha \cap B_\beta$ and $A_\alpha \cap (\mathbb{N} \setminus B) = A_\alpha \cap B_\gamma$ are infinite, and we are done. $\square$

In the next theorem we provide the compact space $\mathscr{A}$ whose space of continuous functions will provide the desired example. This compactum is constructed in ZFC, though we will focus on the case when $\mathfrak{c} = \aleph_2$. This value of the continuum is taken mainly for convenience. The construction is the same as the one performed in [16, 87], that we present in purely topological language. It is an inductive construction of length $\mathfrak{c}$ in which at each successor step we split a couple of disjoint F_σ open sets, and we do this exhaustively. Those successive steps can be interpreted as pull-backs with respect to metrizable quotients, cf. Sect. 3.4.5 for further information about this compactum.

Theorem 2.37 ($\mathfrak{c} = \aleph_2$) *There exists an infinite zero-dimensional compact F-space $\mathscr{A}$ such that no closed G_δ subset of $\mathscr{A}$ contains any $\aleph_2$-Lusin family.*

Proof We construct this compact space as an inverse limit of length $\mathfrak{c}$. So, we shall produce compact spaces $\{K_\alpha : \alpha < \mathfrak{c}\}$ and continuous onto maps $\{\pi_{\beta\alpha} : K_\beta \longrightarrow K_\alpha : \alpha \leq \beta\}$ such that $\pi_{\alpha\alpha}$ is the identity in K_α and $\pi_{\beta\alpha} \circ \pi_{\gamma\beta} = \pi_{\gamma\alpha}$ for all $\alpha < \beta < \gamma$, and then $\mathscr{A}$ will be the limit of the system in the sense that

$$\mathscr{A} = \left\{ (x_\alpha)_{\alpha < \mathfrak{c}} \in \prod_{\alpha < \mathfrak{c}} K_\alpha : \pi_{\beta\alpha}(x_\beta) = x_\alpha \text{ for all } \alpha < \beta < \mathfrak{c} \right\}$$

We fix a partition $c = \bigcup_{\alpha < c} S_i$ into c many subsets such that $|S_\alpha| = c$ and $\alpha \leq \min(S_\alpha)$ for all α. The inductive construction is as follows. Let K_0 be the Cantor set. Once the compact space K_α is constructed, we produce an enumeration

$$\{(V_\beta, W_\beta) : \beta \in S_\alpha\}$$

of all pairs of disjoint open F_σ subsets of K_α. This will be possible because the weight of each K_α will be less than c. If the system has been defined for all ordinals below a given γ, we distinguish two cases. If γ is a limit ordinal, then we take K_γ to be the inverse limit of the preceding system:

$$K_\gamma = \left\{ (x_\alpha)_{\alpha < \gamma} \in \prod_{\alpha < \gamma} K_\alpha : \pi_{\beta\alpha}(x_\beta) = x_\alpha \text{ for all } \alpha < \beta < \gamma \right\}.$$

If $\gamma = \beta + 1$ is a successor, we pick the α such that $\beta \in S_\alpha$, and then define

$$K_\gamma = (K_\beta \setminus \pi_{\beta\alpha}^{-1}(V_\beta)) \times \{0\} \ \cup \ (K_\beta \setminus \pi_{\beta\alpha}^{-1}(W_\beta)) \times \{1\}$$

and then $\pi_{\gamma\delta}(x, i) = \pi_{\beta\delta}(x)$ for $\delta < \gamma$. This finishes the inductive construction. We will denote by $\pi_\alpha : \mathscr{A} \longrightarrow K_\alpha$ the canonical projection.

We start now exploring the properties of $\mathscr{A}$. First, it is clear that $\mathscr{A}$ is zero-dimensional since each K_α along the construction is such. Second, we check that $\mathscr{A}$ is an F-space. If we take V and W two disjoint open F_σ-sets, then they must be of the form $V = \pi_\alpha^{-1}(V')$ and $W = \pi_\alpha^{-1}(W')$ for some disjoint open F_σ sets in K_α for some $\alpha < c$. But then, $(V, W')' = (V_\beta, W_\beta)$ for some $\beta > \alpha$ and

$$\pi_\beta^{-1}\left((K_\beta \setminus \pi_{\beta\alpha}^{-1}(V_\beta)) \times \{0\} \right) \text{ and } \pi_\beta^{-1}\left((K_\beta \setminus \pi_{\beta\alpha}^{-1}(W_\beta)) \times \{1\} \right)$$

are disjoint clopens that separate W and V. Third, we prove that if c is a clopen subset of $\mathscr{A}$, then $\pi_\alpha(c)$ is a closed G_δ set for all $\alpha < c$. Indeed, the set c must be of the form $\pi_\beta^{-1}(b)$ for some clopen subset $b \subset K_\beta$ and some $\beta < c$. So it is enough to show that $\pi_{\beta\alpha}(b)$ is a G_δ for every clopen b of K_β and every $\alpha < \beta < c$. We prove it by induction on β. If $\beta = \gamma + 1$ is a successor, it is easily checked that the one-step map $\pi_{\beta\gamma}$ takes clopen sets onto G_δ-sets. If β is a limit ordinal, there exists indeed $\alpha < \gamma < \beta$ such that $\pi_{\beta\gamma}(b)$ is a clopen set.

Finally, we fix a closed G_δ set F and we prove that F does not contain any $\aleph_2$-Lusin family of clopen subsets of F. So suppose that we have such a family $\mathscr{F}$, and we construct by induction subfamilies $\mathscr{F}_i \subset \mathscr{F}$ and ordinals $\alpha(i) < c$ for $i < \omega_1$ with the following properties:

1. $\mathscr{F}_i \subsetneq \mathscr{F}_j$ and $\alpha(i) < \alpha(j)$ if $i < j$.
2. Each family $\mathscr{F}_i$ has cardinality $\aleph_1$.

3. Each $a \in \mathcal{F}_i$ is determined up to $\alpha(i + 1)$, in the sense that a is of the form $a = \pi_{\alpha(i+1)}^{-1}(a')$ for some clopen set a' of $K_{\alpha(i+1)}$.
4. If b is a clopen subset of $K_{\alpha(i)}$ such that $|\{a \in \mathcal{F} : a \subset \pi_{\alpha(i)}^{-1}(b)\}| \leq \aleph_1$, then $\{a \in \mathcal{F} : a \subset \pi_{\alpha(i)}^{-1}(b)\} \subset \mathcal{F}_{i+1}$

The construction is possible because each K_α has weight less than $\mathfrak{c} = \aleph_2$ so it has at most $\aleph_1$ many clopens. Now, consider $\alpha(\infty) = \sup_{i<\omega_1} \alpha(i)$. We pick $a_\infty \in \mathcal{F} \setminus \bigcup_{i<\omega_1} \mathcal{F}_i$. Write $a_\infty = c_\infty \cap F$ where c_∞ is a clopen subset of $\mathscr{A}$. Now, $\pi_{\alpha(\infty)}(c_\infty)$ is a closed G_δ subset of $K_{\alpha(\infty)}$ which is disjoint from $\pi_{\alpha(\infty)}(a)$ for all $a \in \bigcup_{i<\omega_1} \mathcal{F}_i$, because c_∞ is disjoint from every such a, which is determined up to $\alpha(i) < \alpha(\infty)$. The complement of $\pi_{\alpha(\infty)}(c_\infty)$ is a countable union of clopen sets, so we can conclude that there exists a clopen subset b of K which depends up to $\alpha(\infty)$ such that

$$|\{i < \omega_1 : \text{there is } a \in \mathcal{F}_i \text{ such that } a_\infty \subset b \text{ and } a \cap b = \varnothing\}| = \aleph_1$$

The clopen b must in fact depend up to α_k for some $k < \omega_1$. On the one hand,

$$|\{a \in \mathcal{F} : a \subset b\}| = \aleph_2$$

because of property (4) of the families $\mathcal{F}_i$ since $a_\infty \notin \mathcal{F}_{k+1}$. On the other hand,

$$|\{a \in \mathcal{F} : a \cap b = \varnothing\}| = \aleph_2$$

again by property (4) of the families $\mathcal{F}_i$ because there exist $a \in \mathcal{F}_i$ with $a \subset \mathscr{A} \setminus b$ for many $i > k + 1$.   □

Lemma 2.38 *Let K, L, M be compact spaces and let $f : K \longrightarrow M$ be a continuous map. We denote by $j = f^\circ : C(M) \longrightarrow C(K)$ the composition operator induced by f. Let $\imath : C(M) \longrightarrow C(L)$ be a positive operator of norm one and suppose that $S : C(L) \longrightarrow C(K)$ is an operator with $\|S\| = 1$ and $S\imath = j$. Then S is a positive operator.*

Proof Obviously $S \geq 0$ if and only if $S^*\delta_x \geq 0$ for all $x \in K$, where δ_x is the unit mass at x and $S^* : C(K)^* \to C(L)^*$ is the adjoint operator. Fix $x \in K$. By Riesz theorem we have that $S^*\delta_x = \mu$ is a measure of total variation $\|\mu\| \leq 1$. Let $\mu = \mu^+ - \mu^-$ be the Hahn-Jordan decomposition of μ as the difference of two disjointly supported positive measures, so that $\|\mu\| = \|\mu^+\| + \|\mu^-\|$, with $\mu^+, \mu^- \geq 0$. We have that $\delta_{f(x)} = j^*\delta_x = \imath^*S^*\delta_x = \imath^*\mu$, thus

$$\delta_{f(x)} = \imath^*\mu^+ - \imath^*\mu^- \quad \text{and} \quad \|\delta_{f(x)}\| = 1 \geq \|\imath^*\mu^+\| + \|\imath^*\mu^-\|.$$

Since $\imath$ is a positive operator, $\imath^*\mu^+$ and $\imath^*\mu^-$ are positive measures, so all this implies that the above is the Hahn-Jordan decomposition of $\delta_{f(x)}$, and in particular $\imath^*\mu^- = 0$, hence $\mu^- = 0$.   □

Theorem 2.39 (MA, $\mathfrak{c} = \aleph_2$) *The Banach space $C(\mathscr{A})$ is 1-separably injective but not universally 1-separably injective.*

Proof Since $\mathscr{A}$ is an F-space, $C(\mathscr{A})$ is 1-separably injective by Theorem 2.14. We suppose that $C(\mathscr{A})$ is universally 1-separably injective, and we will derive a contradiction. We pick $\{U_n : n \in \mathbb{N}\}$ a sequence of pairwise disjoint clopen subsets of $\mathscr{A}$, and let $U = \bigcup_n U_n$.

Let $c \subset \ell_\infty$ be the Banach space of convergent sequences, and let $t : c \longrightarrow C(\mathscr{A})$ be the operator given by $t(z_1, z_2, \ldots)(x) = z_n$ if $x \in U_n$ and $t(z_1, z_2, \ldots)(x) = \lim z_n$ if $x \notin U$. If $C(\mathscr{A})$ were universally 1-separably injective, we should have an extension $T : \ell_\infty \longrightarrow C(\mathscr{A})$ of t with $\|T\| = 1$. We shall derive a contradiction from the existence of such operator.

The first observation is that T must be a positive operator because we are in a position to apply Lemma 2.38. It might not be obvious at first glance how we apply the lemma. Let $\alpha\mathbb{N} = \mathbb{N} \cup \{\infty\}$ be the one-point compactification of the natural numbers. The space c of convergent sequences is naturally identified with $C(\alpha\mathbb{N})$ and ℓ_∞ with $C(\beta\mathbb{N})$. The operator $t : c \longrightarrow C(\mathscr{A})$ is thus identified with $f^\circ : C(\alpha\mathbb{N}) \longrightarrow C(\mathscr{A})$ where $f : \mathscr{A} \longrightarrow \alpha\mathbb{N}$ is given by $f(x) = n$ if $x \in U_n$ and $f(x) = \infty$ if $x \notin U$. After this translation, it is clear that we can apply Lemma 2.38, and thus T is positive.

For every $A \subset \mathbb{N}$ we will denote $[A] = \overline{A}^{\beta\mathbb{N}} \setminus \mathbb{N}$. The clopen subsets of $\mathbb{N}^*$ are exactly the sets of the form $[A]$, and we have that $[A] = [B]$ if and only if $(A \setminus B) \cup (B \setminus A)$ is finite.

Let $\mathcal{F}$ be an $\aleph_2$-Lusin family in $\mathbb{N}^*$, which exists by Lemma 2.36. For $F \in \mathcal{F}$ and $0 < \varepsilon < \frac{1}{2}$, let

$$F_\varepsilon = \{x \in \mathscr{A} \setminus U : T(1_A)(x) > 1 - \varepsilon\},$$

where $F = [A]$. This F_ε depends only on F and not on the choice of A because if $[A] = [B]$, then $1_A - 1_B \in c_0$, hence $T(1_A - 1_B) = t(1_A - 1_B)$ vanishes out of U, so $T(1_A)|_{\mathscr{A} \setminus U} = T(1_B)|_{\mathscr{A} \setminus U}$.

CLAIM 1 If $\delta < \varepsilon$ and $F \in \mathcal{F}$ then $\overline{F_\delta} \subset F_\varepsilon$.

CLAIM 2 $F_\varepsilon \cap G_\varepsilon = \varnothing$ for every $F \neq G$.

Proof of Claim 2 Since $F \cap G = \varnothing$ we can choose $A, B \subset \mathbb{N}$ such that $F = [A]$, $G = [B]$ and $A \cap B = \varnothing$. If $x \in F_\varepsilon \cap G_\varepsilon$, $T(1_A + 1_B)(x) > 2 - 2\varepsilon > 1$ which is a contradiction because $1_A + 1_B = 1_{A \cup B}$ and $\|T(1_{A \cup B})\| \leq \|T\|\|1_{A \cup B}\| = 1$. END OF THE PROOF OF CLAIM 2.

For every $F \in \mathcal{F}$, let $\tilde{F}$ be a clopen subset of $\mathscr{A} \setminus U$ such that $\overline{F_{0.2}} \subset \tilde{F} \subset F_{0.3}$. By the preceding claims, this is a disjoint family of clopen sets. As we mentioned above, the key property of $\mathscr{A}$ is that $\mathscr{A} \setminus U$ does not contain any $\aleph_2$-Lusin family.

Therefore we can find $\mathcal{G}, \mathcal{H} \subset \mathcal{F}$ with $|\mathcal{G}| = |\mathcal{H}| = \aleph_2$ such that

$$\overline{\bigcup\{\tilde{G} : G \in \mathcal{G}\}} \cap \overline{\bigcup\{\tilde{H} : H \in \mathcal{H}\}} = \varnothing.$$

Now, for every $n \in \mathbb{N}$ choose a point $p_n \in U_n$. Let $g : \beta\mathbb{N} \longrightarrow \mathscr{A}$ be a continuous function such that $g(n) = p_n$.

CLAIM 3   For $u \in \beta\mathbb{N}$ and $A \subset \mathbb{N}$ one has $T(1_A)(g(u)) = \begin{cases} 1, & \text{if } u \in \overline{A}^{\beta\mathbb{N}}; \\ 0, & \text{if } u \notin \overline{A}^{\beta\mathbb{N}}. \end{cases}$

Proof of Claim 3 It is enough to check it for $u \in \mathbb{N}$. This is a consequence of the fact that T is positive, because if $m \in A, n \notin A$, then $0 \le t(1_m) \le T(1_A) \le t(1_{\mathbb{N}\setminus\{n\}}) \le 1$. END OF THE PROOF OF CLAIM 3.

The function g is one-to-one because

$$\overline{\{p_n : n \in A\}} \cap \overline{\{p_n : n \notin A\}} = \varnothing$$

for every $A \subset \mathbb{N}$, as the function $T(1_A)$ separates these sets. On the other hand, as a consequence of Claim 3 above, for every $F \in \mathcal{F}$ and every ε, $g^{-1}(F_\varepsilon) \cap \mathbb{N}^* = F$, and also $g^{-1}(\tilde{F}) \cap \mathbb{N}^* = F$. But then, for the families $\mathcal{H}$ and $\mathcal{G}$ that we found before, we have

$$\overline{\bigcup \mathcal{G}} \cap \overline{\bigcup \mathcal{H}} \subset g^{-1}\left(\overline{\bigcup\{\tilde{G} : G \in \mathcal{G}\}} \cap \overline{\bigcup\{\tilde{H} : H \in \mathcal{H}\}}\right) \cap \mathbb{N}^* = \varnothing.$$

And this contradicts that $\mathcal{F}$ is an $\aleph_2$-Lusin family in $\mathbb{N}^*$.   □

We do not know whether the space $C(\mathscr{A})$ is universally separably injective, or whether it contains copies of ℓ_∞.

2.5   Injectivity Properties of $C(\mathbb{N}^*)$

In this section we take a closer look at $C(\mathbb{N}^*)$. As usual, $\mathbb{N}^* = \beta\mathbb{N}\setminus\mathbb{N}$ is the growth of the integers in its Stone-Čech compactification. Since ℓ_∞ can be identified with $C(\beta\mathbb{N})$ we also have $C(\mathbb{N}^*) = \ell_\infty/c_0$ in the obvious way and therefore the exact sequence

$$0 \longrightarrow c_0 \longrightarrow \ell_\infty \stackrel{\pi}{\longrightarrow} \ell_\infty/c_0 \longrightarrow 0 \tag{2.8}$$

can be thought as

$$0 \longrightarrow c_0(\mathbb{N}) \longrightarrow C(\beta\mathbb{N}) \stackrel{r}{\longrightarrow} C(\mathbb{N}^*) \longrightarrow 0 \tag{2.9}$$

where r is plain restriction. Most of the injectivity properties of $C(\mathbb{N}^*)$ can be deduced from these representations. Indeed, in view of (2.8) and Proposition 2.11(3), the injectivity of ℓ_∞ and Sobczyk theorem already imply that $C(\mathbb{N}^*)$ is universally separably injective, hence ℓ_∞-upper-saturated.

What about the constants? That $C(\mathbb{N}^*)$ is 1-separably injective follows from the fact that being $\mathbb{N}^*$ a closed subset of the F-space $\beta\mathbb{N}$, it is itself an F-space. See Theorem 2.14, especially the equivalences between (1), (4) and (5). Also, it is clear from (2.9) that $C(\mathbb{N}^*)$ is universally 1-separably injective, according to Borsuk-Dugundji. All these properties of $C(\mathbb{N}^*)$ are implied by the conclusion of the following result we shall prove from scratch.

Theorem 2.40 *Every separable subspace of ℓ_∞/c_0 is contained in a subalgebra of ℓ_∞/c_0 isometrically isomorphic to ℓ_∞. That algebra can be lifted through the quotient homomorphism $\pi : \ell_\infty \longrightarrow \ell_\infty/c_0$ by means of an isometric homomorphism.*

Proof The assertion is a consequence of an analogue result in the category of Boolean algebras: every countable Boolean subalgebra of $\mathbb{P}(\mathbb{N})/\mathrm{fin}$ is contained in a subalgebra isomorphic to $\mathbb{P}(\mathbb{N})$. Those readers that are familiar with the relations of a Boolean algebra with its Stone compact and the corresponding space of continuous functions will have little difficulties in deriving the functional analytic result from the Boolean algebraic one. Anyway, we present a detailed account of the proof.

Let $\mathcal{P}$ be a partition of $\mathbb{N}$ into infinite sets. Associated with $\mathcal{P}$ we define:

$$Y_\mathcal{P} = \{x \in \ell_\infty : x \text{ is constant on every } P \in \mathcal{P}\}$$

$$\mathfrak{Y}_\mathcal{P} = \pi(Y_\mathcal{P}) \subset \ell_\infty/c_0$$

Notice that $Y_\mathcal{P}$ is a subspace of ℓ_∞ (actually a subalgebra: it is closed under products and contains constant functions) isometric to ℓ_∞, and that $\pi : Y_\mathcal{P} \longrightarrow \mathfrak{Y}_\mathcal{P}$ is an isometry and hence $\mathfrak{Y}_\mathcal{P}$ is a subalgebra of ℓ_∞/c_0 isometric to ℓ_∞. We will show that for every separable subspace $S \subset \ell_\infty/c_0$ there exists a partition $\mathcal{P}$ of $\mathbb{N}$ into infinite sets such that $S \subset \mathfrak{Y}_\mathcal{P}$.

Let $\mathbb{P}(\mathbb{N})$ be the Boolean algebra of all subsets of $\mathbb{N}$. Each $A \in \mathbb{P}(\mathbb{N})$ can be identified with its characteristic function $1_A \in \ell_\infty$. Let $\mathbb{P}(\mathbb{N})/\mathrm{fin}$ be the quotient Boolean algebra obtained from $\mathbb{P}(\mathbb{N})$ by the equivalence relation $A \sim B$ if $(A \setminus B) \cup (B \setminus A)$ is finite. The operations of union, intersection and complement and the inclusion relation are defined in $\mathbb{P}(\mathbb{N})/\mathrm{fin}$ as inherited from $\mathbb{P}(\mathbb{N})$ modulo finite sets. Let $\pi' : \mathbb{P}(\mathbb{N}) \longrightarrow \mathbb{P}(\mathbb{N})/\mathrm{fin}$ be the canonical projection. Elements of $\mathbb{P}(\mathbb{N})/\mathrm{fin}$ can be viewed as elements of ℓ_∞/c_0 by identifying each $a = \pi'(A)$ with $x_a = \pi(1_A)$. For a subset $\mathcal{A} \subset \mathbb{P}(\mathbb{N})/\mathrm{fin}$, we define $X_\mathcal{A}$ to be the Banach subalgebra of ℓ_∞/c_0 generated by $\{x_a : a \in \mathcal{A}\}$.

CLAIM 1   For every separable subspace $S \subset \ell_\infty/c_0$ there exists a countable Boolean subalgebra $\mathcal{A} \subset \mathbb{P}(\mathbb{N})/\mathrm{fin}$ such that $S \subset X_\mathcal{A} \subset \ell_\infty/c_0$.

Proof of the claim It is clear that the algebra generated by $\{1_A : A \in \mathbb{P}(\mathbb{N})\}$ is the whole space ℓ_∞. Thus, each vector of ℓ_∞ is in the algebra generated by a countable subset of $\{1_A : A \in \mathbb{P}(\mathbb{N})\}$. Hence, every separable subspace of ℓ_∞ is contained in the algebra generated by a countable subset of $\{1_A : A \in \mathbb{P}(\mathbb{N})\}$. It follows that every separable subspace of ℓ_∞/c_0 is contained in the subalgebra generated by $\{x_a : a \in \mathscr{A}\}$ with $\mathscr{A}$ a countable set. But the Boolean algebra generated by a countable set is countable, so we can assume that $\mathcal{A} = \mathscr{A}$ is a Boolean subalgebra, and the claim follows.

We will also make use of the following standard fact about the Boolean algebra $\mathbb{P}(\mathbb{N})/$ fin:

CLAIM 2 Let $\mathcal{U}$ be a countable set of nonzero elements of $\mathbb{P}(\mathbb{N})/$ fin that is closed under finite intersections. Then there exists a nonzero b in $\mathbb{P}(\mathbb{N})/$ fin such that $b \subset a$ for all $a \in \mathcal{U}$.

Proof of the claim Let us enumerate $\mathcal{U} = \{u_1, u_2, \ldots\}$ and set $v_n = \bigcap_{i \leq n} u_i$. We have $v_1 \supset v_2 \supset \cdots$ and all v_n's are nonzero. Choose sets $V_n \subset \mathbb{N}$ with $v_n = \pi'(V_n)$. All V_n's are infinite and $V_n \setminus V_m$ is finite whenever $n < m$. Inductively, construct a sequence of natural numbers $k_1 < k_2 < \cdots$ such that $k_n \in \bigcap_{i \leq n} V_i$. Set $A = \{k_1, k_2, \ldots\}$ and $a = \pi'(A)$. This is the desired element a. It is nonzero since A is infinite. And $a \subset u$ for $u \in \mathcal{U}$ because $V_n \setminus A$ is finite for all n. This proves the claim.

By Claim 1 it is enough to prove that for every countable subalgebra $\mathcal{A} \subset \mathbb{P}(\mathbb{N})/$ fin there exists a partition $\mathcal{P}$ such that $X_A \subset \mathfrak{Y}_\mathcal{P}$. So we fix such a subalgebra. We take $\{\mathcal{U}_n : n \in \mathbb{N}\}$ a sequence of ultrafilters of the Boolean algebra $\mathcal{A}$ such that for every nonzero $a \in \mathcal{A}$ there exists n with $a \in \mathcal{U}_n$. In other words, $\{\mathcal{U}_n\}$ is a dense sequence in the Stone space of $\mathcal{A}$. For every n, either

1. $\mathcal{U}_n$ is principal, in which case we define $a_n = \min \mathcal{U}_n$, or
2. $\mathcal{U}_n$ is nonprincipal. In this case, by the previous Claim 2 we can pick a nonzero $a_n \in \mathbb{P}(\mathbb{N})/$ fin such that $a_n \subset a$ for all $a \in \mathcal{U}_n$.

The elements a_n defined above are pairwise disjoint in $\mathbb{P}(\mathbb{N})/$ fin: If $n \neq m$, then $\mathcal{U}_n \neq \mathcal{U}_m$, there exists $b \in \mathcal{U}_n$, with $\pi'(\mathbb{N}) \setminus b \in \mathcal{U}_m$, so $a_n \cap a_m \subset b \cap \pi'(\mathbb{N}) \setminus b = \varnothing$. The partition $\mathcal{P} = \{P_n\}$ we are looking for will be such that $a_n = \pi'(P_n)$ for all n. It remains to carefully choose each P_n in the equivalence class a_n, so that they constitute a partition and $X_A \subset \mathfrak{Y}_\mathcal{P}$. Since in any case, $\mathfrak{Y}_\mathcal{P}$ will be a Banach subalgebra, we just have to take care that $x_a \in \mathfrak{Y}_\mathcal{P}$ for all $a \in \mathcal{A}$. We will actually show that

$$ a = \pi'\left(\bigcup_{a \in \mathcal{U}_n} P_n \right). $$

Since every finitely generated Boolean algebra is finite, we can write $\mathcal{A}$ as an increasing union of finite subalgebras $\mathcal{A} = \bigcup_{m=1}^\infty \mathcal{A}_m$. For fixed m, it is easy to choose a partition $\mathcal{P}^m = \{P_n^m\}_{n \in \mathbb{N}}$ of $\mathbb{N}$ with $\pi'(P_n^m) = a_n$ and $a = \pi'\left(\bigcup_{a \in \mathcal{U}_n} P_n^m \right)$

for all $a \in \mathcal{A}_m$: one only has to take care of each of the finitely many atoms (minimal nonzero elements) of $\mathcal{A}_m$. Moreover, if we choose the partitions $\mathcal{P}^m$ inductively one after another, it is possible to do it in such a way that the following conditions hold:

1. $P_k^m = P_k^{m-1}$ for all $k \leq m$,
2. $P_k^m \cap \{0, \ldots, m\} = P_k^{m-1} \cap \{0, \ldots, m\}$ for all k.
3. $\bigcup\{P_k^m : a \in U_k\} = \bigcup\{P_k^{m-1} : a \in U_k\}$ for all $a \in \mathcal{A}_{m-1}$.

The sets $\{P_n = P_n^n : n \in \mathbb{N}\}$ constitute the partition $\mathcal{P}$ that we are looking for. $\square$

Corollary 2.41 ℓ_∞/c_0 is universally 1-separably injective.

Proof It follows from Theorem 2.40, taking into account that ℓ_∞ is 1-injective. $\square$

The subtleties in the proof of Theorem 2.40 are necessary only to construct the "enveloping" subspace isometric to ℓ_∞ in the right position since the lifting of a separable subspace of ℓ_∞/c_0 to ℓ_∞ is nearly trivial. The following result is, formally, a Corollary of Theorem 2.40:

Proposition 2.42 *If S is a separable subspace of ℓ_∞ containing c_0 then there is a contractive projection p on S whose kernel is c_0. When, additionally, S is a subalgebra of ℓ_∞ then p is a unital homomorphism. In any case, $1_S - p$ is a projection onto c_0 of norm at most 2.*

Proof Let us show that if S is a separable subalgebra of ℓ_∞/c_0 then there is a continuous homomorphism $\varphi : S \to \ell_\infty$ such that $\pi \circ \varphi = 1_S$, where $\pi : \ell_\infty \to \ell_\infty/c_0$ is the natural quotient map. From here, the proposition follows.

It is clear that for every $f \in \ell_\infty$ and $\varepsilon > 0$ there is a partition $\mathbb{N} = A_1 \cup \cdots \cup A_k$ and numbers t_i such that

$$\left\| f - \sum_{i=1}^{k} t_i 1_{A_i} \right\| \leq \varepsilon.$$

It follows that S is contained in the closure of the union of a (increasing) sequence of algebras of the form $S_n = \pi(R_n)$, where R_n is the algebra associated to a certain (finite) partition of $\mathbb{N}$. From the viewpoint of ℓ_∞/c_0 we see that each S_n has a basis of idempotents whose sum is 1. Adding some "intermediate" subalgebras if necessary we may an do assume $\dim S_n = n$. Let us construct the required homomorphism

$$\varphi : \bigcup_{n=1}^{\infty} S_n \longrightarrow \ell_\infty$$

by showing that every lifting $\varphi : S_n \to \ell_\infty$ (in the category of unital algebras) extends to a lifting of S_{n+1}. Write $S_n = \text{span}\{u_1, \ldots, u_n\}$ where u_k are idempotents such that $u_1 + \cdots + u_n = 1$. Clearly, $\varphi(u_k) = 1_{A_k}$, where $\mathbb{N} = A_1 \cup \cdots \cup A_n$ (this is the "induction" hypothesis). We may assume $S_{n+1} = \text{span}\{u_1, \ldots, u_{n-1}, v, w\}$

where v and w are idempotents such that $v + w = u_n$. Since v is idempotent there is $V \subset \mathbb{N}$ such that $v = \pi 1_V$. We extend φ to S_{n+1} taking $\varphi(v) = 1_{V \cap A_n}$ (which forces $\varphi(w) = 1_{A_n \setminus V}$). The definition is correct since $V \setminus A_n$ is at most finite (otherwise the decomposition $\pi 1_{A_n} = \pi 1_V + w$ with $w \geq 0$ is impossible). □

We have already shown (several times) that ℓ_∞ / c_0 is not injective. The simplest argument was to observe that the (images of the) characteristic functions of the elements of an almost disjoint family $\mathcal{M}$ of infinite subsets of $\mathbb{N}$ having size $\mathfrak{c}$ generate a subspace isometric to $c_0(\mathfrak{c})$; that ℓ_∞ / c_0 has density character $\mathfrak{c}$ and that therefore it cannot contain any copy of $\ell_\infty(\mathfrak{c})$, which has density character $2^{\mathfrak{c}}$. The above argument is quite rough in a sense: it says that ℓ_∞ / c_0 is uncomplemented in its bidual, a huge superspace. Not being injective, ℓ_∞ / c_0 cannot be complemented in its bidual and therefore it cannot be complemented in any dual space (see [196]). In any case, Amir had shown in [5] that $C(\mathbb{N}^*)$ is not complemented in $\ell_\infty(\mathbb{P}(\mathbb{N}^*)) \approx \ell_\infty(2^{\mathfrak{c}})$, which provides another proof that ℓ_∞ / c_0 is not injective. Amir's proof can be refined in order to get $C(\mathbb{N}^*)$ uncomplemented in a much smaller space. We are indebted to Anatolij Plichko for calling our attention to Amir's paper.

Proposition 2.43 *There exists a Banach space of density character $\mathfrak{c}$ that contains an uncomplemented copy of $C(\mathbb{N}^*)$.*

Proof Following Amir's paper [5], let Σ be a family of subsets of $\mathbb{N}^*$ that contains a basis of open sets of the topology of $\mathbb{N}^*$, and which is closed under complementation, finite union and the closure operation. We can consider the Banach space $B(\Sigma)$, sitting as $C(\mathbb{N}^*) \subset B(\Sigma) \subset \ell_\infty(\mathbb{N}^*)$ defined as the subspace of $\ell_\infty(\mathbb{N}^*)$ generated by the characteristic functions of the elements of Σ. Let also D_Σ be the union of the boundaries of all open sets living in Σ. By [5, Corollary 1], if $C(\mathbb{N}^*)$ is complemented in $B(\Sigma)$, then D_Σ is nowhere dense in $\mathbb{N}^*$. We indicate now how to construct such a family Σ of cardinality $\mathfrak{c}$ and with D_Σ dense in $\mathbb{N}^*$, so that the space $X = B(\Sigma)$ is as stated in the Proposition. For every clopen subset A of $\mathbb{N}^*$, choose $U_A \subset A$ to be an open not closed set. Consider then Σ the least family of subsets of $\mathbb{N}^*$ that contains all clopens A and all open sets U_A and that is closed under complementation, finite union and the closure operation. □

A different proof of Proposition 2.43 can be found in [18]. We do not know whether the space X in the preceding result can be obtained so that $\mathrm{dens}(X/C(\mathbb{N}^*)) = \aleph_1$. By Parovičenko's theorem [40], [245, p. 81], $\mathbb{N}^*$ can be mapped onto any compact space having weight at most $\aleph_1$. Consequently:

Lemma 2.44 *Every Banach space of density character $\aleph_1$ or less is isometric to a subspace of $C(\mathbb{N}^*)$.*

Proof Let X denote a Banach space with $\mathrm{dens}\, X \leq \aleph_1$. Its dual unit ball B_{X^*} in the weak* topology has weight at most $\aleph_1$. Let $\varphi : \mathbb{N}^* \to B_{X^*}$ be the surjective mapping given by Parovičenko theorem. The operator $\varphi^\circ : C(B_{X^*}) \longrightarrow C(\mathbb{N}^*)$ given by

$$(\varphi^\circ f)(u) = f(\varphi(u))$$

is an into isometry. The space X is isometric to a subspace of $C(B_{X^*})$ and this concludes the proof.   □

The following immediate application can be found in [67, Proposition 5.3]:

Corollary 2.45 (CH) $C(\mathbb{N}^*)$ *contains an uncomplemented subspace isometric to* $C(\mathbb{N}^*)$.

Proof By Lemma 2.44, the space in Proposition 2.43 is a subspace of $C(\mathbb{N}^*)$.   □

The argument of Lemma 2.44, together with some interesting applications to the existence of nontrivial twisted sums, can be found in [251]. Lemma 2.44 is actually related to the topic of universal disposition discussed in Chap. 3. Although $C(\mathbb{N}^*)$ cannot be of almost universal disposition (since no C-space can be of almost universal disposition—see the discussion before Theorem 3.34), the compact space $\mathbb{N}^*$ is of "universal co-disposition in the category of compact spaces"— see Definition 5.23 and Corollary 5.24; and the Boolean algebra $\mathbb{P}(\mathbb{N})/\text{fin}$ is of "universal disposition in the category of Boolean algebras". All this can be understood as the real content of Parovičenko theorem.

2.6 Automorphisms of Separably Injective Spaces

Lindenstrauss and Rosenthal proved in [180] that every isomorphism between two infinite codimensional subspaces of c_0 can be extended to an automorphism of c_0.

Definition 2.46 A Banach space is said to be *automorphic* if every isomorphism between two subspaces whose corresponding quotients have the same density character can be extended to an automorphism of the whole space.

Observe that the extension trivially exists when the subspaces are finite dimensional or, in the hypothesis above, finite codimensional. It is clear that Hilbert spaces are automorphic, and in [199] it was proved that also $c_0(\Gamma)$ is automorphic. Lindenstrauss and Rosenthal formulated what has been called *the automorphic space problem*: Does there exist an automorphic space different from $c_0(\Gamma)$ and $\ell_2(\Gamma)$? Different approaches and partial positive answers to the automorphic space problem have been considered and obtained in [17, 19, 63, 67, 199]. There emerged the notion of partially automorphic space, of which we isolate now the following:

Definition 2.47 Let X, Y be Banach spaces.

- We say that X is Y-automorphic if every isomorphism $\tau : A \to B$ between two subspaces of X isomorphic to Y with $\text{dens}(X/A) = \text{dens}(X/B)$ can be extended to an automorphism of X.

- A Banach space X is said to be separably automorphic if it is Y-automorphic for every separable Y.

Observe that X is Y-automorphic if and only if given two embeddings $i, j : Y \to X$ with $\mathrm{dens}(X/i[Y]) = \mathrm{dens}(X/j[Y])$ there is an automorphism τ of X such that $j = \tau \circ i$.

Lindenstrauss and Rosenthal also prove in [180] that ℓ_∞ is separably automorphic (see also [182, Theorem 2.f.12]). The proof can be easily adapted to the general case to obtain that, for every set Γ, the space $\ell_\infty(\Gamma)$ is separably automorphic. We shall see that indeed every universally separably injective space is separably automorphic.

Lemma 2.48 *Let Y be a Banach space isomorphic to its square. Assume that every copy of Y is complemented in X. Then X is Y-automorphic if and only if every complement of Y with the same density character as X contains Y.*

Proof As every copy of Y is complemented in X it is clear that X is Y-automorphic if and only if the complements of copies of Y in X with the same density character are all isomorphic. In fact, using that Y is isomorphic to its square, all the complements with density character equal to $\mathrm{dens}\, X$ must be isomorphic to X. Now, the "if" part is as follows. Let Y_1 be a subspace of X isomorphic to Y. We have $X = Y_1 \oplus Z$ and $Z = Y_2 \oplus A$, with $Y_2 \sim Y$. Hence $X = Y_1 \oplus Y_2 \oplus A \sim Y_2 \oplus A = Z$. The converse is also easy: if $X \sim Y \oplus Z$, with $\mathrm{dens}\, Z = \mathrm{dens}\, X$, then $X \sim Y \oplus Y \oplus Z$ and if X is Y-automorphic, then $Z \sim Y \oplus Z$. □

Corollary 2.49 *Universally separably injective spaces are ℓ_∞-automorphic.*

Proof We apply Lemma 2.48 for $Y = \ell_\infty$. Observe that every copy of ℓ_∞ is complemented as ℓ_∞ is an injective space, and that every complemented subspace of a universally separably injective space is universally separably injective, so it contains ℓ_∞ by Theorem 2.26. □

Now we want to jump from "X is Y-automorphic" to "X is H-automorphic for every subspace H of Y". The obvious result is:

Lemma 2.50 *Let X be Y-automorphic and let $H_1 \subset Y_1 \subset X$ and $H_2 \subset Y_2 \subset X$ be spaces where H_1, H_2 and Y_1, Y_2 are isomorphic to H and Y, respectively. If there is an automorphism of Y transforming H_1 into H_2 then there is an automorphism of X transforming H_1 into H_2 every time $\mathrm{dens}(X/Y_1) = \mathrm{dens}(X/Y_2)$.*

There is an alternative approach to obtain the partially automorphic character of a space: to combine the Y-upper-saturaturation and the fact that every copy of Y is complemented instead of relying on the Y-automorphic character of the space:

Lemma 2.51 *Let E, Y and X be Banach spaces. Suppose that Y is E-automorphic, and that every two copies E_1, E_2 of E inside X are contained in a single complemented copy Y_0 of Y inside X such that $\mathrm{dens}(Y_0/E_1) = \mathrm{dens}(Y_0/E_2)$. Then X is E-automorphic.*

Proof Let $i, j : E \to X$ be two embeddings of a space E into X, with $\mathrm{dens}(X/i[E]) = \mathrm{dens}(X/j[E])$. Obviously, i and j factorize through the inclusion $\omega : \overline{i[E] + j[E]} \longrightarrow X$. By hypothesis, there is a complemented subspace Y_0, which is isomorphic to Y and contains both $i[E]$ and $j[E]$. If τ_0 is an automorphism of Y_0 such that $j = \tau_0 \circ i$ and A is a complement of Y_0 in X, then the automorphism of X is $\tau = \tau_0 \oplus \mathbf{1}_A$. $\quad\square$

Thus, using Theorem 2.26, and the result of Lindenstrauss and Rosenthal asserting that ℓ_∞ is separably automorphic, we can apply Lemma 2.51 for $Y = \ell_\infty$ and E any separable space, to obtain:

Proposition 2.52 *Universally separably injective spaces are separably automorphic.*

Corollary 2.53 *The space ℓ_∞/c_0 is separably automorphic.*

The separably automorphic character of $C(\mathbb{N}^*)$ seems to be connected with the fact that the underlying Boolean algebra has analogous properties; namely, it is "countably automorphic" (every isomorphism between countable Boolean algebras is extended to an automorphism of $\mathbb{P}(\mathbb{N})/\mathrm{fin}$) and every countable Boolean algebra is contained in a copy of $\mathbb{P}(\mathbb{N})$, cf. [77]. The proof of this fact is just the last step (after Claim 2) of the proof of Theorem 2.40. If we are given $\mathcal{A}$ and $\mathcal{A}'$ two isomorphic countable subalgebras of $\mathbb{P}(\mathbb{N})/\mathrm{fin}$, that proof provides two copies of $\mathbb{P}(\mathbb{N})$ of the form $\mathfrak{Y}_\mathcal{P}$ and $\mathfrak{Y}_{\mathcal{P}'}$, and the isomorphism between $\mathcal{A}$ and $\mathcal{A}'$ induces naturally a bijection between the partitions $\mathcal{P}$ and $\mathcal{P}'$, that gives an extended isomorphism between $\mathfrak{Y}_\mathcal{P}$ and $\mathfrak{Y}_{\mathcal{P}'}$. It is not however so clear how to pass from "Boolean-automorphic" to "Banach-automorphic".

Other quotients of ℓ_∞ also have a partially automorphic character. We require a lemma.

Lemma 2.54 *Assume that for $k = 1, 2$ one has pull-back diagrams*

$$
\begin{array}{ccccccccc}
0 & \longrightarrow & A & \overset{i}{\longrightarrow} & X & \overset{q}{\longrightarrow} & X/A & \longrightarrow & 0 \\
 & & \| & & \uparrow{\scriptstyle \Delta_k} & & \uparrow{\scriptstyle \delta_k} & & \\
0 & \longrightarrow & A & \longrightarrow & \mathrm{PB}_k & \underset{q_k}{\longrightarrow} & Y & \longrightarrow & 0
\end{array}
$$

where δ_k are isomorphic embeddings. If there exists isomorphisms $\Theta : X \to X$ and $\theta : \mathrm{PB}_1 \to \mathrm{PB}_2$ such that $\Theta \Delta_1 = \Delta_2 \theta$ and $q_2 \theta = q_1$ then there is an automorphism $\tau : X/A \to X/A$ such that $\tau \delta_1 = \delta_2$.

Proof Observe first that $\theta(A) \subset A$ since $q_2 \theta(a) = q_1(a) = 0$; and then that $\theta(A) = A$ since if $p \in \mathrm{PB}_1$ is such that $\theta(p) \in A$ then $0 = q_2 \theta(p) = q_1(p)$, so $p \in A$. Since Θ is an automorphism of X that extends θ, then also $\Theta(A) = A$. One can then define an automorphism of X/A by $\tau(x + A) = \Theta(x) + A$. It verifies

$$\tau \delta_1 q_1(p) = \Theta(p) + A = q \Delta_2 \theta(p) + A = \delta_2 q_2 \theta(p) = \delta_2 q_1(p)$$

and thus $\tau \delta_1 = \delta_2$. $\quad\square$

The lemma says, in particular, that if the pull-back sequences are isomorphically equivalent and X is PB_k-automorphic then X/A is Y-automorphic. So, it provides relevant information about the automorphic character of quotient spaces. When applied to quotients of ℓ_∞ one gets:

Proposition 2.55

1. If E is a separably injective subspace of ℓ_∞, then ℓ_∞/E is separably automorphic.
2. Let A be any subspace of ℓ_∞ complemented in its bidual and so that ℓ_∞/A is not reflexive. Then ℓ_∞/A is automorphic for all $\mathscr{L}_1$-spaces.
3. For every subspace H of c_0 the space ℓ_∞/H is automorphic for all separable $\mathscr{L}_1$-spaces.

Proof Part (1) follows from a general fact: ℓ_∞/E is universally separably injective, hence separably automorphic. An independent proof is however as follows: Let $\delta_k : Y \to \ell_\infty/E$ be two embeddings of a separable space Y into ℓ_∞/E. Since $\mathrm{Ext}(Y, E) = 0$ then $PB_k \sim E \oplus Y$. Since ℓ_∞/PB_k is isomorphic to $(\ell_\infty/E)/\delta_k[Y]$ one gets that ℓ_∞/PB_k contains ℓ_∞ and this implies that ℓ_∞ is PB_k-automorphic, because Lindenstrauss and Rosenthal [180] proved in fact that ℓ_∞ is Z-automorphic whenever ℓ_∞/Z is not reflexive. So, Lemma 2.54 applies, which proves (1). Assertions (2) and (3) follow the same schema: (2) using Lindenstrauss' lifting (i.e., $\mathrm{Ext}(\mathscr{L}_1, A) = 0$ for every Banach space A complemented in its bidual; see Proposition A.18) and (3) using the identity $\mathrm{Ext}(\mathscr{L}_1, H) = 0$ obtained in [65] (see also [62]).   □

Separably injective spaces are not necessarily separably automorphic as the example of $c_0 \oplus \ell_\infty$ shows: no automorphism can send a complemented copy of c_0 such as $c_0 \oplus 0$ onto an uncomplemented copy such as $0 \oplus c_0$. And automorphic spaces, such as ℓ_2, are not necessarily separably injective. One however has:

Proposition 2.56

1. Every separably automorphic space containing ℓ_1 is separably injective.
2. Every separably automorphic space containing ℓ_∞ is universally separably injective.

Proof Let $i : \ell_1 \to X$ be an into isomorphism. By Proposition 2.5 it is enough to prove that for every closed subspace K of ℓ_1, every operator $K \to X$ extends to ℓ_1. Assume otherwise; let $\delta : K \to \ell_1$ be an into isomorphism and let $t : K \to X$ be an operator that cannot be extended through δ, therefore neither it can be extended to X through $i\delta$. Then, for some $\varepsilon > 0$, the operator $i\delta + \varepsilon t$ is an into isomorphism. If X is separably automorphic, we could find an isomorphism $F : X \longrightarrow X$ such that $Fi\delta = i\delta + \varepsilon t$. But then $\varepsilon^{-1}(T - \mathbf{1}_X)i$ would be an extension of t through δ, and this contradicts our hypothesis. The proof of the second assertion is simpler: every separable subspace of X must be contained in a copy of ℓ_∞ and thus the space is universally separably injective.   □

Many other Banach spaces have been shown to have a partially automorphic character; for instance, Lindenstrauss and Pełczyński prove in [178] that $C[0, 1]$ is H-automorphic for all subspaces H of c_0 while Kalton shows in [155] that $C[0, 1]$ is also ℓ_1-automorphic and in [156] that it is not ℓ_2-automorphic. Another example of separably injective spaces that are also separably automorphic are the $C(K)$ spaces with K Eberlein compact of finite height.

Proposition 2.57

1. If K is an Eberlein compact, then $C(K)$ is c_0-automorphic.
2. Every c_0-upper-saturated WCG-space is separably automorphic.
3. A $C(K)$ space that is c_0-automorphic is also H-automorphic for every subspace H of c_0.
4. Let E be an Eberlein compact. Then $C(E)$ is H-automorphic for every subspace H of c_0.

Proof Assertion (1) follows from Lemma 2.48 for $Y = c_0$. On the one hand, every infinite dimensional complemented subspace of a C-space contains a copy of c_0, cf. (Proposition A.5). On the other hand, if $C(K)$ is WCG, then it has the separable complementation property (every separable subspace is contained in a separable complemented subspace), hence by Sobczyk's theorem, every copy of c_0 is complemented. For assertion (2), we have again that every copy of c_0 is complemented, and since c_0 is automorphic by the Lindenstrauss-Rosenthal theorem, Lemma 2.51 applies. To get (3), assume first that neither $C(K)$ nor H are isomorphic to c_0, otherwise the result is trivial. Now, since c_0 is automorphic, we only need to prove that every subspace H of c_0 contained in $C(K)$ is actually contained in a copy of c_0 contained in $C(K)$. But every separable subspace S of $C(K)$ is contained in a separable subspace $C(T)$ of $C(K)$. This subspace $C(T)$ is H-automorphic [178], and the result follows. Assertion (4) is an immediate consequence of (1) and (2).   □

Assertion (4) is actually a non-separable extension of the main result in [71] asserting that separable Lindenstrauss-Pełczyński spaces are characterized as those which are H-automorphic for all subspaces H of c_0. Concerning the automorphic character of $C(K)$-spaces we obtain from Lemma 6.2 and the proof of Lemma 2.22 that every $C(K)$-space with K a compact of finite height is c_0-upper-saturated. As a consequence, applying Lemma 2.57(2):

Corollary 2.58 *Every $C(K)$-space with K an Eberlein compact of finite height is separably automorphic.*

A generalization of the preceding result was obtained in [17]:

Proposition 2.59 *If K is an Eberlein compact of finite height, the (separably injective) space $C(K)$ is automorphic for all possible subspaces of density character less than $\aleph_\omega$.*

These results are somewhat optimal: there exist separably injective $C(K)$-spaces such as $c_0 \oplus \ell_\infty$ which are not c_0-automorphic; there also exist non-Eberlein compacta of height 3 which are not c_0-automorphic (since they contain complemented and uncomplemented copies of c_0); while Eberlein compacta of infinite height, such as $C[0, 1]$, are not separably automorphic.

See Theorem 5.30 and Sect. 6.4.3 for further information and open problems on partially automorphic spaces.

2.7 Notes and Remarks

2.7.1 Extensions vs. Projections

In this section we take a closer look at the constants implicit in the characterizations given in Proposition 2.5 and we consider the corresponding "quantified" properties and the relationships between the involved constants.

Proposition 2.60

1. If E is λ-separably injective, for every Banach space X and each subspace Y such that X/Y is separable, every operator $t : Y \to E$ admits an extension $T : X \to E$ with $\|T\| \leq 3\lambda\|t\|$.
2. A space E is λ-complemented in every Z such that Z/E is separable if and only if whenever Y is a subspace of X with X/Y separable every operator $t : Y \to E$ admits an extension $T : X \to E$ with $\|T\| \leq \lambda\|t\|$.

Proof

1. We have to follow the trace of λ through the proof of (2) $\Rightarrow$ (3) in Proposition 2.5. With the same notation, consider thus the commutative diagram

$$
\begin{array}{ccccccccc}
0 & \longrightarrow & \ker q & \xrightarrow{\ j\ } & \ell_1 & \xrightarrow{\ q\ } & X/Y & \longrightarrow & 0 \\
& & \phi \downarrow & & \varrho \downarrow & & \| & & \\
0 & \longrightarrow & Y & \longrightarrow & X & \longrightarrow & X/Y & \longrightarrow & 0
\end{array}
$$

Let us construct the true push-out of the couple (ϕ, j) and the corresponding complete diagram

$$
\begin{array}{ccccccccc}
0 & \longrightarrow & \ker q & \xrightarrow{\ j\ } & \ell_1 & \xrightarrow{\ q\ } & X/Y & \longrightarrow & 0 \\
& & \phi \downarrow & & \downarrow \phi' & & \| & & \\
0 & \longrightarrow & Y & \xrightarrow{\ j'\ } & \text{PO} & \longrightarrow & X/Y & \longrightarrow & 0
\end{array}
$$

We can consider without loss of generality that $\|\phi\| = 1$. Let $S : \ell_1 \to E$ be an extension of $t\phi$ with $\|S\| \leq \lambda\|t\phi\| \leq \lambda\|t\|$. By the universal property of the push-out, there exists an operator $L : PO \rightarrow E$ such that $L\phi' = S$ and

$$\|L\| \leq \max\{\|t\|, \|S\|\} \leq \lambda \|t\|.$$

Again by the universal property of the push-out, there is a diagram of equivalent exact sequences

$$
\begin{array}{ccccccccc}
0 & \longrightarrow & Y & \xrightarrow{\ j'\ } & PO & \longrightarrow & X/Y & \longrightarrow & 0 \\
& & \| & & \downarrow{\scriptstyle \gamma} & & \| & & \\
0 & \longrightarrow & Y & \xrightarrow{\ \iota\ } & X & \xrightarrow{\ p\ } & X/Y & \longrightarrow & 0
\end{array}
$$

where the isomorphism γ is defined as $\gamma((y, u) + \Delta) = j(y) + Q(u)$ is such that $\|\gamma\| \leq \max\{\|j\|, \|Q\|\} \leq 1$. Remember that there exists a selection $s : X/Y \rightarrow \ell_1$ for the quotient map $q : \ell_1 \rightarrow X/Y$ which is not necessarily linear but is homogeneous and has $\|s\| = 1$ (that is, $s(\mu u) = \mu s(u)$ and $\|s(u)\| \leq \|u\|$ for all $u \in X/Y$ and scalar μ). The desired extension of t to X is $T = L\gamma^{-1}$, where γ^{-1} comes defined by

$$\gamma^{-1}(x) = (x - Qspx, spx) + \Delta.$$

Notice that γ^{-1} is well defined because $x - Qspx \in Y$ since $p(x - Qspx) = px - pQspx = px - qspx = 0$. Notice also that γ^{-1} is linear because for $x, z \in X$ we can write $\gamma^{-1}(x + z) - \gamma^{-1}(x) - \gamma^{-1}(z)$ as

$$Q(spx + spz - sp(x + z)), sp(x + z) - spx - spz) + \Delta$$

and this is zero because $spx + spz - sp(x + z) \in \ker q$ as s is a selection for q. Finally, one clearly has $\|\gamma^{-1}\| \leq 3$, and therefore $\|T\| \leq 3\lambda$.

2. The complementation of E is achieved simply considering t as the identity on E. The other implication is contained in the proof of the implication (4) $\Rightarrow$ (3) in Proposition 2.5: if $p' : PO \rightarrow E$ is a projection with norm at most λ, since $\|t'\| \leq 1$, the composition $p't' : X \rightarrow E$ yields an extension of t with norm at most λ. $\qquad\square$

We do not know if the bound 3 appearing in Proposition 2.60(1) is sharp. Observe that if every operator $t : Y \rightarrow E$ can be extended preserving the norm to any superspace X such that $\dim X/Y = 1$, then E is 1-injective, as can be seen by transfinite induction. Thus one cannot replace 3 by 1. The following example shows that at least 2 is required. See Example 2.4:

Example 2.61 The projection constant of $\ell_\infty^c(\aleph_1)$ in $\ell_\infty^c(\aleph_1)_+$ is 2.

Proof Each element of $\ell_\infty^c(\aleph_1)_+$ can be written as $\lambda + f$, with $f \in \ell_\infty^c(\aleph_1)$. The map $\lambda + f \mapsto f$ is a projection of norm 2, so the projection constant is at most 2. To

see the reversed inequality, let $p : \ell_\infty^c(\aleph_1)_+ \longrightarrow \ell_\infty^c(\aleph_1)$ be any linear projection
and take $\phi = p(1)$, so that $p(\lambda + f) = \lambda\phi + f$. Take any i such that $\phi(i) = 0$. Then
$\|1 - 21_i\| = 1$ but $\|p(1 - 21_i)\| = 2$ since $p(1 - 21_i)(i) = -2$. $\square$

Moreover, a λ-separably injective space E is not necessarily λ-complemented
in every superspace Z such that Z/E is separable. Let us therefore consider the
following "one dimensional" version of Proposition 2.5: A Banach space E is said
to enjoy property (c_λ) when for every Banach space X and each subspace Y such
that $\dim X/Y = 1$, every operator $t : Y \to E$ extends to an operator $X \to E$ with
norm at most $\lambda\|t\|$. Lemma 2.33 says:

Proposition 2.62 *A Banach space E has property (c_λ) if and only if given a family*
$B(x_\alpha, r_\alpha)$ *of mutually intersecting balls whose centers lie on a separable subspace*
there exists a point p such that $\|x_\alpha - p\| \le \lambda r_\alpha$.

Every λ-separably injective space has property (c_λ), although it is not clear if
there is a function f so that a space with property (c_λ) is $f(\lambda)$-separably injective.
Kalton is able to show in [155, Theorem 5.2] that such is the case of c_0:

Lemma 2.63 *Let Y be a closed subspace of a Banach space X such that X/Y is*
separable. Let $\tau : Y \to c_0$ be a norm one operator. If for every $x \in X$ there is
an extension $\tau_x : Y + [x] \to c_0$ with norm at most λ then there is an extension
$T : X \to c_0$ with norm at most λ.

Of course the result contains extra information only for $\lambda < 2$. As it is clear from
Propositions 2.32 or 2.34, some break occurs at $\lambda = 2$. Moreover, all C-spaces
have property (c_2) since they actually have the following property: for every family
$B(x_\alpha, r_\alpha)$ of mutually intersecting balls $\bigcap_\alpha B(x_\alpha, 2r_\alpha) \ne \varnothing$. Indeed, every Banach
space has the property that $\bigcap_\alpha B(x_\alpha, 2r_\alpha + \varepsilon) \ne \varnothing$ for every family $B(x_\alpha, r_\alpha)$ of
mutually intersecting balls [116, p.198]. To deduce from here that a 2 is enough in C-
spaces, Lindenstrauss [177] reasons as follows: in a $C(K)$ space $\bigcap_\alpha B(f_\alpha, t_\alpha) \ne \varnothing$
if and only if for every $k_0 \in K$

$$\limsup_{k \to k_0} \sup_\alpha (f_\alpha(k) - t_\alpha) \le \liminf_{k \to k_0} \inf_\alpha (f_\alpha(k) + t_\alpha).$$

From here it is clear that if the inequality holds for all $t_\alpha + \varepsilon$ then it also holds for
t_α. Observe that Proposition 2.32 actually shows that a space with property (c_λ) for
$\lambda < 2$ and containing almost isometric copies of c_0 has density character $\mathfrak{c}$.

2.7.2 Complex Separably Injective Spaces

Although these notes deal with real Banach spaces we will make a few remarks
on injective-like complex Banach spaces. First of all one can consider (universally)
separably injective complex spaces just assuming that the underlying field in the
definitions is $\mathbb{C}$. Then Sects. 2.1, 2.3 and 2.4 apply verbatim to complex spaces, with

the sole exception that the characterization of 1-separable injectivity by intersection properties of balls has to be reformulated. The new property required here is the *weak intersection property*, introduced by Hustad [137] as follows: a family of balls $\{B(x_\alpha, r_\alpha)\}_\alpha$ in a Banach space X said to be weakly intersecting if for every norm one $f \in X^*$ the balls $\{B(f(x_\alpha), r_\alpha)\}_\alpha$ in the scalar field have nonempty intersection. All what remains is to replace "mutually intersecting balls" (real case) by "weakly intersecting balls" (complex case); see Proposition 2.30.

In general, given a real vector space X one can "change" the scalar field just taking $X \otimes_{\mathbb{R}} \mathbb{C}$, which is a complex vector space by the very definition. Observe that there is a natural embedding of X into $X \otimes_{\mathbb{R}} \mathbb{C}$ given by $x \mapsto x \otimes 1$ and that every $z \in X \otimes_{\mathbb{R}} \mathbb{C}$ has a unique decomposition $z = x + iy$, where $x, y \in X$. If, besides, X is a real Banach space, then $X \otimes_{\mathbb{R}} \mathbb{C}$ can be equipped with a variety of norms, making it a complex Banach space, which is called a *complexification* of X when its norm is reasonable in the sense that $\|x \otimes \zeta\| = \|x\| \cdot |\zeta|$ for every $x \in X, \zeta \in \mathbb{C}$.

Examples of such can be found in [3, 41, 182, 241]. Unfortunately, those complexifications that are suitable for some purposes may be not for others, and the interested reader may peruse the paper [200] to get an idea of the situation. Those readers familiar with tensor products of Banach spaces will guess that for what these notes are concerned (namely, the extension of operators) the most convenient norm on $X \otimes_{\mathbb{R}} \mathbb{C}$ is that arising from the *injective tensor product* $X \check{\otimes}_{\mathbb{R}} \mathbb{C}$. Without entering into any details, the injective norm in $X \otimes_{\mathbb{R}} Y$ is given by

$$\|u\|_\varepsilon = \sup\{|u(x^* \otimes y^*)| : \|x^*\|, \|y^*\| \le 1\},$$

where x^* and y^* are real-linear functionals on X and Y, respectively (see [83]). Needless to say, every real-linear functional on $\mathbb{C}$ has the form $\zeta = \alpha + i\beta \mapsto s\alpha + t\beta$ for some fixed $s, t \in \mathbb{R}$ and the norm of such functionals is just $\|(s, t)\|_2 = \sqrt{s^2 + t^2}$. Thus, the injective norm of $z = x + iy = x \otimes 1 + y \otimes i$ in $X \otimes_{\mathbb{R}} \mathbb{C}$ is

$$\begin{aligned}
\|z\|_\varepsilon &= \sup \{|sx^*(x) + tx^*(y)| : \|x^*\|, s^2 + t^2 \le 1\} \\
&= \sup \{|x^*(sx + ty)| : \|x^*\| \le 1, s^2 + t^2 \le 1\} \\
&= \sup \{\|sx + ty\|_X : s^2 + t^2 \le 1\}. \tag{2.10}
\end{aligned}$$

Let us denote by $X_{\mathbb{C}}$ the complexification of X associated to the just defined norm. Observe that this is not the same complexification as in, say, [181, p. 81]. The basic property of this construction is the following: if $u : Y \to X$ is a linear isometry between two real Banach spaces, then $u \otimes \mathbf{1}_{\mathbb{C}} : Y_{\mathbb{C}} \longrightarrow X_{\mathbb{C}}$ is again an isometry. Another pleasant feature of this norm is that if $X = C(K)$, then $X_{\mathbb{C}} = C(K, \mathbb{C})$, with the sup norm. It follows that if the real space X is a Lindenstrauss space, then $X_{\mathbb{C}}$ is a (complex) Lindenstrauss space: if $X^* = L_1(\mu, \mathbb{R})$, then $X_{\mathbb{C}}^* = L_1(\mu, \mathbb{C})$. One has

Proposition 2.64 *Let E be a real Banach space E and $\lambda \ge 1$. Then E is (universally) λ-separably injective, as a real Banach space if and only if $E_{\mathbb{C}}$ is (universally) λ-separably injective, as a complex Banach space.*

Proof Observe that the inclusion map $E \to E_{\mathbb{C}}$ given by $x \mapsto x \otimes 1$ has a contractive real-linear left-inverse $\Re : E_{\mathbb{C}} \longrightarrow E$ given by $\Re(x + iy) = x$. Actually this map is nothing different from the tensorization of the "real part" map $\mathbb{C} \to \mathbb{R}$ with the identity on E. Anyway is trivial to check that $\|x\|_E \leq \|x + iy\|_{E_{\mathbb{C}}}$ in view of (2.10).

Suppose E is (universally) λ-separably injective, as a real Banach space. Let X be a complex Banach space and $t : Y \to E_{\mathbb{C}}$ a complex-linear operator, where Y is a closed subspace of X. Then $\Re(t) : Y \to E$ is a real-linear operator with $\|\Re(t)\| \leq \|t\|$. If $\tau : X \to E$ is a real-linear extension of $\Re(t)$, then the map $T : X \to E_{\mathbb{C}}$ defined by

$$T(x) = \frac{\tau(x) - i\tau(ix)}{2}$$

is a complex-linear extension of t. This establishes the "only if" part.

To prove the converse, let us assume that $E_{\mathbb{C}}$ is (universally) λ-separably injective, as a complex Banach space. Let Y be a subspace of a real Banach space X, and let $t : Y \to E$ be a real-linear operator. Consider $Y_{\mathbb{C}}$ as a complex subspace of $X_{\mathbb{C}}$ and the complex operator $t_{\mathbb{C}} : Y_{\mathbb{C}} \longrightarrow E_{\mathbb{C}}$ defined as

$$t_{\mathbb{C}}(x + iy) = t(x) + it(y)$$

If this operator extends to a complex operator $T : X_{\mathbb{C}} \longrightarrow E_{\mathbb{C}}$ then the "restriction" of $\Re(T)$ to X is a real-linear extension of t.   □

It follows, for instance, that a compact space K is an F-space if and only if the complex space $C(K, \mathbb{C})$ is 1-separably injective.

In the opposite direction, every complex space is *also* a real space. One has:

Lemma 2.65 *A complex Banach space is (universally) separably injective if and only if its underlying real space is (universally) separably injective.*

Proof First, we suppose that E is a complex (universally) separably injective Banach space. Let Y be a subspace of a real Banach space X, and let $t : Y \to E$ be a real-linear operator.

Consider $Y_{\mathbb{C}}$ as a complex subspace of $X_{\mathbb{C}}$ and the complex operator $\tau : Y_{\mathbb{C}} \longrightarrow E$ defined as

$$\tau(x + iy) = t(x) + it(y)$$

If this operator extends to a complex operator $T : X_{\mathbb{C}} \longrightarrow E$ then the "restriction" of T to X is a real-linear extension of t. For the converse implication, assume now that the underlying real space of E is (universally) separably injective Let X be complex Banach space, Y a complex subspace of X and $t : Y \longrightarrow E$ be a complex operator.

If $\tau : X \longrightarrow E$ a real-linear extension of t, it is easy to check that the formula

$$T(x) = \frac{\tau(x) - i\tau(ix)}{2}$$

defines a complex operator $T : X \longrightarrow E$ that extends t. $\qquad\qquad\square$

The preceding proof shows that if E is (universally) λ-separably injective as a real space, then so is as a complex space. However, when E is a complex (universally) λ-separably injective, the proof gives only that E is (universally) $\lambda\sqrt{2}$-separably injective as a real Banach space. No more can be expected: $\mathbb{C}$ is 1-injective in the complex domain, while, being isometric to ℓ_2^2, it is only $\sqrt{2}$-separably injective as a real space.

Chapter 3
Spaces of Universal Disposition

In this chapter we deal with Banach spaces of universal disposition and almost universal disposition. These notions were introduced in the sixties by Gurariy, who constructed the (unique, up to isometries) separable Banach space of almost universal disposition for finite dimensional spaces in [118]. Spaces of universal disposition for separable Banach spaces are interesting for us because they are 1-separably injective (Theorem 3.5). More yet, the only way we know of obtaining separably injective p-Banach spaces is to construct p-Banach spaces of universal disposition (see Sect. 3.4.3).

For this reason, we treat first Banach spaces of universal disposition. We provide, for every separable Banach space X, natural isometric embeddings of X into new Banach spaces $\mathscr{S}^{\omega_1}(X)$ and $\mathscr{U}^{\omega_1}(X)$ that are of universal disposition for separable Banach spaces. The space $\mathscr{S}^{\omega_1}(X)$ is 1-separably injective while the space $\mathscr{U}^{\omega_1}(X)$ is universally 1-separably injective. The procedure we present for the construction of the spaces $\mathscr{S}^{\omega_1}(X)$ and $\mathscr{U}^{\omega_1}(X)$ is rather flexible and, when performed with the appropriate input data, is able to produce a wide variety of examples; in particular, the "genuine" Gurariy space $\mathscr{G}$, the p-Gurariy spaces, the Kubiś space, the countable ultrapowers of Gurariy space (to be treated in Chap. 4), and also new spaces such as $\mathscr{F}^{\omega_1}(X)$, which is of universal disposition for finite dimensional spaces but not for separable spaces; after several ad-hoc fine tuning, the procedure could also produce the so called $\mathscr{L}_\infty$-envelopes [69] of Banach spaces, although this topic will not be presented here. Moreover, the embedding of X into $\mathscr{S}^{\omega_1}(X)$ has the additional property that every operator from X into any 1-separably injective space can be extended to $\mathscr{S}^{\omega_1}(X)$ without increasing the norm. The embedding of X into $\mathscr{U}^{\omega_1}(X)$ enjoys the analogous property when the target space is universally 1-separably injective. A similar construction is worked out for finite dimensional spaces. Our study of the main properties of those Banach spaces includes uniqueness, sizes and universality. The corresponding results for p-Banach spaces are treated in Sect. 3.4.3.

© Springer International Publishing Switzerland 2016
A. Avilés et al., *Separably Injective Banach Spaces*, Lecture Notes in Mathematics 2132, DOI 10.1007/978-3-319-14741-3_3

We then study spaces of almost universal disposition, Sect. 3.3, which historically came first and spurred great interest very soon. Their (often surprising) properties and generalizations have been studied by many authors and are still object of intense research. It is remarkable that spaces of almost universal disposition are never complemented in M-spaces; in particular, they cannot be injective. Recall that in Chap. 2 we presented Benyamini's M-space which is a Grothendieck space and 5-separably injective but is not complemented in any C-space (Example 2.24). Such example is not universally separably injective and therefore it cannot be 1-separably injective if we assume the continuum hypothesis. In this context, the spaces $\mathscr{S}^{\omega_1}$ and $\mathscr{U}^{\omega_1}$ constructed in this chapter provide the first examples of 1-separably injective spaces which are no complemented in any M-space, let alone in a C-space, in striking contrast with what is known for injective spaces.

3.1 Spaces of (Almost) Universal Disposition

Before introducing the main definitions of the chapter, let us fix some notations and conventions which apply specially here. By an isometry we mean a linear map which preserves the norm. Also, given $\varepsilon \in [0, 1)$, we say that $u : X \to Y$ is an ε-isometry if

$$(1 - \varepsilon)\|x\| \leq \|u(x)\| \leq (1 + \varepsilon)\|x\|$$

for every $x \in X$. In general we do not assume surjectivity unless explicitly stated. However we say that two spaces are isometric (respectively, ε-isometric) if there exist a surjective isometry (respectively, ε-isometry) between them.

In [118] Gurariy introduces the notions of spaces of universal and almost universal disposition as follows:

Definition 3.1 Let $\mathfrak{J}$ be a class of isometries between Banach spaces.

- A Banach space U is said to be of almost universal disposition for $\mathfrak{J}$ if, given isometries $u : A \to U$ and $\iota : A \to B$ such that $\iota \in \mathfrak{J}$, and $\varepsilon > 0$, there is an ε-isometry $u' : B \to U$ such that $u = u'\iota$.
- A Banach space U is of universal disposition for $\mathfrak{J}$ if, given isometries $u : A \to U$ and $\iota : A \to B$ such that $\iota \in \mathfrak{J}$, there is an isometry $u' : B \to U$ such that $u = u'\iota$.
- If $\mathfrak{M}$ is a class of Banach spaces, the space U is said to be of (almost) universal disposition for $\mathfrak{M}$ if it is of (almost) universal disposition for the isometries acting between spaces in $\mathfrak{M}$.

We are particularly interested in spaces of universal disposition for the classes of separable and finite dimensional Banach spaces that we denote by $\mathfrak{S}$ and $\mathfrak{F}$, respectively. We will not consider spaces of almost universal disposition for any other class than $\mathfrak{F}$, and so we will, in this context, omit the words "for finite dimensional spaces" during the chapter.

The first thing one has to know about spaces of universal disposition is that they are $\mathscr{L}_\infty$-spaces. Precisely:

Lemma 3.2 *All Banach spaces of almost universal disposition are Lindenstrauss spaces.*

Proof Recall that Lindenstrauss spaces are exactly the $\mathscr{L}_{\infty,1+}$-spaces. Suppose U is of almost universal disposition and let A be a finite dimensional subspace of U. Fix any $\varepsilon > 0$ and take n large enough so that there is an ε-isometry $\imath : A \to \ell_\infty^n$. Let B the Banach space whose underlying linear space is ℓ_∞^n and the unit ball is the closed convex hull of the set

$$\{x \in \ell_\infty^n : \|x\| \leq 1 - \varepsilon\} \bigcup \{\imath(a) : a \in A, \|a\| \leq 1\}.$$

Then the formal identity is an ε-isometry from B onto ℓ_∞^n and $\imath$ is an isometry from A to B. By the very definition there is an ε-isometry $v : B \to U$ such that $v(\imath(a)) = a$ for every $a \in A$. Clearly, $v[B]$ is a subspace of U that is $(1 + \varepsilon)^2$-isometric to ℓ_∞^n and contains A. □

3.1.1 The Basic Construction

Let us consider an isometry $u : A \to B$ and an operator $t : A \to E$. Our method is based in the push-out construction, which we will use in order to get an extension of t through u at the cost of embedding E in a larger space as it is showed in the diagram

$$
\begin{array}{ccc}
A & \xrightarrow{u} & B \\
{\scriptstyle t}\big\downarrow & & \big\downarrow{\scriptstyle t'} \\
E & \xrightarrow{u'} & PO
\end{array}
$$

where $t'u = u't$. It is important to realize that u' is again an isometry and that t' is a contraction or an isometry if t is; see Lemma A.19. Given specific families of isometries and norm one operators, we will iterate the push-out scheme up to get the desired space. More precisely, once set a starting Banach space E, the input data for the construction are:

- A class $\mathfrak{M}$ of Banach spaces.
- A family $\mathfrak{J}$ of isometries acting between the elements of $\mathfrak{M}$.
- A family $\mathfrak{L}$ of norm one E-valued operators defined on elements of $\mathfrak{M}$.

For any operator $s : X \to Y$, we establish $\mathrm{dom}(s) = X$ and $\mathrm{cod}(s) = Y$. Notice that the codomain of an operator is usually larger than its range, and that the unique codomain of the elements of $\mathfrak{L}$ is E. To avoid complications we will assume that $\mathfrak{J}$ and $\mathfrak{L}$ are sets.

Set $\Gamma = \{(u,t) \in \mathfrak{J} \times \mathfrak{L} : \mathrm{dom}\, u = \mathrm{dom}\, t\}$ and consider the Banach spaces of summable families $\ell_1(\Gamma, \mathrm{dom}\, u)$ and $\ell_1(\Gamma, \mathrm{cod}\, u)$. We have an obvious isometry

$$\oplus \mathfrak{J} : \ell_1(\Gamma, \mathrm{dom}\, u) \longrightarrow \ell_1(\Gamma, \mathrm{cod}\, u)$$

defined by $(x_{(u,t)})_{(u,t)\in\Gamma} \longmapsto (u(x_{(u,t)}))_{(u,t)\in\Gamma}$; and a contractive operator

$$\Sigma \mathfrak{L} : \ell_1(\Gamma, \mathrm{dom}\, u) \longrightarrow E,$$

given by $(x_{(u,t)})_{(u,t)\in\Gamma} \longmapsto \sum_{(u,t)\in\Gamma} t(x_{(u,t)})$. Observe that the notation is slightly imprecise since both $\oplus \mathfrak{J}$ and $\Sigma \mathfrak{L}$ depend on Γ. We can form their push-out diagram

$$
\begin{array}{ccc}
\ell_1(\Gamma, \mathrm{dom}\, u) & \xrightarrow{\ \oplus \mathfrak{J}\ } & \ell_1(\Gamma, \mathrm{cod}\, u) \\
{\scriptstyle \Sigma \mathfrak{L}}\downarrow & & \downarrow \\
E & \xrightarrow{\quad \imath \quad} & \mathrm{PO}
\end{array}
$$

We obtain in this way an isometric enlargement of E such that for every $t : A \to E$ in $\mathfrak{L}$, the operator $\imath t$ can be extended to an operator $t' : B \to \mathrm{PO}$ through any embedding $u : A \to B$ in $\mathfrak{J}$ provided $\mathrm{dom}\, u = \mathrm{dom}\, t = A$. In the next step we keep the family $\mathfrak{J}$ of isometries, replace the starting space E by PO and $\mathfrak{L}$ by a family of norm one operators $\mathrm{dom}\, u \to \mathrm{PO}$, $u \in \mathfrak{J}$, and proceed again.

This construction can be iterated until any countable or uncountable ordinal. Depending on the choice of the families, essentially on $\mathfrak{M}$ and $\mathfrak{J}$, we will produce different spaces of universal disposition.

3.1.2 Some Specific Spaces of Universal Disposition

Let us proceed to present some specific constructions in detail. We set a starting Banach space X. To fix the inconvenient that the class of "all separable Banach spaces" is not a set, we may take $\mathfrak{M} = \tilde{\mathfrak{S}}$ as the set of all closed subspaces of $C(\Delta)$, $\mathfrak{J}$ the set of all isometries with domain and codomain in $\tilde{\mathfrak{S}}$, and $\mathfrak{L}$ the set of all isometries $S \to X$, with $S \in \tilde{\mathfrak{S}}$.

We want to define, for every ordinal α, a Banach space $\mathscr{S}^\alpha(X)$, depending on X, in such a way that the family $(\mathscr{S}^\alpha(X))_\alpha$ forms a directed system of Banach spaces (see Appendix A.6 in the Appendix). As before, one should take into account that the class of "all ordinals" is not a set, and to be true one should first fix an ordinal β and then consider the set of all ordinals $\alpha \leq \beta$. We use transfinite induction. We start with $\mathscr{S}^0(X) = X$. The inductive step is as follows. Suppose we have constructed the directed system $(\mathscr{S}^\alpha(X))_{\alpha<\beta}$, including the corresponding linking maps $\iota_{(\alpha,\gamma)} : \mathscr{S}^\alpha(X) \longrightarrow \mathscr{S}^\gamma(X)$ for $\alpha < \gamma < \beta$. To define $\mathscr{S}^\beta(X)$ and the maps $\iota_{(\alpha,\beta)} : \mathscr{S}^\alpha(X) \longrightarrow \mathscr{S}^\beta(X)$ we consider separately two cases, as usual: if β is a limit ordinal, then we take $\mathscr{S}^\beta(X)$ as the direct limit of the system $(\mathscr{S}^\alpha(X))_{\alpha<\beta}$ and $\iota_{(\alpha,\beta)} : \mathscr{S}^\alpha(X) \longrightarrow \mathscr{S}^\beta(X)$ the natural inclusion map. Otherwise $\beta = \alpha + 1$ is a successor ordinal and we construct $\mathscr{S}^\beta(X)$ applying the basic construction of Sect. 3.1.1 with the following data: $\mathscr{S}^\alpha(X)$ is the starting space, $\mathfrak{J}$ keeps being the set of all isometries acting between the elements of $\tilde{\mathfrak{G}}$ and $\mathfrak{L}_\alpha$ is the family of all isometries $t : S \to \mathscr{S}^\alpha(X)$, where $S \in \tilde{\mathfrak{G}}$.

We then set $\Gamma_\alpha = \{(u,t) \in \mathfrak{J} \times \mathfrak{L}_\alpha : \mathrm{dom}\, u = \mathrm{dom}\, t\}$ and make the push-out

$$
\begin{array}{ccc}
\ell_1(\Gamma_\alpha, \mathrm{dom}\, u) & \xrightarrow{\oplus \mathfrak{J}_\alpha} & \ell_1(\Gamma_\alpha, \mathrm{cod}\, u) \\
{\scriptstyle \Sigma \mathfrak{L}_\alpha} \downarrow & & \downarrow \\
\mathscr{S}^\alpha(X) & \longrightarrow & \mathrm{PO}
\end{array}
\tag{3.1}
$$

thus obtaining $\mathscr{S}^{\alpha+1}(X) = \mathrm{PO}$. The embedding $\iota_{(\alpha,\beta)}$ is the lower arrow in the above diagram; by composition with $\iota_{(\alpha,\beta)}$ we get the embeddings $\iota_{(\gamma,\beta)} = \iota_{(\alpha,\beta)}\iota_{(\gamma,\alpha)}$, for all $\gamma < \alpha$.

We will also consider the following two variations of this construction: Fix a starting Banach space X and an ordinal α.

- The space $\mathscr{F}^\alpha(X)$, obtained taking as input data: $\mathfrak{M} = \tilde{\mathfrak{F}}$ the family of all finite dimensional subspaces of $C(\Delta)$, $\mathfrak{J}$ the set of all isometries between elements of $\tilde{\mathfrak{F}}$ and $\mathfrak{L}$ all X-valued isometries defined on elements of $\tilde{\mathfrak{F}}$.
- The space $\mathscr{U}^\alpha(X)$, obtained taking as initial data: $\mathfrak{M} = \tilde{\mathfrak{G}}$ all closed subspaces of $C(\Delta)$, $\mathfrak{J} = \mathfrak{J}^\infty$ all isometries from the elements of $\tilde{\mathfrak{G}}$ into ℓ_∞, so that $\mathrm{cod}\, u = \ell_\infty$ for every $u \in \mathfrak{J}^\infty$, and $\mathfrak{L}$ all isometries from the elements of $\tilde{\mathfrak{G}}$ to X.

Continuing the process until the first uncountable ordinal ω_1, one obtains very interesting creatures:

Proposition 3.3 *Let X be a Banach space.*

1. *The spaces $\mathscr{S}^{\omega_1}(X)$ and $\mathscr{U}^{\omega_1}(X)$ are of universal disposition for separable Banach spaces.*
2. *The space $\mathscr{F}^{\omega_1}(X)$ is of universal disposition for finite dimensional Banach spaces.*

Proof

1. We write the proof for $\mathscr{S}^{\omega_1} = \mathscr{S}^{\omega_1}(X)$. The case $\mathscr{U}^{\omega_1}(X)$ is analogous and we leave it to the reader. We must show that if $v : A \to B$ and $\ell : A \to \mathscr{S}^{\omega_1}$ are isometries and B is separable, then there is an isometry $L : B \to \mathscr{S}^{\omega_1}$ such that $Lv = \ell$. We may and do assume $A, B \in \tilde{\mathfrak{G}}$ so that $v \in \mathfrak{J}$. On the other hand there is $\alpha < \omega_1$ such that $\ell[A] \subset \mathscr{S}^\alpha$ and we may consider that ℓ is one of the

operators in $\mathfrak{L}_\alpha$. Therefore ℓ has an extension ℓ' making the following square commutative:

$$
\begin{array}{ccc}
A & \xrightarrow{\;v\;} & B \\[2pt]
\ell \downarrow & & \downarrow \ell' \\[2pt]
\mathscr{S}^\alpha & \xrightarrow{\;I_{(\alpha,\alpha+1)}\;} & \mathscr{S}^{\alpha+1}
\end{array}
$$

Actually ℓ' is the composition of the inclusion $j_{(v,\ell)}$ of $B = \operatorname{cod} v$ into the (v,ℓ)-th coordinate of $\ell_1(\Gamma_\alpha, \operatorname{cod} u)$ with the right descending arrow in the diagram

$$
\begin{array}{ccc}
\ell_1(\Gamma_\alpha, \operatorname{dom} u) & \xrightarrow{\;\oplus\mathfrak{J}\;} & \ell_1(\Gamma_\alpha, \operatorname{cod} u) \\[2pt]
\Sigma\mathfrak{L}_\alpha \downarrow & & \downarrow \\[2pt]
\mathscr{S}^\alpha & \longrightarrow & \mathrm{PO} = \mathscr{S}^{\alpha+1}
\end{array}
\qquad (3.2)
$$

We known that ℓ' is contractive, but we still have to prove that it is isometric. We have

$$
\mathrm{PO} = \left(\mathscr{S}^\alpha \oplus_1 \ell_1(\Gamma_\alpha, \operatorname{cod} u) \right) / \Delta
$$

with

$$
\Delta = \left\{ \left(\sum_{(u,t)\in\Gamma_\alpha} tx_{(u,t)}, \; - \sum_{(u,t)\in\Gamma_\alpha} ux_{(u,t)} \right) : (x_{(u,t)}) \in \ell_1(\Gamma_\alpha, \operatorname{dom} u) \right\}.
$$

Thus, for $b \in B$ we have $\ell'(b) = (0, j_{(v,\ell)}b) + \Delta$ and

$$
\|\ell'(b)\|_{\mathrm{PO}} = \operatorname{dist}((0, j_{(v,\ell)}b), \Delta) = \inf_{a\in A} \{\|\ell(a)\|_{\mathscr{S}^\alpha} + \|b - v(a)\|_B\} = \|b\|_B
$$

since both ℓ and v preserve the norm.

2. The proof is completely analogous to the previous one just considering B finite dimensional. $\square$

The space $\mathscr{S}^{\omega_1}(X)$ confirms the conjecture of Gurariy that spaces of universal disposition exist. We will later show that it does not depend on the initial space X and that, under CH, coincides with the Fraïssé limit in the category of separable Banach spaces and isometries constructed by Kubiś [169]; and also with any "countable" ultrapower of the Gurariy space, see Sect. 3.3.1 and Chap. 4.

3.2 Properties of Spaces of Universal Disposition

3.2.1 Operator Extension Properties

The following Lemma gathers the main connections between "universal disposition" and "injectivity" properties. The basic idea is that if one can extend isometries, then one can extend operators with the same bound.

Lemma 3.4 *Let E be a Banach space and $\lambda \geq 1$ a constant.*

1. *Suppose that for every pair of isometries $u : A \to B$ and $v : A \to E$ there exists an operator $V : B \to E$ such that $Vu = v$ with $\|V\| \leq \lambda$. Then E is λ-injective.*
2. *If condition (1) holds whenever B is separable, then E is λ-separably injective.*
3. *If condition (1) holds whenever A is separable and $B = \ell_\infty$, then E is universally λ-separably injective.*
4. *If condition (1) holds when B is finite dimensional, then E is an $\mathscr{L}_\infty$-space.*

Proof The proof is the same in all cases. Let $t : A \to E$ have norm one and make the push-out:

$$
\begin{array}{ccc}
A & \xrightarrow{\;u\;} & B \\
{\scriptstyle t}\downarrow & & \downarrow{\scriptstyle t'} \\
\overline{t[A]} & \xrightarrow{\;u'\;} & \mathrm{PO}
\end{array}
\qquad\qquad (3.3)
$$

By Lemma A.19, u' is an into isometry, and the hypothesis yields and operator $t'' : \mathrm{PO} \to E$ such that $t''u'$ is the inclusion of $\overline{t[A]}$ into E, with $\|t''\| \leq \lambda$. Taking $T = t''t'$ concludes the proof. $\qquad\square$

The second part of the preceding Lemma yields as immediate consequence:

Theorem 3.5 *A space of universal disposition for separable Banach spaces is 1-separably injective.*

In particular, for every Banach space X, the spaces $\mathscr{S}^{\omega_1}(X)$ and $\mathscr{U}^{\omega_1}(X)$ are 1-separably injective. And from the third part we obtain:

Proposition 3.6 *The space $\mathscr{U}^{\omega_1}(X)$ is universally 1-separably injective.*

The following result establishes the partial automorphic character of the spaces of universal disposition. In practice, the second part holds only under **CH**, since the minimum size of a Banach space of universal disposition is the continuum (see Proposition 3.10).

Proposition 3.7 *Let E be a space of universal disposition for separable spaces.*

1. *Given a separable Banach space B and a subspace $A \subset B$, every isomorphic embedding $t : A \to E$ extends to an isomorphic embedding $T : B \to E$ with $\|T\| = \|t\|$ and $\|T^{-1}\| = \|t^{-1}\|$.*
2. *Consequently, if $\operatorname{dens} E \leq \aleph_1$, then E is separably automorphic; namely, any isomorphism between two subspaces of E can be extended to an automorphism of E.*

Proof

1. Let u denote the inclusion of A into B and assume, without loss of generality, that $\|t\| = 1$. We follow the same notation as in Lemma 3.4. Looking at Diagram 3.3 we have $\|t'\| = 1$ and u' is isometric, so there is an isometric embedding $t'' :$ PO $\to E$ such that $t''u'$ is the inclusion of $A' = t[A]$ into E. Now $T = t''t'$ is the extension of t we wanted. Clearly, $\|T\| = \|t\| = 1$. On the other hand, by Lemma A.19, one has $\|(t')^{-1}\| \leq \max\{1, \|t^{-1}\|\}$ hence $\|T^{-1}\| = \|t^{-1}\|$.
2. It suffices to show that if Y is a separable subspace of E, every isomorphic embedding $\varphi_0 : Y \to E$ extends to an automorphism of E. This is proved through the obvious back-and-forth argument: write $E = \bigcup_{\alpha < \omega_1} E_\alpha$ as an ω_1-increasing sequence of separable subspaces starting with $E_0 = Y$. Consider the embedding $\varphi_0 : E_0 \to E$. By Part 1, let $\psi_1 : \varphi_0[E_0] + E_1 \to E$ be an extension of $\varphi_0^{-1} : \varphi_0[E_0] \to E$, with $\|\psi_1\| = \|\varphi_0^{-1}\|$ and $\|\psi_1^{-1}\| = \|\varphi_0\|$. Notice that $\operatorname{ran} \psi_1 = E_0 + \psi_1[E_1]$. Let φ_2 be the extension of ψ_1^{-1} to $E_0 + \psi_1[E_1] + E_2$ provided by Part (1) and so on. Proceeding by transfinite induction one gets a couple of endomorphisms φ and ψ such that $\psi\varphi = \varphi\psi = 1_E$, with $\|\varphi\| = \|\varphi_0\|$ and $\|\psi\| = \|\varphi_0^{-1}\|$ and $\varphi = \varphi_0$ on Y. $\qquad\square$

The canonical inclusion of X into the spaces $\mathscr{S}^{\omega_1}(X)$ and $\mathscr{U}^{\omega_1}(X)$ constructed above enjoys the following additional properties:

Proposition 3.8

1. *Every operator from X into a 1-separably injective space can be extended to $\mathscr{S}^{\omega_1}(X)$ with the same norm.*
2. *Every operator from X into a 1-universally separably injective space extends to $\mathscr{U}^{\omega_1}(X)$ with the same norm.*

Proof We write the proof of the first part, the other is almost the same. Clearly, it suffices to see that for each $\alpha < \omega_1$, every norm one operator τ from $\mathscr{S}^\alpha = \mathscr{S}^\alpha(X)$ to a 1-separably injective Banach space E extends to $\mathscr{S}^{\alpha+1} = \mathscr{S}^{\alpha+1}(X)$ without increasing the norm. Iterating the process until ω_1 one gets a norm one extension $\tau_{\omega_1} : \mathscr{S}^{\omega_1}(X) \to E$.

Let us first assume that τ is an isometry and consider the composition $\tau \circ (\Sigma\mathfrak{L}_\alpha) :$ $\ell_1(\Gamma_\alpha, \operatorname{dom} u) \longrightarrow \mathscr{S}^\alpha \longrightarrow E$. For each $(u, t) \in \Gamma_\alpha$, the "component" $\tau \circ t$ is an operator (actually an isometry) from the separable Banach space $\operatorname{dom} u = \operatorname{dom} t$ to E and there is an extension $T_{(u,t)} : \operatorname{cod} u \to E$ through u with $\|T_{(u,t)}\| = 1$. The coproduct of all these operators yields a norm one operator $T : \ell_1(\Gamma_\alpha, \operatorname{cod} u) \longrightarrow$

E such that $T \circ (\oplus \mathfrak{J}) = \tau \circ (\Sigma \mathfrak{L}_\alpha)$ and the universal property of the push-out construction yields a norm one operator $\tilde{\tau} : \mathscr{S}^{\alpha+1} \longrightarrow E$ extending τ.

If τ is not an isometry, take any isometry $\tau_1 : \mathscr{S}^\alpha \longrightarrow \ell_\infty(I)$, define $\tau_2 : \mathscr{S}^\alpha \longrightarrow E \oplus_\infty \ell_\infty(I)$ by $\tau_2(x) = (\tau(x), \tau_1(x))$ and apply the preceding reasoning to the isometry τ_2 to obtain a norm one extension $\tilde{\tau}_2 : \mathscr{S}^{\alpha+1} \longrightarrow E \oplus_\infty \ell_\infty(I)$. Killing the second coordinate of $\tilde{\tau}_2$ gives the required extension. $\square$

3.2.2 Sizes

It is clear that, when the starting space X is separable, the Banach spaces appearing in Proposition 3.3 have density character $\mathfrak{c}$ since each of them is the union of an ω_1-sequence formed by Banach spaces of density $\mathfrak{c}$. One may wonder if there are smaller examples. A juxtaposition of Theorem 3.5 and Proposition 2.32 shows that any Banach space of universal disposition for separable spaces must have density character at least $\mathfrak{c}$. As for finite dimensional spaces, observe that "universal disposition" for the single isometry $\imath : t \in \mathbb{R} \mapsto (t, 0) \in \ell_1^2$ destroys the (Gâteaux) differentiability of the norm at every point of the unit sphere. Since the norm of any separable Banach space has to be Gâteaux differentiable on a dense G_δ (a classical result by Mazur [195]) we see that no Banach space of universal disposition for finite dimensional spaces can be separable.

Actually it is impossible to reduce the size of spaces of universal disposition for finite dimensional spaces. Let us see why.

Lemma 3.9 *Let X be a Banach space containing a 2-dimensional Hilbert subspace H. Assume that for every isometric embedding into a 3-dimensional space $i : H \to F$ there exists an isometric embedding $j : F \to X$ such that $j \circ i$ is the inclusion $H \subset X$. Then, the density character of X is at least the continuum.*

Proof Let S be the positive part of the sphere of H. Given $u \in S$, we define a norm on $H \oplus \mathbb{R}$ by the formula

$$\|(x, \lambda)\|_u = \max \left\{ \|x\|_2, |\lambda| + |(x|u)| \right\},$$

where $(\cdot|\cdot)$ denotes the usual scalar product on H. Note that $\|(0, 1)\|_u = 1$ and $\|\cdot\|_u$ extends the Euclidean norm $\|\cdot\|_2$ of H, where $x \in H$ is identified with $(x, 0)$. By hypothesis, for each $u \in S$ we can find $e_u \in X$ such that the map $i_u : H \times \mathbb{R} \to X$, defined by $i_u(x, \lambda) = x + \lambda e_u$, is an isometric embedding with respect to $\|\cdot\|_u$.

Fix $u, v \in S$ such that $u \neq v$ and let $\|\cdot\|$ denote the norm of X. Fix $\mu > 0$ and let $w = \mu u \in H \subset X$. Then

$$\|e_u - e_v\| \geq \|e_u + w\| - \|e_v + w\| = \|(\mu u, 1)\|_u - \|(\mu u, 1)\|_v.$$

Finally, observe that $\|(\mu u, 1)\|_u = 1 + \mu$ and

$$\|(\mu u, 1)\|_v = \max\left\{\mu, 1 + \mu|(u|v)|\right\} = \mu,$$

whenever μ is big enough, because $|(u|v)| < 1$ (recall that u, v are distinct, positive vectors of the sphere of H). Thus, we conclude that $\|e_u - e_v\| \geq 1$ whenever $u \neq v$, which shows that the density of X is at least $|S| = \mathfrak{c}$.   □

It is immediate from that:

Proposition 3.10 *The density character of any space of universal disposition for finite dimensional spaces is at least the continuum.*

This last result will get full sense in Sect. 3.2.4 where it will be shown that there exist spaces which are of universal disposition for finite dimensional spaces that are not of universal disposition for separable spaces.

3.2.3   Universality

Definition 3.11 A Banach space is universal for a given class $\mathfrak{M}$ if it contains isometric copies of all elements of $\mathfrak{M}$.

One has the following relation between universal disposition and universality:

Lemma 3.12 *Let $\aleph$ be a cardinal. A Banach space of universal disposition for spaces of density character strictly lesser than $\aleph$ is universal for spaces of density character at most $\aleph$.*

Proof Assume that U is a space of universal disposition for Banach spaces of density character strictly lesser than $\aleph$ and let X have density $\aleph$. Let us write $X = \bigcup_{\gamma < \aleph} X_\gamma$ as an increasing family of subspaces X_γ of X of density (strictly) lesser than $\aleph$. There is no loss of generality assuming that X_0 has dimension 1. Take an isometric embedding $f_0 : X_0 \to U$ and use the universal disposition of U to get an isometric extension $f_1 : X_1 \to U$. A transfinite iteration of the extension process produces an isometric embedding $f : X \to U$.   □

An immediate consequence is:

Proposition 3.13

1. *A Banach space of universal disposition for finite dimensional spaces is universal for separable spaces.*
2. *A Banach space of universal disposition for separable spaces is universal for the class of Banach spaces having density character $\aleph_1$ or less.*

Must a space of universal disposition for finite-dimensional spaces contain an isometric copy of each Banach space of density $\aleph_1$ or less? Our guess is no.

3.2.4 Uniqueness

In this section we face the problem of comparing the different spaces of universal disposition we have encountered. Since being of (almost) universal disposition is a demanding property, it can be very difficult to distinguish between those spaces that fulfill it, unless they have different density characters. In fact, it can be impossible: we shall prove later that all separable Banach spaces of almost universal disposition are isometric (see Sect. 3.3.5; another result in this line is Theorem 4.29).

Let us begin by showing that there are (at least) two spaces of universal disposition for finite dimensional spaces whose density character is the continuum. We need the following elementary lemma which rests on the peculiarities of c_0.

Lemma 3.14 *A c_0-valued operator defined on a finite dimensional Banach space admits a (compact) extension with the same norm to any superspace.*

Proof Let $F \subset X$ be a finite dimensional subspace of a Banach space X, and let $\tau : F \to c_0$ be a norm one operator. Write $\tau = (\tau_n)$ as a sequence of functionals. Then (τ_n) is pointwise null and since F is finite dimensional, the sequence (τ_n) is actually norm null. Thus, any sequence of Hahn-Banach extensions will also be norm null, and the operator they define is a compact extension of τ having the same norm. $\square$

Now we can use the idea behind the proof of Proposition 3.8 to show that there are spaces of universal disposition for finite dimensional Banach spaces which are not of universal disposition for separable spaces.

Proposition 3.15 *The space $\mathscr{F}^{\omega_1}(c_0)$, which is of universal disposition for finite dimensional spaces, is not isomorphic to any space of universal disposition for separable spaces.*

Proof It follows from Lemma 3.14 that the embedding $X \to \mathscr{F}^{\omega_1}(X)$ has the property that every operator $X \to c_0$ can be extended to $\mathscr{F}^{\omega_1}(X)$. Therefore $\mathscr{F}^{\omega_1}(c_0)$ contains c_0 complemented, and thus it cannot be 1-separably injective in any equivalent norm; see Proposition 2.31. $\square$

This suggests that quite plausibly there is a continuum of mutually non-isomorphic spaces of universal disposition for finite dimensional spaces, even under CH.

Regarding spaces of universal disposition for separable Banach spaces, the situation is more subtle and depends on what axioms of set theory are assumed. Our next result implies, among other things, that under CH there is no dependence on the initial separable space X in the constructions appearing in Proposition 3.3(1).

Proposition 3.16 (CH) *Up to isometries, there is a unique space of universal disposition for separable spaces, having density character the continuum.*

Proof Let U and V be spaces of universal disposition for separable spaces and with density character $\aleph_1$. Let us write $U = \bigcup_{\alpha < \omega_1} X_\alpha$ and $V = \bigcup_{\beta < \omega_1} Y_\beta$

as increasing ω_1-sequences of separable subspaces. Pick β_1 such that there is an isometric embedding $\varphi_0 : X_0 \to Y_{\beta_1}$. Let $\psi_1 : Y_{\beta_1} \to X$ be an isometric extension of φ_0^{-1}. As ψ_1 has separable range there is $\alpha_2 < \omega_1$ such that ran $\psi_1 \subset X_{\alpha_2}$. Let $\varphi_2 : X_{\alpha_2} \to Y$ be an isometric extension of ψ_1^{-1}. A transfinite iteration of the process produces an isometry from U onto V. $\square$

Corollary 3.17 (CH) *If X is separable, the spaces $\mathscr{S}^{\omega_1}(X)$ and $\mathscr{U}^{\omega_1}(X)$ are isometric to $\mathscr{S}^{\omega_1}(\mathbb{R})$.*

The continuum hypothesis cannot be omitted in Proposition 3.16, as witnessed by the following result. Note that the hypothesis is consistent by a result of Brech and Koszmider [49, Theorem 1.3].

Proposition 3.18 *Assume that no Banach space of density character $\mathfrak{c}$ is universal for all Banach spaces with density character $\mathfrak{c}$. Then there exist $\mathfrak{c}^+$ many non-isomorphic spaces having density character $\mathfrak{c}$ that are of universal disposition for separable spaces*

Proof We proceed by transfinite induction. To make the induction start, form the space $S_1 = \mathscr{S}^{\omega_1}(\mathbb{R})$. Take, by hypothesis, a Banach space X_1 with density character $\mathfrak{c}$ not contained in S_1 and form then $S_2 = \mathscr{S}^{\omega_1}(X_1 \oplus S_1)$. Take a new Banach space X_2 with density character $\mathfrak{c}$ not contained in S_2 and continue in this way.

Let $\beta < \mathfrak{c}^+$, and assume that we have already constructed nonisomorphic Banach spaces $\{S_\alpha : \alpha < \beta\}$ of density $\mathfrak{c}$ and of universal disposition for $\mathfrak{S}$. Then, consider Y to be the ℓ_1-sum of all spaces S_α with $\alpha < \beta$. Since this space cannot be universal, find X a space which is not contained in Y, and then define $S_\beta = \mathscr{S}^{\omega_1}(X)$. $\square$

Although spaces of universal disposition need not be all of them isometric, we will show that those obtained by the procedure of iterated push-out as described in Sect. 3.1.2 are all isometric in some cases. So, while the continuum hypothesis is really necessary in Proposition 3.16, it is not in its Corollary 3.17, at least in the part concerning $\mathscr{S}^{\omega_1}(X)$.

Let $\mathfrak{M}$ be a class of Banach spaces. We say that an isometry $u : A \to B$ is an $\mathfrak{M}$-cell if it fits in an isometric push-out square

where $R, S \in \mathfrak{M}$. Recall that, according to our conventions, the arrows in isometric push-out squares are all isometries; see Appendix A.4.1.

The following definition isolates the relevant properties of the spaces constructed in Sect. 3.1.2.

Definition 3.19 Let $\mathfrak{M}$ be a class of Banach spaces. We will say that a Banach space Z is tightly $\mathfrak{M}$-filtered if there is an ordinal λ and a family of subspaces $\{Z_i : i \leq \lambda\}$ such that:

1. $Z_i \subset Z_j$ if $i < j$,
2. $Z_\lambda = Z$,
3. $Z_i = \overline{\bigcup_{j<i} Z_j}$ for every limit ordinal i,
4. Each inclusion $Z_i \longrightarrow Z_{i+1}$ is an $\mathfrak{M}$-cell.

We will say that Z is exhaustively tightly $\mathfrak{M}$-filtered if, in addition, the density of Z is the continuum and

5. For every $R \subset Z$ and any isometry $\phi : R \rightarrow S$ with $R, S \in \mathfrak{M}$, there exist continuum many ordinals $i < \lambda$ for which Z_i contains R and an isometry $\varphi_i : S \rightarrow Z_{i+1}$ depending on i so that the diagram

$$
\begin{array}{ccc}
R & \xrightarrow{\ \phi\ } & S \\
\downarrow & & \downarrow{\scriptstyle\varphi} \\
Z_i & \longrightarrow & Z_{i+1}
\end{array}
$$

is an isometric push-out square, where the unlabelled arrows denote the corresponding canonical inclusions.

Let us show that the "iterated push-out spaces" we constructed in Sect. 3.1.2 are in most cases of the type just described.

Lemma 3.20 *Let $\xi \leq \mathfrak{c}$ be an ordinal having uncountable cofinality.*

1. *If $\mathrm{dens}(X) \leq \aleph_1$, then $\mathscr{S}^\xi(X)$ is exhaustively tightly $\mathfrak{S}$-filtered.*
2. *If X is separable, the space $\mathscr{F}^\xi(X)$ is exhaustively tightly $\mathfrak{F}$-filtered.*

Proof We prove only (1) because (2) is completely analogous. Just to simplify notation in this proof, we will construct the family of subspaces $\{Z_i : i \leq \lambda\}$ as a well ordered family $\mathfrak{Z}$ of subspaces of $\mathscr{S}^\xi(X)$ without writing an explicit enumeration on an ordinal. Since $\mathrm{dens}(X) \leq \aleph_1$ it is clear that X is tightly $\mathfrak{S}$-filtered, with ordinal $\lambda = 1$ if X is separable; and with $\lambda = \omega_1$ if X is not separable. Let $\mathfrak{Z}'$ be the family of all subspaces Z_i of the filtration of X. The family $\mathfrak{Z}'$ will constitute the initial segment of $\mathfrak{Z}$. The space $\mathscr{S}^\xi(X)$ was defined as the inductive limit of a chain of spaces $\{\mathscr{S}^\alpha : \alpha < \xi\}$, that we view now as subspaces of the final limit space. We had $\mathscr{S}^0 = X$. For each $\alpha < \xi$, we will define a well ordered chain of subspaces $\mathfrak{Z}_\alpha$, and finally we will declare

$$\mathfrak{Z} = \mathfrak{Z}' \cup \bigcup_{\alpha<\xi} \mathfrak{Z}_\alpha \cup \{\mathscr{S}^\xi(X)\}.$$

Let us see how $\mathfrak{Z}_\alpha$ is constructed. Recall the push-out Diagram 3.2:

$$
\begin{array}{ccc}
\ell_1(\Gamma_\alpha, \operatorname{dom} u) & \longrightarrow & \ell_1(\Gamma_\alpha, \operatorname{cod} u) \\
\downarrow & & \downarrow \\
\mathscr{S}^\alpha & \longrightarrow & \mathscr{S}^{\alpha+1}.
\end{array}
\tag{3.4}
$$

The push-out space $\mathscr{S}^{\alpha+1}$ can be explicitly described as

$$
\mathscr{S}^{\alpha+1} = R(\Gamma_\alpha)/N(\Gamma_\alpha),
$$

where, for $\Upsilon \subset \Gamma_\alpha$, we denote

$$
R(\Upsilon) = \mathscr{S}^\alpha \oplus_1 \ell_1(\Upsilon, \operatorname{cod} u),
$$
$$
N(\Upsilon) = \overline{\operatorname{span}} \left\{ (t(x), -[u(x)]_{(u,t)}) : x \in \operatorname{dom} u, (u,t) \in \Upsilon \right\}
$$

Here, $[y]_\gamma$ means the element of $\ell_1(\Gamma_\alpha, \operatorname{cod} u)$ whose only nonzero coordinate is y at place γ. Because all u and t are isometries, we easily check that $R(\Upsilon) \cap N(\Upsilon') = N(\Upsilon \cap \Upsilon')$. Thus, one has the following subspaces of $\mathscr{S}^{\alpha+1}$:

$$
Z(\Upsilon) = \frac{R(\Upsilon)}{N(\Gamma)} = \frac{R(\Upsilon)}{N(\Upsilon \cap \Gamma_\alpha)} = \frac{R(\Upsilon) + N(\Gamma_\alpha)}{N(\Gamma_\alpha)} = \frac{R(\Upsilon)}{N(\Gamma_\alpha)} \subset \mathscr{S}^{\alpha+1}.
$$

In order to define the family $\mathfrak{Z}_\alpha$ we enumerate $\Gamma_\alpha = \{(u_\gamma, t_\gamma) : \gamma < \zeta_\alpha\}$ on an ordinal ζ_α, and consider the initial segments

$$
\Upsilon_{\alpha,\beta} = \{(u_\gamma, t_\gamma) : \gamma < \beta\}, \quad \text{and} \quad \mathfrak{Z}_\alpha = \{Z(\Upsilon_{\alpha,\beta}) : \beta < \zeta_\alpha\}.
$$

We need to check all properties in Definition 3.19. It is clear that the family $\mathfrak{Z}$ can be enumerated on an ordinal λ so that Properties (1), (2) and (3) hold: we just need to take the sum of ordinals $\lambda = \omega_1 + \sum_{\alpha < \xi} \zeta_\alpha$. As for (4), let us verify that the inclusion $Z(\Upsilon_{\alpha,\beta}) \longrightarrow Z(\Upsilon_{\alpha,\beta+1})$ is an $\mathfrak{S}$-cell. Recalling that

$$
Z(\Upsilon_{\alpha,\beta+1}) = \frac{\mathscr{S}^\alpha \oplus_1 \ell_1(\Upsilon_{\alpha,\beta}, \operatorname{cod} u) \oplus_1 \operatorname{cod} u_\beta}{N(\Upsilon_{\alpha,\beta+1})},
$$
$$
Z(\Upsilon_{\alpha,\beta}) = \frac{\mathscr{S}^\alpha \oplus_1 \ell_1(\Upsilon_{\alpha,\beta}, \operatorname{cod} u)}{N(\Upsilon_{\alpha,\beta})} = \frac{\mathscr{S}^\alpha \oplus_1 \ell_1(\Upsilon_{\alpha,\beta}, \operatorname{cod} u)}{N(\Upsilon_{\alpha,\beta+1})}
$$

we have an obvious push-out square of isometries

$$
\begin{array}{ccc}
\operatorname{dom} u_\beta & \xrightarrow{\ u_\beta\ } & \operatorname{cod} u_\beta \\
{\scriptstyle t_\beta}\downarrow & & \downarrow{\scriptstyle g} \\
Z(\Upsilon_{\alpha\beta}) & \longrightarrow & Z(\Upsilon_{\alpha,\beta+1})
\end{array}
$$

In order to verify Property (5), the preceding reasoning shows that each pair $(u, t) \in \Gamma_\alpha$ can be inserted into an isometric push-out diagram

$$
\begin{array}{ccc}
\operatorname{dom} u & \xrightarrow{\ u\ } & \operatorname{cod} u \\
{\scriptstyle t}\downarrow & & \downarrow \\
Z_i & \longrightarrow & Z_{i+1}
\end{array}
$$

where Z_{i+1} is the "immediate successor" of Z_i along $\mathfrak{Z}$. Since ξ has uncountable cofinality, we have $\mathscr{S}^\xi(X) = \bigcup_{\alpha < \xi} \mathscr{S}^\alpha$ (otherwise, we would need to take the closure of the union). Thus, Property (5) would be immediate after another formal make-up: make sure that each Γ_α contains $\mathfrak{c}$ many copies of each pair (u, t), replacing Γ_α by $\Gamma_\alpha \times \mathfrak{c}$. Or, if the reader prefers, relabel the push-outs "repeating them" as needed: a formal trick could be to compose each u with inclusions inside universal spaces $C[0, 1] \oplus \ell_p$, for a set of $\mathfrak{c}$ many p's which are disjoint for different u's. $\qquad\square$

The following definitions and notations will be helpful. Suppose that the class $\mathfrak{M}$ is stable under (surjective) isometries and that Z is tightly $\mathfrak{M}$-filtered so that for each $i < \lambda$ one has an isometric push-out square

$$
\begin{array}{ccc}
R & \xrightarrow{\ \phi\ } & S \\
\downarrow & & \downarrow{\scriptstyle \varphi} \\
Z_i & \longrightarrow & Z_{i+1}
\end{array}
$$

witnessing that the inclusion of Z_i into Z_{i+1} is an $\mathfrak{M}$-cell. Then, letting $S_i = \varphi[S]$ and then $R_i = Z_i \cap S_i$ we obtain and "equivalent" isometric push-out square

$$
\begin{array}{ccc}
R_i & \longrightarrow & S_i \\
\downarrow & & \downarrow \\
Z_i & \longrightarrow & Z_{i+1}
\end{array}
$$

in which all arrows are inclusions that we consider fixed in the remainder of this section.

With these conventions, a subset $\Gamma \subset \lambda$ is said to be *saturated* when for every $\alpha \in \Gamma$ one has

$$
R_\alpha \subset \overline{\operatorname{span}}\{S_\beta : \beta \in \Gamma, \beta < \alpha\}
$$

Given a saturated Γ let us write

$$E(\Gamma) = \overline{\mathrm{span}}\{S_\gamma : \gamma \in \Gamma\}.$$

Lemma 3.21 *Let $\Gamma \subset \lambda$ and $\delta \in \lambda \setminus \Gamma$ such that both Γ and $\Gamma \cup \{\delta\}$ are saturated. Then*

$$
\begin{array}{ccc}
R_\delta & \longrightarrow & S_\delta \\
\downarrow & & \downarrow \\
E(\Gamma) & \longrightarrow & E(\Gamma \cup \{\delta\})
\end{array}
$$

is an isometric push-out diagram.

Proof As it is explained in Appendix A.4.2 (the "push-out made with inclusions"), the "isomorphic part" of the assertion holds since $E(\Gamma) \cap S_\delta = R_\delta$ and $\overline{E(\Gamma) + S_\delta} = E(\Gamma \cup \{\delta\})$. We will need however the "isometric" assertion, for which it is necessary to show that, given $x \in E(\Gamma)$ and $s \in S_\delta$ then

$$\inf\{\|x + r\| + \|s - r\| : r \in R_\delta\} \le \|x + s\|.$$

We prove that the statement of the Lemma holds for $\Gamma_\xi := \Gamma \cap \xi$ for all ξ, and we prove it by induction on ξ. For $\xi \le \delta + 1$, what we have to do is to prove the lemma assuming that $\Gamma \subset \delta$. In that case we have a diagram of inclusions

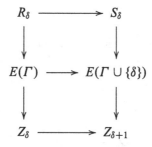

The inclusion $R_\delta \subset E(\Gamma)$ follows from saturation of $E(\Gamma \cup \{\delta\})$. The fact that the lower square is an isometric push-out diagram follows from the fact the whole big square is an isometric push-out diagram.

Assume now that $\xi > \delta$ and ξ is a limit ordinal. This case is a mere limit process: Given $x \in E(\Gamma_\xi)$, $s \in S_\delta$ and $\varepsilon > 0$, we can find $\nu < \xi$ and $x' \in E(\Gamma_\nu)$ such that $\|x' - x\| < \varepsilon/3$ and, by inductive hypothesis, $r \in R_\delta$ such that $\|x' + r\| + \|s - r\| < \|x + s'\| + \varepsilon/3$. The remaining case is that $\xi = \eta + 1 > \delta + 1$. In the nontrivial case, $\eta \in \Gamma$ and $\Gamma_\xi = \Gamma_\eta \cup \{\eta\}$. We start again with $x \in E(\Gamma_\xi)$, $s \in S_\delta$ and $\varepsilon > 0$, and we want to find $r \in R_\delta$ such that $\|x + r\| + \|s - r\| < \|x + s\| + \varepsilon$. Since $E(\Gamma_\xi) = \overline{E(\Gamma_\eta) + S_\eta}$, by a limit argument as before, we can suppose that x is of the form $x = s_\eta + x_\eta$ with $s_\eta \in S_\eta$ and $x_\eta \in E(\Gamma_\eta)$. Notice that now $x_\eta + s_\delta \in Z_\eta$, so since we have a push-out diagram

$$
\begin{array}{ccc}
R_\eta & \longrightarrow & S_\eta \\
\downarrow & & \downarrow \\
Z_\eta & \longrightarrow & Z_{\eta+1}
\end{array}
$$

for $x_\eta + s \in Z_\eta$ and $s_\eta \in S_\eta$, we can find $r_\eta \in R_\eta$ such that

$$\|s_\eta + x_\eta + s\| > \|x_\eta + s + r_\eta\| + \|s_\eta - r_\eta\| - \varepsilon/2.$$

Since $\eta \in \Gamma$ and Γ is saturated, we have that $R_\eta \subset E(\Gamma_\eta)$, hence $x_\eta + r_\eta \in E(\Gamma_\eta)$ and $s \in S_\delta$. So we can now use the inductive hypothesis to find $r \in R_\delta$ such that

$$\|x_\eta + r_\eta + s\| > \|s + r\| + \|x_\eta + r_\eta - r\| - \varepsilon/2.$$

Combining, we get

$$\|s_\eta + x_\eta + s\| > \|s + r\| + \|x_\eta + r_\eta - r\| + \|s_\eta - r_\eta\| - \varepsilon,$$

and applying the triangle inequality,

$$\|s_\eta + x_\eta + s\| > \|s + r\| + \|x_\eta + s_\eta - r\| - \varepsilon,$$

which is exactly what we wanted. $\qquad\qquad\square$

All these efforts lead to:

Proposition 3.22 *All exhaustively tightly $\mathfrak{M}$-filtered spaces are isometric.*

Proof Let $Z, \tilde{Z}$ be two exhaustively tightly $\mathfrak{M}$-filtered spaces, and Let $\{Z_i : i < \lambda\}$ and $\{\tilde{Z}_i : i < \tilde{\lambda}\}$ be the corresponding towers of subspaces and $E(\Gamma)$ and $\tilde{E}(\tilde{\Gamma})$ the subspaces of Z and $\tilde{Z}$ respectively defined for $\Gamma \subset \lambda$ or $\tilde{\Gamma} \subset \tilde{\lambda}$. We define inductively re-enumerations $\lambda = \{\gamma_i : i < \mathfrak{c}\}$ and $\tilde{\lambda} = \{\tilde{\gamma}_i : i < \mathfrak{c}\}$ such that, if we call $\Gamma_i = \{\gamma_j : j < i\}$ and $\tilde{\Gamma}_i = \{\tilde{\gamma}_j : j < i\}$, then all these sets will be saturated and we will have a long commutative diagram with isometries ϕ_i as

$$
\begin{array}{ccccc}
E(\Gamma_0) & \longrightarrow & E(\Gamma_1) & \longrightarrow & \cdots \\
\downarrow{\scriptstyle\phi_0} & & \downarrow{\scriptstyle\phi_1} & & \\
\tilde{E}(\tilde{\Gamma}_0) & \longrightarrow & \tilde{E}(\tilde{\Gamma}_1) & \longrightarrow & \cdots
\end{array}
$$

which shows that Z and $\tilde{Z}$ are isometric as they are the respective unions of the two long rows of the diagram. We define γ_i, $\tilde{\gamma}_i$ and ϕ_i by induction. Formally, it is better to think that a step i we define Γ_i, $\tilde{\Gamma}_i$ and ϕ_i. This way, the limit step of the inductive procedure is obvious. At the successive step, we are given $\phi_i : E(\Gamma_i) \longrightarrow \tilde{E}(\tilde{\Gamma}_i)$ and we need to define γ_i and $\tilde{\gamma}_i$ and an isometry $\phi_{i+1} : E(\Gamma^i \cup \{\gamma_i\}) \longrightarrow \tilde{E}(\tilde{\Gamma}^i \cup \{\tilde{\gamma}_i\})$ extending ϕ_i. We distinguish the cases when i an even or an odd ordinal. If i is even, we declare $\gamma_i = \min(\lambda \setminus \Gamma_i)$. Using property (5) of the definition of exhaustively tightly $\mathfrak{M}$-filtered, we can find $\tilde{\gamma}_i \in \tilde{\lambda} \setminus \tilde{\Gamma}_i$ such that $\tilde{R}_{\tilde{\gamma}_i} = \phi_i[R_{\gamma_i}]$ and there is an isometry $\psi : S_{\gamma_i} \longrightarrow \tilde{S}_{\tilde{\gamma}_i}$ that coincides with ϕ_i on R_{γ_i}. By Lemma 3.21, both diagrams below

$$
\begin{array}{ccc}
R_{\gamma_i} \longrightarrow S_{\gamma_i} & & \tilde{R}_{\tilde{\gamma}_i} \longrightarrow \tilde{S}_{\tilde{\gamma}_i} \\
\downarrow \qquad\quad \downarrow & \text{and} & \downarrow \qquad\quad \downarrow \\
E(\Gamma_i) \longrightarrow E(\Gamma \cup \{\gamma_i\}) & & E(\tilde{\Gamma}_i) \longrightarrow \tilde{E}(\tilde{\Gamma} \cup \{\tilde{\gamma}_i\})
\end{array}
$$

are push-out diagrams. So the isometries can be extended to the desired isometry $\phi_{i+1} : E(\Gamma \cup \{\gamma_i\}) \longrightarrow \tilde{E}(\tilde{\Gamma} \cup \{\tilde{\gamma}_i\})$. In the odd case we proceed similarly but the other way around, we start choosing $\tilde{\gamma}_i = \min(\tilde{\xi} \setminus \tilde{\Gamma}_i)$ and then we choose γ_i.   □

A direct consequence of Lemma 3.20 and Proposition 3.22 is the fact that iterated push-out spaces are isometric in many cases:

Theorem 3.23 *Let $\xi \leq \mathfrak{c}$ be an ordinal of uncountable cofinality.*

1. $\mathscr{S}^\xi(X) \approx \mathscr{S}^{\omega_1}(\mathbb{R})$ *for every Banach space X of density at most $\aleph_1$.*
2. $\mathscr{F}^\xi(X) \approx \mathscr{F}^{\omega_1}(\mathbb{R})$ *for every separable Banach space X.*

From now on we write $\mathscr{S}^{\omega_1}$ and $\mathscr{F}^{\omega_1}$ to denote the isometric type of any of the spaces appearing in the first and second parts of the preceding Theorem. Of course, we know from Proposition 3.15 that $\mathscr{S}^{\omega_1}$ is not isomorphic to $\mathscr{F}^{\omega_1}$: actually the proof gives that the later space is a subspace but not a quotient of the former.

The following result formally upgrades the "universal disposition character" of these spaces:

Theorem 3.24

1. The space $\mathscr{S}^{\omega_1}$ is of universal disposition for all $\mathfrak{S}$-cells $A \to B$ such that $\mathrm{dens}(A) < \mathfrak{c}$.
2. The space $\mathscr{F}^{\omega_1}$ is of universal disposition for all $\mathfrak{F}$-cells $A \to B$ such that $\mathrm{dens}(A) < \mathfrak{c}$.

Proof

1. Let $\kappa < \mathfrak{c}$ be an infinite cardinal below the continuum, and let $\kappa^+ \leq \mathfrak{c}$ be its successor cardinal. Just imitating the proof of Theorem 3.3 one gets that $\mathscr{S}^{\kappa^+}(\mathbb{R})$ is of universal disposition for all $\mathfrak{S}$-cells $A \to B$ such that $\mathrm{dens}(A) \leq \kappa$. But by Theorem 3.23, this space is actually isometric to $\mathscr{S}^{\omega_1}$.
2. Replace $\mathfrak{S}$ by $\mathfrak{F}$ and $\mathscr{S}$ by $\mathscr{F}$ in the proof of Part 1.   □

Omitting the "exhaustiveness" condition (5) in Definition 3.19 one obtains the following result, whose proof is very similar to that of Theorem 3.23, using the auxiliary subspaces $E(\Gamma)$ and back-and-forth. We shall not repeat it here (cf. [16, Theorem 16]).

Theorem 3.25 *Any tightly $\mathfrak{S}$-filtered Banach space of universal disposition for all $\mathfrak{S}$-cells $A \to B$ such that* $\mathrm{dens}(A) < \mathfrak{c}$ *is isometric to* $\mathscr{S}^{\omega_1}$.

The proofs of Theorems 3.23 and 3.24 follow ideas from [16], where the main result was a restricted version of Theorems 3.24 and 3.25. We notice that, although it is assumed in [16, Theorem 16] that $\mathfrak{c}$ is regular, this assumption was only used in the *existence* part of the proof, where by the previously exposed arguments it is anyway unnecessary.

3.3 Spaces of Almost Universal Disposition

In this section we adapt the main construction to obtain Gurariy's separable space of almost universal disposition. Recall that a Banach space U is of almost universal disposition if given isometries $\imath : A \to B, u : A \to U$ with B finite dimensional and $\varepsilon > 0$ there is an ε-isometry $u' : B \to U$ such that $u\imath = u$. Spaces of almost universal disposition provide a general framework for the constructions of the preceding sections (since "universal disposition" implies "almost universal disposition") and, moreover, several properties of spaces of universal disposition rely in the end in the "almost" character. For instance, we shall prove later that no space of almost universal disposition is complemented in a C-space; and we will see in Chap. 4 that ultraproducts of spaces of almost universal disposition are of universal disposition for separable Banach spaces (which leads to an alternative construction for the spaces of this chapter). Finally, there is a wide variety of spaces of almost universal disposition in the literature; some of them will be reviewed in Sect. 3.4.1.

3.3.1 Construction of Separable Spaces of Almost Universal Disposition

Let us show now how the basic method explained above can produce the Gurariy space. To keep our spaces separable we fix a countable system of isometries $\mathfrak{J}_0$ having the following density property: given an isometry $w : A \to B$ between finite dimensional spaces, and $\varepsilon > 0$, there is $u \in \mathfrak{J}_0$, and surjective ε-isometries

$\alpha : A \to \operatorname{dom} u$ and $\beta : B \to \operatorname{cod} u$ making the square

$$
\begin{array}{ccc}
A & \xrightarrow{\ w\ } & B \\[2pt]
{\scriptstyle\alpha}\big\downarrow & & \big\downarrow{\scriptstyle\beta} \\[2pt]
\operatorname{dom} u & \xrightarrow{\ u\ } & \operatorname{cod} u
\end{array}
\tag{3.5}
$$

commutative.

Although there is no need to be that specific let us indicate one possible choice for $\mathfrak{J}_0$. Let us say that a norm on $\mathbb{R}^n$ is "rational" if its unit ball is the convex hull of finitely many points whose coordinates are rational numbers. A "rational" (normed) space is just $\mathbb{R}^n$ furnished with a rational norm. Consider the family of those isometries f whose codomain is a rational normed space $(\mathbb{R}^n, \|\cdot\|)$, its domain is $\mathbb{R}^m$ for some $m \le n$ equipped with a not necessarily rational norm and have the form

$$
f(x_1, \ldots, x_m) = (x_1, \ldots, x_m, 0, \ldots, 0).
$$

An obvious compactness argument shows that these into isometries are "dense" amongst all finite dimensional isometries. Once $\mathfrak{J}_0$ has been fixed, set $\mathfrak{F}_0 = \operatorname{dom} \mathfrak{J}_0$.

Let now X be a separable Banach space. We define an increasing sequence of Banach spaces

$$
G^0 \longrightarrow G^1 \longrightarrow G^2 \longrightarrow \cdots
$$

depending on X, as follows. We start with $G^0 = X$. Assuming G^n has been defined we get G^{n+1} from the basic construction explained in Sect. 3.1.1 just taking $\mathfrak{L}_n$ as a countable set of G^n-valued contractions with domains in $\mathfrak{F}_0$ such that, for every $\varepsilon > 0$, and every ε-isometry $s : F \to G^n$, with $F \in \mathfrak{F}_0$, there is $t \in \mathfrak{L}_n$ such that $\|s - t\| < \varepsilon$. We consider the index set $\Gamma_n = \{(u, t) \in \mathfrak{J}_0 \times \mathfrak{L}_n : \operatorname{dom} u = \operatorname{dom} t\}$ and the push-out diagram

$$
\begin{array}{ccc}
\ell_1(\Gamma_n, \operatorname{dom} u) & \xrightarrow{\ \oplus \mathfrak{J}_0\ } & \ell_1(\Gamma_n, \operatorname{cod} u) \\[2pt]
{\scriptstyle\Sigma\mathfrak{L}_n}\big\downarrow & & \big\downarrow \\[2pt]
G^n & \longrightarrow & \mathrm{PO}
\end{array}
\tag{3.6}
$$

Then we set $G^{n+1} = \mathrm{PO}$. The linking map $G^n \to G^{n+1}$ is given by the lower arrow in the push-out diagram.

Proposition 3.26 *Let X be a separable Banach space. The space $G^\omega(X) = \varinjlim_n G^n$ is a separable Banach space of almost universal disposition.*

Proof Let $w : A \to B$ and $s : A \to G^\omega$ be isometries, with B a finite dimensional space and fix $\varepsilon > 0$. Choose $u \in \mathfrak{J}_0$, as in (3.5). Clearly, for m large enough there is a contractive ε-isometry $t : \operatorname{dom} u \to G^m$ satisfying $\|s - t\alpha\| < \varepsilon$. Let $t' : \operatorname{cod} u \to G^{m+1}$ be the extension provided by Diagram 3.6, so that t is a contractive ε-isometry such that $t'u = t$. Therefore $t'\beta$ is a contractive 2ε-isometry satisfying $\|s - t'\beta w\| \le \varepsilon$. The following perturbation result ends the proof.

Lemma 3.27 *A Banach space U is of almost universal disposition if and only if, given isometries $u : A \to U$ and $\imath : A \to B$ with B finite dimensional, and $\varepsilon > 0$, there is an ε-isometry $u' : B \to U$ such that $\|u - u'\imath\| \le \varepsilon$.*

Proof If $\{b_i : 1 \le i \le n\}$ is a basis for B, then for every $\varepsilon > 0$ there is δ (depending on ε and the basis) such that if $t : B \to U$ is linear map with $\|t(b_i)\| \le \delta$ for every $1 \le i \le n$, then $\|t\| \le \varepsilon$.   □

We shall prove later that there is a unique separable Banach space of almost universal disposition, up to isometries; see Sect. 3.3.5. A weaker (and easier) result which suffices for most purposes is contained in Proposition 3.29. We now study a number of structural properties which do not depend on this fact.

3.3.2 Operator Extension Properties

Let us observe that another way to say that a Banach space U is of almost universal disposition is: given $\varepsilon > 0$ and an isometry $g : A \to B$, with B finite dimensional and A a subspace of U, there is an ε-isometry $g : B \to U$ such that $f(g(x)) = x$, for all $x \in A$.

This (trivially) equivalent formulation will make (surprisingly) easier to prove in Sect. 3.3.5 the main isometric properties of the Gurariy space, in particular its uniqueness. We will also need in Chap. 4 the next two results, which already appear in Gurariy's work [118], in order to prove that ultrapowers of the Gurariy space have nice *transitivity* properties (see Proposition 4.16).

Lemma 3.28 *Let U be a Banach space of almost universal disposition for finite dimensional spaces and let $g : A \to B$ be a linear embedding, where B is finite dimensional and A is a subspace of U. Then for each $\varepsilon > 0$ there is an embedding $f : B \to U$ such that $f(g(a)) = a$ for every $a \in A$ and $\|f\| \le (1 + \varepsilon)\|g^{-1}\|$ and $\|f^{-1}\| \le (1 + \varepsilon)\|g\|$.*

Proof Let us make push-out with the canonical embedding $g[A] \to B$ and $h = g^{-1}/\|g^{-1}\|$ as follows:

$$
\begin{array}{ccc}
g(A) & \longrightarrow & B \\
\Big\downarrow{\scriptstyle h} & & \Big\downarrow{\scriptstyle h'} \\
A & \overset{\imath}{\longrightarrow} & \mathrm{PO}
\end{array}
$$

According to Lemma A.19 in the Appendix, $\|h'\| \leq 1$ and, since $\|h\| = 1$, the embedding ι is an isometry. Now, let $\tau : PO \to U$ be an ε-isometry such that $\tau \iota (a) = a$, for every $a \in A$. Then, $f = \|g^{-1}\| \tau h$ is an embedding which obviously verifies $fg(a) = a$, for all $a \in A$, and $\|f\| \leq (1 + \varepsilon)\|g^{-1}\|$. On the other hand, again by Lemma A.19, we have $\|h'^{-1}\| \leq \max\{1, \|h^{-1}\|\}$ then, $\|h'^{-1}\| \leq 1$ since $\|h^{-1}\| = \|g\|\|g^{-1}\| \geq 1$, and therefore

$$\|f^{-1}\| = \|h^{-1}\tau^{-1}\|/\|g^{-1}\| \leq \|g\|/(1 - \varepsilon). \quad \square$$

Proposition 3.29 *Let U and V be separable Banach spaces of almost universal disposition. Suppose that $\varphi_0 : A \to B$ is a linear isomorphism, where A and B are finite dimensional subspaces of U and V, respectively. Then, for each $\varepsilon > 0$, there is a surjective isomorphism $\varphi : U \to V$ extending φ_0 and such that $\|\varphi\| \leq (1+\varepsilon)\|\varphi_0\|$ and $\|\varphi^{-1}\| \leq (1 + \varepsilon)\|\varphi_0^{-1}\|$.*

In particular U and V are almost isometric: for every $\varepsilon > 0$ there is a surjective ε-isometry from U onto V.

Proof The result follows from a simple back-and-forth argument analogous to the one used in the proof of Proposition 3.7(2): let (ε_n) be a sequence of positive integers such that $\prod_n (1 + \varepsilon_n) \leq 1 + \varepsilon$ and write $U = \bigcup_n U_n$ where (U_n) is an increasing sequence of finite dimensional subspaces of U starting with $U_0 = A$. Also, let (V_n) be an increasing sequence of finite dimensional subspaces of V such that $V = \bigcup_n V_n$, with $V_0 = B$.

Let $\varphi_0 : A \to B$ be an isomorphic embedding. By Lemma 3.28, let $\psi_1 : V_1 \to U$ be an extension of $\varphi_0^{-1} : \varphi_0[U_0] \to U$, with $\|\psi_1\| \leq (1 + \varepsilon_1)\|\varphi_0^{-1}\|$ and $\|\psi_1^{-1}\| \leq (1 + \varepsilon_1)\|\varphi_0\|$. Let then $\varphi_2 : \psi_1[V_1] + U_2 \to V$ be an extension of $\psi_1^{-1} : \psi_1[V_1] \to V$ such that such that $\|\varphi_2\| \leq (1 + \varepsilon_2)\|\psi_1^{-1}\|$ and $\|\varphi_2^{-1}\| \leq (1 + \varepsilon_2)\|\psi_1\|$ provided by Lemma 3.28. Proceeding by induction one gets a couple of endomorphisms φ and ψ such that $\psi\varphi = 1_U, \varphi\psi = 1_V$, with $\|\varphi\| \leq (1+\varepsilon)\|\varphi_0\|$ and $\|\psi\| \leq (1+\varepsilon)\|\varphi_0^{-1}\|$ and $\varphi_{|A} = \varphi_0$. $\quad \square$

The canonical embedding of X into $G^\omega(X)$ provided by the first step of the construction enjoys the analogous "approximate" property to the universal property of the embedding $X \to \mathscr{S}^{\omega_1}(X)$ appearing in Proposition 3.8:

Proposition 3.30

1. *Every norm one operator from X into a Lindenstrauss space admits, for every $\varepsilon > 0$, an extension to $G^\omega(X)$ of norm at most $1 + \varepsilon$.*
2. *If X is a separable Lindenstrauss space, then one can construct $G^\omega(X)$ so that there is a contractive projection from $G^\omega(X)$ onto X.*

Proof

1. Given $\varepsilon > 0$ we fix a sequence (ε_n) such that $\prod(1 + \varepsilon_n) \leq 1 + \varepsilon$. Now, let $\mathscr{L}$ be a Lindenstrauss space and $\tau : X \to \mathscr{L}$ be a norm one operator.

Look at the diagram

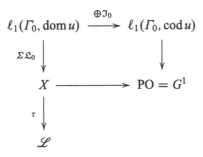

and consider the composition $\tau \circ \Sigma \mathfrak{L}_0$. Since $\mathscr{L}$ is a Lindenstrauss space, for each fixed $(u, t) \in \Gamma_0$, the restriction of $\tau \circ \Sigma \mathfrak{L}_0$ to the corresponding "coordinate" maps $\operatorname{dom} u = \operatorname{dom} t$ into a finite dimensional subspace of $\mathscr{L}$ and so it is contained in a $(1 + \varepsilon_1)$-isomorph of some finite dimensional ℓ_∞^n. Therefore, it can be extended to $\operatorname{cod} u$ through u with norm at most $(1 + \varepsilon_1)$. The coproduct of all these extensions yields thus an extension $T : \ell_1(\Gamma_0, \operatorname{cod} u) \to \mathscr{L}$ of $\tau \circ \Sigma \mathfrak{L}_0$ with norm at most $(1 + \varepsilon_1)$. The push-out property of $PO = G^1$ yields therefore an operator $\tau_1 : G^1 \to \mathscr{L}$ that extends τ with norm at most $(1 + \varepsilon_1)$. Iterating the process ω times, working with $(1 + \varepsilon_n)$ at step n, one gets an extension $\tau_\omega : G^\omega \to \mathscr{L}$ of τ with norm at most $\prod(1 + \varepsilon_n) \leq 1 + \varepsilon$.

2. We shall establish a slightly more general result, namely, that if $\mathscr{L}$ is a separable Lindenstrauss space and $\tau : X \to \mathscr{L}$ is surjective, then one can construct $G^\omega(X)$ so that τ admits an extension $\tau_\omega : G^\omega(X) \to \mathscr{L}$ with $\|\tau_\omega\| = \|\tau\|$. Taking $\tau = \mathbf{1}_X$ gives (2).

The proof is a re-examination of the preceding one, using induction. We shall construct $G^n(X)$ as in Sect. 3.3.1 together with an operator $\tau_n : G^n(X) \to \mathscr{L}$ extending τ, with $\|\tau_n\| = \|\tau\|$. These operators must be compatible in the sense that the restriction of τ_{n+1} to $G^n(X)$ agrees with τ_n. The initial step is obvious: we take $G^0(X) = X$ and $\tau_0 = \tau$.

Now suppose we have constructed $\tau_n : G^n(X) \to \mathscr{L}$ extending τ, with $\|\tau_n\| = \|\tau\|$. Let us observe that if $\pi : Y \to Z$ is a surjective operator acting between Banach spaces and $\bigcup_k Z_k$ is dense in Z, then $\bigcup_k \pi^{-1}[Z_k]$ is dense in Y. This is just the open mapping theorem. By a result of Michael and Pełczyński one can write $\mathscr{L} = \overline{\bigcup_k L_k}$, where L_k is isometric to ℓ_∞^k and so 1-injective; see [198, Theorem 1.1] or [173, Theorem 3.2]. Set $U_k = \tau_n^{-1}[L_k]$ so that $\bigcup_k U_k$ is dense in $G^n(X)$ since τ_n is onto. Take care here to achieve that all the operators in $\mathfrak{L}_n$ have range in $\bigcup_k U_k$, which can be done since every operator in $\mathfrak{L}_n$ has finite dimensional domain and $\bigcup_k U_k$ is dense in $G^n(X)$. After that, look

at the diagram

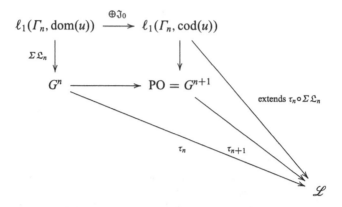

and let us argue as before. For each $(u, t) \in \Gamma_n$ the restriction of $\Sigma \mathcal{L}_n$ to dom $u =$ dom t into some U_k and composing with τ_n one falls in $L_k \approx \ell_\infty^k$. Therefore it can be extended to $\mathrm{cod}(u)$ without increasing the norm. The coproduct of all these extensions gives the arrow on the left of the diagram, still with the same norm as τ_n. Finally the universal property of the push-out construction yields τ_{n+1} and shows that it is an extension of τ_n with the same norm; see Appendix A.4.1. This completes the inductive step and we are done.   □

Corollary 3.31 *Let U be a separable Banach space of almost universal disposition. Then for every separable Lindenstrauss space $\mathcal{L}$ and every $\varepsilon > 0$ there is an ε-isometry $u : \mathcal{L} \to U$ and linear map $\pi : U \to \mathcal{L}$ such that $\pi \circ u = \mathbf{1}_{\mathcal{L}}$, with $\|\pi\| \leq 1 + \varepsilon$. In particular U is not isomorphic to a complemented subspace of any M-space.*

The first part is a juxtaposition of Propositions 3.29 and 3.30. The second part follows from the fact, proved in [34], that there exist separable Lindenstrauss spaces that are not complemented subspaces of any M-space. We shall prove later that the Corollary is true even with $\varepsilon = 0$ (see Corollary 3.39).

3.3.3 The Structure of a Space of Almost Universal Disposition

In this section and the next one we present some remarkable features of the spaces of almost universal disposition. We begin with the following observation.

Lemma 3.32 *Let X be a separable subspace of a space of almost universal disposition U. Then there is another separable subspace X′ containing X having the following property: given isometric embeddings $u : A \to X$ and $\imath : A \to B$ with*

B finite dimensional, and $\varepsilon > 0$, there is an ε-isometric embedding $u' : B \to X'$. such that $\|u - u'\iota\| \le \varepsilon$.

Proof Let $\mathfrak{J}_0$ and $\mathfrak{F}_0$ be as in Sect. 3.3.1, that is, $\mathfrak{J}_0$ is a "dense" countable family of isometries between finite dimensional spaces and $\mathfrak{F}_0 = \operatorname{dom}\mathfrak{J}_0$. As X is separable we may select a countable set of injective operators $\mathfrak{L}$ from the spaces in $\mathfrak{F}_0$ to X with the following density property: for every injective operator $f : E \to X$ and every $\varepsilon > 0$ there is $f' : E \to X$ in $\mathfrak{L}$ such that $\|f - f'\| < \varepsilon$.

Suppose now that we are given $f \in \mathfrak{L}$ and $v : E \to F$ in $\mathfrak{J}_0$ with $\operatorname{dom}(f) = \operatorname{dom}(v)$. Then we may use Lemma 3.28 to obtain, for each integer n, an injective operator $f'_n : F \to U$ such that $f = f'_n v$, with $\|f'_n\| \le (1 + \frac{1}{n})\|f^{-1}\|$ and $\|(f'_n)^{-1}\| \le (1 + \frac{1}{n})\|f\|$. Let X' be the least closed subspace of U containing X and the range of all these f'_n. Clearly, X' is separable. Let us check that X' has the property we claimed. Take isometries $u : A \to U$ and $\iota : A \to B$ with B finite dimensional and fix $\varepsilon > 0$. Take an "approximation" of ι in $\mathfrak{J}_0$, that is, an isometry $v : E \to F$ in $\mathfrak{J}_0$ fitting in a commutative square

$$
\begin{array}{ccc}
E & \xrightarrow{\;v\;} & F \\
\iota\downarrow & & \downarrow s \\
A & \xrightarrow[\;\iota\;]{} & B
\end{array}
$$

in which the vertical arrows are bijective δ-isometries for small δ. Consider the composition $ut : E \to U$ and take $f \in \mathfrak{L}$ such that $\|f - ut\| \le \delta$. Taking δ small enough guarantees that any such a f has to be an almost-isometry which can be extended to an almost-isometry $f' : F \to X'$. The composition $u' = f's^{-1}$ is an ε-isometry and $\|u - u'\iota\| < \varepsilon$—for δ sufficiently small. $\qquad\square$

Which leads to:

Proposition 3.33 *A Banach space is of almost universal disposition if and only if every separable subspace is contained in a separable subspace of almost universal disposition.*

Proof The "if" part is obvious. As for the converse just note that the preceding Lemma allows us to reproduce the construction explained in Sect. 3.3.1 inside any space of almost universal disposition. Let X be a separable subspace of U. Applying the Lemma recursively we obtain an increasing sequence of separable subspaces (X_n), starting $X_0 = X$, having the following property: if $u : A \to X_m$ and $\iota : A \to B$ are isometric embeddings, then for every $\varepsilon > 0$ there is an ε-isometry $u' : B \to X_{m+1}$ such that $\|u - u'\iota\| < \varepsilon$. By Lemma 3.27 the closure of $\bigcup_n X_n$ in U is a separable space of almost universal disposition. $\qquad\square$

3.3.4 Spaces of Almost Universal Disposition Are Not Complemented in Any C-Space

One of the main open problems about injective spaces is to know if every injective space must be isomorphic to a C-space. We already know that there exist separably injective spaces which are not isomorphic to a complemented subspace of any C-space. We now show that even (universally) 1-separably injective spaces can have this pathological behavior, in contrast with the fact that every 1-injective space is isometric to a $C(K)$-space for some extremely disconnected compact space K.

We begin with the elementary observation that no C-space can be of almost universal disposition. The idea is that the metric of $C(K)$ "knows" how large is the support of a given function. Consider the pair of isometries

$$\imath : \mathbb{K} \to \ell_\infty^2, \quad \imath(t) = (t, 0); \qquad u : \mathbb{K} \to C(K), \quad u(t) = t1_K.$$

Let $u' : \ell_\infty^2 \longrightarrow C(K)$ be any operator extending u and look at $u'(0, 1) = f$. One has $\|f\|_\infty = \sigma f(k)$ for some $k \in K$ and some $\sigma \in \mathbb{K}$ with $|\sigma| = 1$. Clearly

$$\|u'(0, 1)\|_\infty = \|f\|_\infty$$

$$\|u'(\sigma, 1)\|_\infty = 1 + \|f\|_\infty$$

and these numbers cannot be simultaneously close to 1. In particular u' cannot be an ε-isometry for $\varepsilon < \frac{1}{2}$.

Of course this naive argument does not apply to the $C_0(L)$-spaces, let alone to M-spaces. However one has the following result, which is a formal consequence of Corollary 3.31:

Theorem 3.34 *Banach spaces of almost universal disposition are not isomorphic to complemented subspaces of M-spaces. In particular they are not injective.*

Proof Suppose U is a space of almost universal disposition and there is an into isomorphism $e : U \to M$ and an operator $\pi : M \to U$ such that $\pi e = 1_U$, where M is an M-space. We are going to reach a contradiction.

Let us first remark that every separable subspace of a Banach lattice is contained in a separable sublattice: this follows from the continuity of the lattice operations. We known that U contains almost isometric copies of all separable Banach spaces. Take G_0 a Gurariy subspace of U and let M_0 be the least sublattice of M containing $e[G_0]$ and U_0 the closure of $\pi[M_0]$ in U. Proposition 3.33 provides a Gurariy subspace $G_1 \subset U$ containing U_0. Now, replace G_0 by G_1 and go ahead. In this way we obtain a sequence of arrows organized as follows:

$$G_0 \longrightarrow U_0 \longrightarrow G_1 \longrightarrow U_1 \longrightarrow G_2 \longrightarrow \cdots$$

$$e \downarrow \qquad \pi \uparrow \qquad e \downarrow \qquad \pi \uparrow \qquad e \downarrow$$

$$e[G_0] \longrightarrow M_0 \longrightarrow e[G_1] \longrightarrow M_1 \longrightarrow e[G_2] \longrightarrow \cdots$$

Unlabeled arrows are just inclusions. Look at the space

$$V = \overline{\bigcup_n G_n} = \overline{\bigcup_n U_n}. \tag{3.7}$$

It is pretty clear that the normed space $\bigcup_n G_n$ is of almost universal disposition and Lemma 3.27 implies that so is its closure V.

On the other hand e embeds the Gurariy space $V = \overline{\bigcup_n U_n}$ in $L = \overline{\bigcup_n M_{n+1}}$ while the restriction of π to L is left inverse to e and so V is (isomorphic to a subspace) complemented in L. Finally, L is a (separable) sublattice of M hence it is itself an M-space. A contradiction. $\qquad \Box$

Corollary 3.35 *None of the spaces $\mathscr{S}^{\omega_1}(X)$, $\mathscr{U}^{\omega_1}(X)$ or $\mathscr{F}^{\omega_1}(X)$, for whichever initial X, are complemented subspaces of an M-space, let alone of a C-space.*

Gelfand's representation theorem states that every complex commutative C^*-algebra is isometrically isomorphic to the algebra of all complex continuous functions on some compact space, with the sup norm. Thus, Theorem 3.34 and the arguments contained in Sect. 2.7.2 show that no complex space of almost universal disposition is isomorphic to a complemented subspace of a commutative complex C^*-algebra. The commutativity is essential since a beautiful result of Lusky [191] establishes that every separable complex Lindenstrauss space is a 1-complemented subspace of the so-called CAR algebra—a certain separable, noncommutative, C^*-algebra.

3.3.5 Isometric Uniqueness of the Gurariy Space

This section is devoted to prove Lusky's result [185] that any two separable Banach spaces of almost universal disposition are isometric. Our exposition follows a more recent approach due to Kubiś and Solecki [170] who found a new proof of the isometric uniqueness of the Gurariy based in the next lemma, which can be regarded as a way to turn almost isometries into isometries.

Lemma 3.36 *Let X and Y be normed spaces and $f : X \to Y$ an ε-isometry, with $\varepsilon \in (0, 1)$. Let $i : X \to X \oplus Y$ and $j : Y \to X \oplus Y$ be the canonical inclusions. Then there is a norm on $X \oplus Y$ such that $\|f \circ j - i\| \le \varepsilon$ and both i and j are isometries.*

Proof Set

$$\|(x, y)\| = \inf \{ \|x_0\|_X + \|y_1\|_Y + \varepsilon\|x_2\|_X : (x, y) = (x_0, 0) + (0, y_1) + (x_2, -f(x_2)) \}.$$

It is easily seen that this formula defines a norm on $X \oplus Y$. Let us check that $\|(x, 0)\| = \|x\|_X$ for all $x \in X$. The inequality $\|(x, 0)\| \leq \|x\|_X$ is obvious. As for the converse, suppose $x = x_0 + x_2$ and $y_1 = f(x_2)$. Then

$$\begin{aligned}
\|x_0\|_X + \|y_1\|_Y + \varepsilon\|x_2\|_X &= \|x_0\|_X + \|f(x_2)\|_Y + \varepsilon\|x_2\|_X \\
&\geq \|x_0\|_X + (1 - \varepsilon)\|x_2\|_X + \varepsilon\|x_2\|_X \\
&\geq \|x\|_X,
\end{aligned}$$

as required. Next we prove that $\|(0, y)\| = \|y\|_Y$ for every $y \in Y$. That $\|(0, y)\| \leq \|y\|_Y$ is again obvious. To prove the reversed inequality assume $x_0 + x_2 = 0$ and $y = y_1 - f(x_2)$,

$$\begin{aligned}
\|x_0\|_X + \|y_1\|_Y + \varepsilon\|x_2\|_X &= \|x_2\|_X + \|y_1\|_Y + \varepsilon\|x_2\|_X \\
&\geq \|y_1\|_Y + (1 + \varepsilon)\|x_2\|_X \\
&\geq \|y_1\|_Y + \|f(x_2)\|_Y \\
&\geq \|y\|_Y.
\end{aligned}$$

To end, let us estimate $\|j \circ f - i\|$. We have

$$\|j \circ f - i\| = \sup_{\|x\| \leq 1} \|j(f(x)) - i(x)\| = \sup_{\|x\| \leq 1} \|(-x, f(x))\| \leq \varepsilon$$

and we are done.   □

We are about seeing that the previous lemma, innocent as it seems, is the crucial tool for proving all isometric properties of separable spaces of almost universal disposition. We first need to introduce a new definition: an operator $f : X \to Y$, which is not assumed to be surjective, is called a *strict ε-isometry*, with $\varepsilon \in [0, 1)$, if

$$(1 - \varepsilon)\|x\|_X < \|f(x)\|_Y < (1 + \varepsilon)\|x\|_X$$

for every nonzero $x \in X$. Note that when X is finite dimensional, every strict ε-isometry is an ε'-isometry for some $\varepsilon' < \varepsilon$.

Lemma 3.37 *Let U be a space of almost universal disposition for finite dimensional spaces and let $f : X \to Y$ be a strict ε-isometry, where X and Y are finite dimensional Banach spaces, with $X \subset U$ and $\varepsilon \in (0, 1)$. Then for each $\delta > 0$ there exists a δ-isometry $g : Y \to U$ such that $\|g(f(x)) - x\| < \varepsilon\|x\|$ for all $x \in X$.*

Proof Choose $0 < \varepsilon' < \varepsilon$ such that f is an ε'-isometry. Shrinking δ if necessary, we may assume that $(1 + \delta)\varepsilon' < \varepsilon$. Let Z denote the direct sum $X \oplus Y$ equipped with the norm given by Lemma 3.36 and let $i : X \to Z$ and $j : Y \to Z$ denote the canonical inclusions, so that $\|j \circ f - i\| < \varepsilon'$. Let $h : Z \to U$ be a δ-isometry such that $h(i(x)) = x$ for $x \in X$. Then $g = h \circ j$ is a δ-isometry from Y into U and for every nonzero $x \in X$,

$$\|x - g(f(x))\| = \|h(i(x)) - h(j(f(x)))\|$$
$$\leq (1 + \delta)\|i(x) - j(f(x))\|_Z \leq (1 + \delta)\varepsilon'\|x\| < \varepsilon\|x\|,$$

as required.   □

We are now ready for the proof of the uniqueness.

Theorem 3.38 *Let U and V be separable spaces of almost universal disposition and $\varepsilon \in (0, 1)$. Let $f : X \to V$ be a strict ε-isometry, where X is a finite dimensional subspace of U. Then there exists a bijective isometry $h : U \to V$ such that $\|h(x) - f(x)\| \leq \varepsilon\|x\|$ for every $x \in X$. In particular, U and V are isometrically isomorphic.*

Proof Set $X_0 = X$, $Y_0 = f[X]$, $f_0 = f$ and $\varepsilon_0 = \varepsilon$. Take $\varepsilon_1 < \varepsilon_0$, by Lemma 3.37 there exists a strict ε_1-isometry $g_0 : Y_0 \to U$ (which is also ε_0-strict isometry) such that $\|g_0 f_0 x - x\| \leq \varepsilon_0\|x\|$ for all $x \in X_0$. Set now $X_1 = g_0[Y_0]$, again by Lemma 3.37 there is a strict ε_1-isometry $f_1 : X_1 \to V$ such that $\|f_1 g_0(y) - y\| \leq \varepsilon_1\|y\|$, for all $y \in Y_0$. The next step is clear: fix $Y_1 = f_1[X_1]$, which obviously contains Y_0, take $\varepsilon_2 < \varepsilon_1$ and get a strict ε_2-isometry $g_1 : Y_1 \to X_2$ such that $\|g_1 f_1 x - x\| < \varepsilon_1\|x\|$, for all $x \in X_1$. In this way, we will construct inductively sequences of linear operators $f_n : X_n \to Y_n$ and $g_n : Y_n \to X_{n+1}$ between finite dimensional spaces such that each f_n is a strict ε_n-isometry, each g_n is a strict ε_{n+1}-isometry, $X_n \subset X_{n+1}$, $Y_n \subset Y_{n+1}$ and the spaces $\bigcup_n X_n$ and $\bigcup_n Y_n$ are dense in U and V, respectively. Furthermore the next conditions are verified for every n:

1. $\|g_n f_n(x) - x\| < \varepsilon_n\|x\|$, for $x \in X_n$
2. $\|f_{n+1} g_n(y) - y\| < \varepsilon_{n+1}\|y\|$, for $y \in Y_n$.

Let us see now that $(f_n(x))_n$ is a Cauchy sequence. Fix $n \in \mathbb{N}$ and $x \in X_n$ with $\|x\| = 1$. Using (2), we get

$$\|f_{n+1} g_n f_n(x) - f_n(x)\| < \varepsilon_{n+1} \cdot \|f_n(x)\| \leq \varepsilon_{n+1} \cdot (1 + \varepsilon_n),$$

and using (1),

$$\|f_{n+1} g_n f_n(x) - f_{n+1}(x)\| \leq \|f_{n+1}\| \cdot \|g_n f_n(x) - x\| < (1 + \varepsilon_{n+1}) \cdot \varepsilon_n.$$

These inequalities give

$$\|f_n(x) - f_{n+1}(x)\| < (\varepsilon_n + \varepsilon_{n+1})^2.$$

Given now $x \in \bigcup_n X_n$, define $h(x) = \lim_{n \geq m} f_n(x)$, where m is such that $x \in X_m$. Then h is an ε_n-isometry for every $n \in \mathbb{N}$, hence it is an isometry. Consequently, it extends to an isometry on U, which we denote also by h. Furthermore, assuming that the sequence (ε_n) has been properly chosen, we get

$$\|f(x) - h(x)\| \leq \sum_{n=0}^{N} (\varepsilon_n + \varepsilon_{n+1})^2 + \|f_{n+1}(x) - h(x)\| < \varepsilon.$$

It remains to see that h is a bijection. To this end, we check as before that $(g_n(y))_{n \geq m}$ is a Cauchy sequence for every $y \in Y_m$. Once this is done, we obtain an isometry g defined on V. Conditions (1) and (2) tell us that $g \circ h = \mathbf{1}_U$ and $h \circ g = \mathbf{1}_V$. This completes the proof.   □

Let us denote by $\mathscr{G}$ the isometric type of any separable Banach space of almost universal disposition and call it the Gurariy space.

Corollary 3.39 *The Gurariy space contains an isometric copy of every separable Banach space. Moreover, it contains a copy of each separable Lindenstrauss which is the range of a contractive projection. The Gurariy space is not complemented in any M-space, let alone in a C-space.*

Universality follows now from the fact that all the spaces $G^\omega(X)$ are isometric for every separable Banach space X. Yet, using Lemma 3.36, one can easily get a direct proof of the fact that $\mathscr{G}$ contains an isometric copy of every separable Banach space just proceeding analogously to the proof of Theorem 3.38. The second part follows from the second part of Proposition 3.30 and Theorem 3.38.

3.4 Notes and Remarks

3.4.1 Other Spaces of Almost Universal Disposition in the Literature

A number of authors constructed spaces of almost universal disposition with rather special properties. Let us mention a few of them.

Motivated by the uniqueness of the Gurariy space, Lusky showed in [188] that for every cardinal $\aleph \geq \mathfrak{c}$ there exist two Banach spaces of almost universal disposition with density character $\aleph$ which are not isomorphic: one of them has weakly* sequentially compact dual ball and the other not. Certainly, the spaces $\mathscr{F}^{\omega_1}$ and $\mathscr{S}^{\omega_1}$, which are even of universal disposition, provide similar counter-examples. However, what it is proven in [188, Theorem 1.4] is that every infinite dimensional Lindenstrauss space X is isometric to the range of a contractive projection defined on a space of almost universal disposition G_X such that $\operatorname{dens}(X) = \operatorname{dens}(G_X)$ and $\operatorname{dens}(X^*) = \operatorname{dens}(G_X^*)$. Besides, the unit ball of G_X^* is weakly* sequentially compact if (and only if) so is that of X^*. Thus for instance, taking first $X = c_0(\mathfrak{c})$ and then $X = \ell_\infty$ one gets the announced examples.

Under the so called "diamond axiom" $\diamondsuit$ (which implies CH), Shelah constructed in [231] a Banach space of almost universal disposition $\mathscr{S}_\diamond$ with density character $\aleph_1$ and no uncountable biorthogonal sequences. The space $\mathscr{S}_\diamond$ is of almost universal disposition, hence a Lindenstrauss space, since it admits a representation $\mathscr{S}_\diamond = \bigcup_{i<\omega_1} \mathscr{G}_i$ in which each $\mathscr{G}_i$ is a Gurariy space and the chain is increasing and continuous. Among the strange properties of $\mathscr{S}_\diamond$ one encounters

1. [143, Corollary 4.5] $\mathscr{S}_\diamond$ is a subspace of ℓ_∞. Moreover, every subspace and quotient of $\mathscr{S}$ is isometric to a subspace of ℓ_∞ under any equivalent renorming.
2. Among any $\aleph_1$ elements of $\mathscr{S}_\diamond$, one of them belongs to the closure of the convex hull of the others.
3. Every quotient of $\mathscr{S}_\diamond$ by a nonseparable subspace is separable.
4. Every operator $T : \mathscr{S}_\diamond \to \mathscr{S}_\diamond$ has the form $\lambda \mathbf{1}_{\mathscr{S}_\diamond} + S$ with S an operator having separable range.
5. [143, Corollary 4.4] For every equivalent norm on $\mathscr{S}_\diamond$, its dual $\mathscr{S}_\diamond{}^*$ has a separable boundary (i.e., a set $\mathscr{B}$ such that for every element x there is some $f \in \mathscr{B}$ such that $|f(x)| = \|x\|$).

We do not know if $\mathscr{S}_\diamond$ is of universal disposition. Property (1) implies that it cannot be of universal disposition for separable spaces since in that case it would contain isometric copies of all spaces having density character $\aleph_1$; and certainly, say, $\ell_2(\aleph_1)$ is not a subspace of ℓ_∞.

Shelah space $\mathscr{S}_\diamond$ is related to an earlier counter-example of Kunen who constructed, under CH, a compact space $\mathscr{K}$, so that $C(\mathscr{K})$ enjoys properties (1), (2) and (5) above (see [143, Corollary 4.5], [202, Theorem 7.1] and [143, Corollary 4.4], respectively). The papers [113, 114] contain additional information about the spaces $\mathscr{S}_\diamond$ and $C(\mathscr{K})$.

López-Abad and Todorcevic [184] pushed these ideas into the set-theoretic forcing setting to produce a variety of pairs of Banach spaces X and Y such that:

- X is of almost universal disposition.
- Y is a c_0-saturated predual of $\ell_1(\omega_1)$.
- Y is isometric to the quotient of X by a (separable) subspace isometric to Gurariy space $\mathscr{G}$.

This last condition implies that X is, in some sense, close to Y and both spaces must share many Banach space properties which are relevant in this setting. Amongst the most shinning examples of [184] we find the following:

- The weak topologies of X and Y are hereditarily Lindelöf (in all finite powers) but no equivalent norm on X and Y has weak* sequentially compact dual ball.
- Neither X nor Y have uncountable ω-independent sequences.
- X and Y have no uncountable biorthogonal sequences, but they have uncountable ε-biorthogonal sequences for each $\varepsilon > 0$.

3.4.2   Transitivity and Universal Disposition

Recall that a Banach space X is said to be isotropic if given two points x, y in the unit sphere there is a linear, surjective isometry of the space sending one to the other. The classical, still open, problem of rotations of Mazur is to determine whether a separable isotropic Banach space must be isometric to a Hilbert space. Isotropic spaces have been called "transitive" by most authors. We will keep the term "isotropic" for the previous classical notion and will use "transitive" for the next generalization:

Definition 3.40 Let $\mathfrak{M}$ be a class of Banach spaces. We will say that a Banach space X is $\mathfrak{M}$-transitive, if given two subspaces A, B of X in $\mathfrak{M}$, any surjective isometry between A and B can be extended to a surjective isometry of X.

It is a question proposed by Kubiś whether every space of universal disposition with respect to a class $\mathfrak{M}$ must be $\mathfrak{M}$-transitive. Garbulińska and Kubiś show in [101]:

Proposition 3.41 *Assume κ is an infinite regular cardinal and let $\mathfrak{M}_\kappa$ be the class of those Banach spaces having density character strictly smaller than κ. Let U be a Banach space of universal disposition for $\mathfrak{M}_\kappa$ having density character κ. Then U is $\mathfrak{M}_\kappa$-transitive.*

Proof Let $f : X \to Y$ be a surjective isometry between two subspaces of U with density character less than κ. By the regularity of κ, we may assume that there exist two continuous chains $\{U_\alpha\}_{\alpha < \kappa}$ and $\{V_\alpha\}_{\alpha < \kappa}$ of subspaces of U such that $U_0 = X$ and $V_0 = Y$ and $\bigcup_{\alpha < \kappa} U_\alpha = U = \bigcup_{\alpha < \kappa} V_\alpha$. Now, construct inductively isometric embeddings $f_\xi : U_{\alpha(\xi)} \to V_{\beta(\xi)}$ and $g_\xi : V_{\beta(\xi)} \to U_{\alpha(\xi+1)}$ so that

1. $f_0 = f$, and for each $\xi < \kappa$:
2. $g_\xi \circ f_\xi$ is the canonical inclusion $U_{\alpha(\xi)} \subset U_{\alpha(\xi+1)}$,
3. $f_{\xi+1} \circ g_\xi$ is the canonical inclusion $V_{\beta(\xi)} \subset V_{\beta(\xi+1)}$.

The limit steps are not a problem because of the continuity of the chain. The limit operators $f_\kappa : U \to U$ and $g_\kappa : U \to U$ are bijective linear isometries because $f_\kappa \circ g_\kappa = 1_U$ and $g_\kappa \circ f_\kappa = 1_U$. Finally, it is clear that f_κ extends f, which completes the proof. $\square$

Corollary 3.42 *A Banach space of density character $\aleph_1$ that is of universal disposition for separable spaces is $\mathfrak{S}$-transitive.*

Observe that CH is a consequence of the other hypotheses of the Corollary. In particular, under CH, the space $\mathscr{S}^{\omega_1}(X) \approx \mathscr{U}^{\omega_1}(X)$ constructed in Sect. 3.1.2 is $\mathfrak{S}$-transitive. We do not know however if a Banach space of density character $\aleph_1$ that is of universal disposition for finite dimensional spaces is necessarily $\mathfrak{F}$-transitive, so not even under CH we know if the spaces $\mathscr{F}^{\omega_1}(X)$ are $\mathfrak{F}$-transitive.

There is an approximate variant of the notion of an isotropic space. Namely, one says that X is almost isotropic if, given $x, y \in X$ with $\|x\| = \|y\| = 1$, and $\varepsilon > 0$, there is an isometric automorphism ϕ of X such that $\|y - \phi(x)\| < \varepsilon$. Observe that this is stronger than the requirement that there is a surjective ε-isometry sending x to y. The corresponding generalization of almost $\mathfrak{F}$-transitive space can also be considered: when given two finite dimensional isometric subspaces F, G of X via an isometry $\delta : F \to G$ and $\varepsilon > 0$, there is an isometric automorphism ϕ of X such that $\|\delta(x) - \phi(x)\| < \varepsilon$ for all $x \in F$. It is a direct consequence of Theorem 3.38 that Gurariy space is almost $\mathfrak{F}$-transitive:

Corollary 3.43 *Let* $\phi : A \to B$ *be a surjective isometry between finite dimensional subspaces of* $\mathscr{G}$. *Then, for each* $\varepsilon > 0$ *there is a surjective* ϕ' *isometry of* $\mathscr{G}$ *such that* $\|\phi(a) - \phi'(a)\| \leq \varepsilon\|a\|$ *for every* $a \in A$. *In particular* $\mathscr{G}$ *is almost isotropic.*

No C-space can be almost isotropic (unless the underlying compact is a singleton). Indeed, if $f \in C(K)$ is normalized and vanishes at some point of K, then $\|f - \phi(1_K)\|_\infty \geq 1$ for every surjective isometry ϕ of $C(K)$, by the Banach-Stone Theorem.

On the other hand, there are isotropic M-spaces and almost isotropic separable M-spaces, necessarily isomorphic to $C(\Delta)$; see [51]. Even more surprising is the result, proved by Rambla and Kawamura (independently and almost simultaneously; see [161, 219]), that if P_* denotes the pseudo-arc with a point deleted, then the complex space $C_0(P_*)$ is almost isotropic for the sup norm. We refer the reader to [50, 117] for two complementary surveys on "Mazur rotations problem". Two very recent outstanding papers on this topic are [84, 94].

3.4.3 *p-Banach Spaces of Almost Universal Disposition*

In this section we review some results on "separably injective" p-Banach spaces. The situation can be summarized by saying that, while in Banach spaces the notion of a space of "universal disposition" leads to separably injective spaces whose structure is "very different" to the "typical examples" (such as C-spaces), in p-Banach spaces the only "separably injective" objects we know are those which are constructed as spaces of universal disposition for the class of all separable p-Banach spaces. To some extent, this bad behaviour arises from the fact that, while the ground field is injective in the category of Banach spaces, it is not in the category of p-Banach spaces for $0 < p < 1$; see Proposition 3.54 below. To avoid repetitions in this section we fix once and for all the number $p \in (0, 1]$ and we work in the category of p-normed spaces and linear, bounded operators.

Let us begin by giving the definitions that one can expect.

Definition 3.44 A p-Banach space E is said to be:

- Injective if for every p-Banach space X and every subspace Y of X, every operator $t : Y \to E$ can be extended to an operator $T : X \to E$. If this can be achieved with

$\|T\| \leq \lambda \|t\|$ for some fixed $\lambda \geq 1$, then E is said to be a λ-injective p-Banach space.

- Universally separably injective or universally λ-separably p-injective if the preceding condition holds when Y is separable.
- Separably injective or λ-separably injective if the preceding condition holds when X is separable.
- Locally λ-injective if for every finite dimensional p-Banach space X and every subspace Y of X, every operator $t : Y \to E$ has an extension $T : X \to E$ with $\|T\| \leq \lambda \|t\|$. Finally, we say that E is locally injective if it is locally λ-injective for some λ.

Contrarily to what happens in Banach spaces, there are very few separably injective p-Banach spaces and no universally separably injective p-Banach spaces at all:

Proposition 3.45 *If* $0 < p < 1$, *the category of p-Banach spaces has no universally separably injective object, apart from zero.*

Proof Let E be a nonzero p-Banach space. Let us see that E cannot be universally separably injective. Let $\aleph$ be the density character of E and let μ denote Haar measure on the product of a family of $2^{\aleph}$ copies of $\mathbb{T}$, the unit circle. Kalton (see [153, p. 163, at the end of Sect. 3]) showed that for $\aleph = \aleph_0$, and Popov [217, Theorem 1] in full generality, that there is no nonzero operator from $L_p(\mu)$ to E. Thus, if we fix some nonzero $e \in E$ and we consider the subspace $\mathbb{K}$ of constant functions in $L_p(\mu)$, then the operator $\lambda \in \mathbb{K} \mapsto \lambda e \in E$ cannot be extended and E is not universally separably injective.   □

Now, even if a good portion of the results in Chap. 2 about separably injective Banach spaces extend straightforwardly to the category of p-Banach spaces, we have no example at hand to apply them. One can construct one as a p-Banach space of universal disposition for separable p-Banach spaces. Before going into this issue, let us made a detour into "universality" and "almost universal disposition".

In [152, Theorem 4.1(a)] was stated without proof that for $0 < p < 1$ there exists a separable p-Banach space which is "universal" for the class of all separable p-Banach spaces. This result also appears mentioned in [221, Theorem 3.2.8] but the proof only gives "universality with respect to ε-isometries". Such a space was finally provided in [58]:

Proposition 3.46 *For every* $p \in (0, 1]$ *there exists a unique separable p-Banach space* $\mathscr{G}_p$ *of almost universal disposition for finite dimensional p-Banach spaces, up to isometries. This space contains an isometric copy of every separable p-Banach space.*

The construction is roughly the same as in Sect. 3.3.1, working with ℓ_p-sums instead of ℓ_1-sums. One starts with a separable p-Banach space X and arrives at an "enveloping" p-Banach space $G_p(X)$, which is of almost universal disposition for finite dimensional p-Banach spaces. Here, the key point is that push-outs of p-

Banach spaces exist and have the expected properties. After that, one proves that any two p-Banach spaces of almost universal disposition are isometric, from where it follows that these spaces contain isometric copies of each separable p-Banach space. The proof is similar to that presented in Sect. 3.3.5 and the key point is a version of Lemma 3.36 for p-normed spaces. At the end it turns out that Theorem 3.38 is true for p-Banach spaces for $p \in (0, 1]$.

Starting with any separable p-Banach space X, considering the universal space $\mathscr{G}_p$ instead of $C(\Delta)$, and using the push-out of p-Banach spaces at every step $\alpha < \omega_1$, one can construct the spaces $\mathscr{S}_p^{\omega_1}(X)$ and $\mathscr{F}_p^{\omega_1}(X)$ which contain X, have density character $\mathfrak{c}$, and are of universal disposition for separable and finite dimensional p-Banach spaces, respectively. Actually, the following result was established in [58]:

Proposition 3.47 *There exists a p-Banach space of density character the continuum which is of universal disposition for separable p-Banach spaces, contains isometric copies of every p-Banach space of density character up to $\aleph_1$ and, under the continuum hypothesis, is unique up to isometries.*

The following result also appears in [58]:

Proposition 3.48 *Every p-Banach space of universal disposition for finite dimensional p-Banach spaces has density character at least the continuum.*

It is interesting to remark that the general ideas leading to the proof of Theorem 3.23 apply verbatim to p-Banach spaces, so we have the following generalization.

Theorem 3.49 *Let $\xi \leq \mathfrak{c}$ be an ordinal of uncountable cofinality.*

1. $\mathscr{S}_p^{\xi}(X) \approx \mathscr{S}_p^{\omega_1}(\mathbb{R})$ *for every p-Banach space X of density at most $\aleph_1$.*
2. $\mathscr{F}_p^{\xi}(X) \approx \mathscr{F}_p^{\omega_1}(\mathbb{R})$ *for every separable p-Banach space X.*

We do not know, however, if $\mathscr{F}_p^{\omega_1}(\mathbb{R})$ is isometric or isomorphic to $\mathscr{S}_p^{\omega_1}(\mathbb{R})$ when $0 < p < 1$ although we suspect that, just as in the case $p = 1$, these spaces are not linearly isomorphic.

Returning to injectivity-like properties, we have the following.

Proposition 3.50

1. *Every p-Banach space of almost universal disposition, in particular $\mathscr{G}_p$ and $\mathscr{F}_p^{\omega_1}$, is locally λ-injective for each $\lambda > 1$.*
2. *Spaces of universal disposition for separable p-Banach spaces are 1-separably injective.*

Proof We prove (1). The proof of (2) is easier. Assume U is of almost universal disposition for finite dimensional p-Banach spaces. Let X be a finite dimensional p-Banach space, Y a subspace of X and $t : Y \to U$ an operator with $\|t\| = 1$. We will prove that, for each $\varepsilon > 0$, there is an extension $T : X \to U$ with $\|T\| \leq 1 + \varepsilon$.

Consider the push-out diagram

$$
\begin{array}{ccc}
Y & \longrightarrow & X \\
{\scriptstyle t}\downarrow & & \downarrow{\scriptstyle t'} \\
\overline{t[Y]} & \stackrel{\imath}{\longrightarrow} & \mathrm{PO}
\end{array}
$$

where the unlabeled arrow is the inclusion and we consider $t[Y]$ as a (closed) subspace of U. As $\imath$ is an isometry, for each $\varepsilon > 0$, there is an ε-isometry $u : \mathrm{PO} \to U$ such that $\imath(t(y)) = u(t(y))$ for all $y \in Y$. Then $u \circ t'$ is an extension of t to X with quasinorm at most $1 + \varepsilon$.   □

Besides this, the characterizations of Proposition 2.5 extend straightforwardly to separably p-injective spaces:

Proposition 3.51 *For a p-Banach space E the following properties are equivalent.*

1. *E is separably injective.*
2. *Every operator from a subspace of ℓ_p into E extends to ℓ_p.*
3. *For every p-Banach space X and each subspace Y such that X/Y is separable, every operator $t : Y \to E$ extends to X.*
4. *If Z is a p-Banach space containing E and Z/E is separable, then E is complemented in Z.*

The last condition can be rephrased by saying that for every separable p-Banach space S one has $\mathrm{Ext}(S, E) = 0$ in the category of p-Banach spaces.

The stability properties in Proposition 2.11 also extend straightforwardly:

Proposition 3.52 *Let $0 \to Y \to X \to Z \to 0$ be a short exact sequence of p-Banach spaces.*

1. *If Y and Z are separably (respectively, locally) injective, then so is X.*
2. *If X and Y are separably (respectively, locally) injective, then so is Z.*

In particular, direct products (and complemented subspaces) of separably injective p-Banach spaces are separably injective. Moreover, since the proof of the vector valued version of Sobczyk's theorem (Proposition 2.12) presented in Chap. 2 uses the triangle inequality only in the last part of the proof, we can add the following:

Proposition 3.53 *If E is a separably λ-injective p-Banach space, then $c_0(E)$ is $\Delta\lambda(1 + \lambda)^+$-separably injective.*

Here, Δ is the "modulus of concavity" of E, that is, the least constant C for which the inequality $\|x + y\| \leq C(\|x\| + \|y\|)$ holds in E. One result however marks the difference:

Proposition 3.54 *If $p < 1$, then all locally injective p-Banach spaces have trivial dual.*

Proof A quasi-Banach space has nontrivial dual if and only if it contains a complemented subspace of dimension one. Since the local injectivity passes to complemented subspaces, it suffices to check that the ground field $\mathbb{K}$ is not locally injective in the category of p-Banach spaces when $p < 1$. Let us consider the "diagonal" embedding $\delta : \mathbb{K} \to \ell_p^n$ given by $\delta(1) = n^{-1/p} \sum_{i=1}^n e_i$. An obvious symmetrization argument shows that any extension to ℓ_p^n of the identity on $\mathbb{K}$ must have norm greater that the obvious one $\Sigma : \ell_p^n \to \mathbb{K}$ given by $\Sigma(x) = n^{1/p-1} \sum_i x_i$. But $\|\Sigma\| \geq n^{1/p-1}$ goes to infinity with n.   □

3.4.4 Fraïssé Limits

Let us briefly introduce the notion of a Fraïssé limit. Although the definition of Fraïssé limits in "abstract" categories is harder (see, e. g., [136]), the following adaptation to "concrete" categories of quasi-Banach spaces suffices for our present purposes.

Fix $p \in (0, 1]$ and let $\mathbf{C}$ be a subcategory of the "isometric" category of p-Banach spaces. This means that the objects of $\mathbf{C}$ are certain p-Banach spaces, while the morphisms $L_{\mathbf{C}}(A, B)$ between two objects A, B of $\mathbf{C}$ consist of certain isometries $A \to B$. We assume for the sake of simplicity that if a subspace of an object of $\mathbf{C}$ is also an object of $\mathbf{C}$, then the inclusion is a morphism in $\mathbf{C}$. In particular we may consider any reasonable "class" of p-Banach spaces as a category just taking the isometries as morphisms.

Definition 3.55 A p-Banach space $\mathscr{F}$ is said to be a Fraïssé limit for $\mathbf{C}$ if the following conditions hold.

1. If X is an object in $\mathbf{C}$ there exists a subspace $H \subset \mathscr{F}$ which is in $\mathbf{C}$ and a morphism $X \to H$.
2. Given a subspace $H \subset \mathscr{F}$, an object X in $\mathbf{C}$ and a morphism $f : H \to X$, there exists a subspace $G \subset \mathscr{F}$ containing H and a morphism $g : X \to G$ such that $g \circ f$ is the inclusion of H into G.

In [169, Theorem 6.3] Kubiś establishes the existence, under CH, of Fraïssé limits corresponding to the category $\mathfrak{S}$ of separable Banach spaces and linear isometric embeddings and shows that these spaces are isometric provided they have density character $\aleph_1$. Fraïssé limits can be characterized—in ZFC!—as spaces of universal disposition for most natural subcategories; amongst them, the most interesting for us are $\mathfrak{F}_p$ (finite dimensional p-Banach spaces) and $\mathfrak{S}_p$ (separable p-Banach spaces).

Proposition 3.56 *Fix $p \in (0, 1]$.*

- *A p-Banach space is of universal disposition for separable p-Banach spaces if and only if it is a Fraïssé limit for $\mathfrak{S}_p$.*

- *A p-Banach space is of universal disposition for finite-dimensional p-Banach spaces if and only if it is a Fraïssé limit for $\mathfrak{F}_p$.*

Proof Only one proof is necessary. We first prove that "Fraïssé" implies "universal disposition". Let $\mathscr{F}$ be a Fraïssé limit for any of the previous classes, denoted $\mathfrak{M}$. Let $u : A \to B$ and $t : A \to \mathscr{F}$ be isometries, with B in $\mathfrak{M}$. Form the push-out diagram

$$
\begin{array}{ccc}
A & \xrightarrow{\ u\ } & B \\
{\scriptstyle t}\downarrow & & \downarrow{\scriptstyle t'} \\
t[A] & \xrightarrow{\ u'\ } & \text{PO}
\end{array}
$$

By property (2), there is an isometry $g : \text{PO} \to \mathscr{F}$ such that $g \circ u'$ is the canonical embedding of $t[A]$ into $\mathscr{F}$. Therefore $gt' : B \to \mathscr{F}$ is the desired into isometry since $gt'u = gu't = t$.

As for the converse, let U be a space of universal disposition for $\mathfrak{M}$. That U satisfies (1) is obvious. To check (2), let X be an object in $\mathfrak{M}$, and let $u : H \to X$ be an isometry, where H is a subspace of U. Let $I \to U$ be any isometric extension of the inclusion of H into U. Then $I[X]$ is an object in $\mathfrak{M}$ and $I \circ u : H \to X \to I[X]$ is the inclusion of H into $I[X]$.   □

3.4.5   Similar Constructions in Other Categories

The procedure of constructing spaces by long transfinite exhaustive sequences of push-outs that we used to produce the Banach spaces $\mathscr{M}^{\omega_1}(X)$ can be repeated for mathematical objects other than Banach spaces, whenever we have analogous notions for *isometric embeddings* and *push-outs* that behave in the right way. We shall briefly discuss the case of compact spaces, because of its connection to Theorem 2.39.

When moving from Banach spaces to compact spaces, the arrows are usually reversed. So we look at pull-backs of compact spaces, rather than push-outs. The pull-back of two continuous functions $f : K_1 \longrightarrow L$ and $g : K_2 \longrightarrow L$ is constructed as $\text{PB} = \{(x, y) \in K_1 \times K_2 : f(x) = g(y)\}$, and the pull-back diagram is obtained taking $\text{PB} \longrightarrow K_1$ and $\text{PB} \longrightarrow K_2$ the restrictions of the two coordinate projections.

Let us denote this time by $\mathfrak{M}$ the class of metrizable compacta. A continuous surjection $f : K \longrightarrow L$ will be called an $\mathfrak{M}$-cell if there exists a push-out diagram of

continuous surjection of the form

where R and S are metrizable.

Lemma 3.57 *A continuous surjection $f : K \longrightarrow L$ between compact spaces is an $\mathfrak{M}$-cell if and only if the following two conditions hold:*

1. *There exist countably many continuous functions $\{g_n\} \subset C(K)$ such that, for all $x, y \in K$, if $f(x) = f(y)$ and $g_n(x) = g_n(y)$ for all n, then $x = y$. (This is equivalent to say that $C(K)/C(L)$ is a separable Banach space.)*
2. *For every closed G_δ subset F of K, we have that $f(F)$ is a closed G_δ subset of L.*

Proof Suppose first that f is an $\mathfrak{M}$-cell, so K is the push-out of the diagram, with arrows $\pi : K \longrightarrow S$, $\tilde{\pi} : L \longrightarrow R$, $f' : S \longrightarrow R$. So we can assume as above that $K = \{(x, y) \in L \times S : \tilde{\pi}(x) = f'(y)\}$. The continuous functions for condition (1) can be obtained by considering countably many continuous functions $\tilde{g}_n : S \longrightarrow \mathbb{R}$ that separate points and taking $g_n = \tilde{g}_n \circ \pi$. Now, if $F \subset K$ is a closed G_δ set, then $f(F) = \tilde{\pi}^{-1}(f\pi(F))$ and $f\pi(F)$ is a compact subset of the metrizable compact space R, hence it is G_δ closed. For the converse, for every n and every rational $q \in \mathbb{Q}$, since we are assuming that $F_{nq} = f(g_n^{-1}(-\infty, q])$ is a closed G_δ set, it is a zero set, so we can find a continuous function $g_{nq} \in C(L)$ such that $g_{nq}^{-1}(0) = F_{nq}$. Then, the push-out diagram is obtained by taking $\pi : K \longrightarrow \mathbb{R}^{\mathbb{N} \cup \mathbb{N} \times \mathbb{Q}}$ as $\pi(x) = (g_n(x), g_{nq}(x))$, $S = \pi(K)$, $\tilde{\pi} : L \longrightarrow \mathbb{R}^{\mathbb{N} \times \mathbb{Q}}$ as $\tilde{\pi}(y) = (\pi_{nq}(y))$, $R = \tilde{\pi}(L)$, and $f' : S \longrightarrow R$ in the obvious way $f'(s_n, s_{nq}) = (s_{nq})$. $\square$

All the definitions that we used for Theorem 3.23 can be now transferred to compact spaces, being careful of reversing the direction of arrows. Thus, a compact space is tightly $\mathfrak{M}$-filtered if there exists an ordinal λ and an inverse system of continuous surjections $\pi_{\beta\alpha} : K_\beta \longrightarrow K_\alpha$ for $\alpha < \beta \leq \lambda$, satisfying:

1. $\pi_{\beta\alpha} \circ \pi_{\gamma\beta} = \pi_{\gamma\beta}$ if $\alpha < \beta < \gamma \leq \lambda$.
2. $K_\xi = K$.
3. If $\gamma \leq \lambda$ is a limit ordinal, then K_γ is the limit of the inverse system below γ. That is to say, for all $x, y \in K$, if $\pi_\alpha(x) = \pi_\alpha(y)$ for all $\alpha < \gamma$, then $\pi_\gamma(x) = \pi_\gamma(y)$.
4. If $\alpha < \lambda$, then $\pi_{\alpha+1,\alpha} : K_{\alpha+1} \longrightarrow K_\alpha$ is an $\mathfrak{M}$-cell.

Moreover, if K has weight $\mathfrak{c}$, we say that K is exhaustively tightly $\mathfrak{M}$-filtered, if the above have the additional property that for every couple of continuous surjections $h : K \longrightarrow R$ and $h' : S \longrightarrow R$, with S, R metrizable, there exists $\mathfrak{c}$

many ordinals $\alpha < \lambda$ for which there is a push-out diagram

$$
\begin{array}{ccc}
K_{\alpha+1} & \xrightarrow{\;\;\pi_{\alpha+1,\alpha}\;\;} & K_\alpha \\[4pt]
{\scriptstyle u}\big\downarrow & & {\scriptstyle v}\big\downarrow \\[4pt]
S & \xrightarrow{\hspace{2cm}} & R
\end{array}
$$

such that $h = v \circ \pi_{\alpha+1,\alpha} \circ \pi_{\lambda,\alpha+1}$ and $h' = u \circ \pi_{\lambda,\alpha+1}$.

An exercise of translation of the proof of Theorem 3.23 to this language yields that there exists a unique (up to homeomorphism) compact space K which is exhaustively tightly $\mathfrak{M}$-filtered. We can observe that such K must be a zero-dimensional F-space. This is because, if we take two disjoint closed G_δ sets $F, G \subset K$, then we can apply the exhaustiveness condition to a function $h : K \longrightarrow [-1, 1]$ which is constant equal to -1 on F and constant equal to 1 on G, and $h' : [0, 1] \times \{-1, 1\} \longrightarrow [-1, 1]$ given by $h'(z, t) = t \cdot z$, and in this way F and G can be separated by clopen sets. This compactum is also the Stone compact of the Boolean algebra that one obtains repeating the same analogous procedures in the category of Boolean algebras, cf. [16].

If we revisit Theorem 2.39 in this light, we see that the compact space that is constructed in that proof is tightly $\mathfrak{M}$-filtered, as it is produced as an inverse limit in which each successor step is an $\mathfrak{M}$-cell. The construction is done by a kind of exhaustive procedure but instead of making sure that all possible pull-backs are represented along the construction as in condition (4), we dealt only with the particular kind of pull-backs that separate pairs of disjoint closed G_δ sets. This was done to make the presentation simpler, but one can see that it does not make any difference, and the compact space used in Theorem 2.39 is precisely the unique exhaustively tightly $\mathfrak{M}$-filtered compact space.

Finally, we can observe that the Boolean algebra $\mathbb{P}(\mathbb{N})/\mathrm{fin}$ is of universal disposition for countable subalgebras, because it contains all countable subalgebras and is *countably automorphic* (cf. paragraph after Corollary 2.53). This gives an alternative way of getting objects of universal disposition in this category without using iterated push-outs. It is unclear if we have, in the category of Banach spaces, a similar space of universal disposition for separable subspaces with a natural definition that does not involve a transfinite inductive construction. The first guess would be ℓ_∞/c_0, but C-spaces cannot work (cf. Sect. 3.3.4).

3.4.6 Sources

As we mentioned before, the notion of a Banach space of (almost) universal disposition was introduced by Gurariy in [118], where $\mathscr{G}$ appeared for the first time. Lemma 3.28 and Proposition 3.29 as well as the observations opening Sects. 3.2.2 and 3.3.4 are taken from Gurariy's paper. The "uniqueness" of Gurariy space was stablished by Lusky in [185] by methods which are completely different than those of Sect. 3.3.5, where we have followed the approach of Kubiś and Solecki in [170]. Anyway, [185] is a fine paper and contains additional information on the structure of the isometry group of $\mathcal{G}$: for instance, that given two smooth points on the unit sphere there is a surjective isometry of $\mathcal{G}$ sending one into the other. See [96] for more precise results on the structure of the isometry group of $\mathcal{G}$.

Lusky's uniqueness result unified two lines of research: Pełczyński and Wojtaszczyk had proved in 1971 that the family of separable Lindenstrauss spaces has a maximal member, namely there is a separable Lindenstrauss space $\mathcal{L}$ such that for every separable Lindenstrauss space X and each $\varepsilon > 0$, there is an operator $u : X \to \mathcal{L}$ such that $\|x\| \leq \|u(x)\| \leq (1 + \varepsilon)\|x\|$ and a contractive projection of $\mathcal{L}$ onto the range of u; see [215, Theorem 4.2]. One year later Wojtaszczyk [248] himself proved that such a space can be constructed as a space of almost universal disposition. Lusky approached Gurariy space by means of triangular matrices, a technique promoted by Lazar and Lindenstrauss (cf. [173]) in the study of separable Lindenstrauss spaces.

Most of the material included in Sects. 3.1 and 3.2 appeared in [19]. We have used isometries instead of contractions in Sect. 3.1.2 in order to simplify the proofs of the last part of Sect. 3.2.4. Proposition 3.10, which solves a problem posed in [19], is due to Ben Yaacov and Henson [30] and the relevant example in the proof of Lemma 3.9 was constructed by Haydon and taken from [58]. The ideas around Theorem 3.23 come from [16]; however Theorems 3.24 and 3.25 are formally stronger than the main result of [16] for Banach spaces.

The results of Sect. 3.3.3 are from [101]. Theorem 3.34 can be atributed to Henson and Moore [134, Theorem 6.8]. The proof is a simplification of the argument in [19]. At the end of the day everything depends on the fact, due to Benyamini and Lindenstrauss, that there are isometric ℓ_1-preduals which are not complemented in any C-space; see [34, Corolaries 1 and 2].

Section 3.4.3 is based on [58] and complements it. A forerunner of $\mathcal{G}_p$ appeared without further explanations in Kalton's paper [153]. For related results in other areas we refer the reader to [168, 205, 206].

Chapter 4
Ultraproducts of Type $\mathscr{L}_\infty$

The Banach space ultraproduct construction is perhaps the main bridge between model theory and the theory of Banach spaces and its ramifications. Ultraproducts of Banach spaces, even at a very elementary level, proved very useful in local theory, the study of Banach lattices, and also in several nonlinear problems, such as the uniform and Lipschitz classification of Banach spaces. We refer the reader to Heinrich's survey paper [126] and Sims' notes [234] for two complementary accounts. Traditionally, the main investigations about Banach space ultraproducts have focused on the isometric theory, reaching a quite coherent set of results very early, as can be seen in [132]. We will review some results on the isometric theory of ultraproducts in Sect. 4.7.4, but most of the Chapter is placed in the isomorphic context.

Indeed, in this Chapter we study ultraproducts of type $\mathscr{L}_\infty$, in particular ultraproducts of $\mathscr{L}_\infty$ spaces. The leading idea is that, if one starts with a family of Banach spaces that have a certain property at the "finite dimensional level", then its ultraproducts "must" have that property at the "separable level". Thus, it will be shown that ultraproducts of Banach spaces are universally separably injective as soon as they are $\mathscr{L}_\infty$-spaces despite of the facts that the starting spaces do not need to be of type $\mathscr{L}_\infty$ or that ultraproducts are never injective. We analyze ultraproducts of C-spaces in some detail and provide alternative approaches to several results and constructions in Chap. 3 by studying ultrapowers of the Gurariy space. We also address the Henson-Moore problem of when two given $\mathscr{L}_\infty$-spaces have isomorphic ultrapowers.

The exposition is basically self-contained, with the exception of Sect. 4.6, which uses a few basic results from "model theory".

© Springer International Publishing Switzerland 2016
A. Avilés et al., *Separably Injective Banach Spaces*, Lecture Notes
in Mathematics 2132, DOI 10.1007/978-3-319-14741-3_4

4.1   Ultraproducts of Banach Spaces

In this Section we present the Banach space ultraproduct construction based on the notion of convergence along a maximal filter. We also define the set-theoretic ultraproduct, which historically came first, as we shall need it from time to time. The topological ultracoproduct (a related construction which applies to families of compact spaces) will appear in Sects. 4.3.1 and 4.7.2.

4.1.1   Ultrafilters

A family $\mathcal{U}$ of subsets of a given set I is said to be a filter if it is closed under finite intersection, does not contain the empty set and, one has $A \in \mathcal{U}$ provided $B \subset A$ and $B \in \mathcal{U}$. An ultrafilter on I is a filter which is maximal with respect to inclusion. An ultrafilter $\mathcal{U}$ on I is said to be fixed if there is $a \in I$ such that $\{a\} \in \mathcal{U}$. By maximality one then has $\mathcal{U} = \{A \subset I : a \in A\}$. Otherwise $\mathcal{U}$ is called free. If X is a topological space, $f : I \to X$ is a function, and $x \in X$, one says that $f(i)$ converges to x along $\mathcal{U}$ (written $x = \lim_\mathcal{U} f(i)$ in short) if whenever V is a neighborhood of x in X the set $f^{-1}(V) = \{i \in I : f(i) \in V\}$ belongs to $\mathcal{U}$. The obvious compactness argument shows that if X is compact and Hausdorff, and $\mathcal{U}$ is an ultrafilter on I, then for every function $f : I \to X$ there is a unique $x \in X$ such that $x = \lim_\mathcal{U} f(i)$. The following Definition isolates the property of ultrafilters that makes ultraproducts interesting.

Definition 4.1 An ultrafilter $\mathcal{U}$ on a set I is countably incomplete if there is a sequence (I_n) of subsets of I such that $I_n \in \mathcal{U}$ for all n, and $\bigcap_{n=1}^\infty I_n = \varnothing$.

Throughout this Chapter all ultrafilters will be assumed to be countably incomplete unless otherwise stated. Notice that $\mathcal{U}$ is countably incomplete if and only if there is a function $n : I \to \mathbb{N}$ such that $n(i) \to \infty$ along $\mathcal{U}$ (equivalently, there is a family $\varepsilon(i)$ of strictly positive numbers converging to zero along $\mathcal{U}$). Indeed, if (I_n) is a sequence witnessing that $\mathcal{U}$ is countably incomplete, for which we may assume $I_1 = I$, then the required function can be defined as $n(i) = \max\{n : i \in I_n\}$. And conversely, if $n : I \to \mathbb{N}$ goes to ∞ along $\mathcal{U}$, then the sets $I_n = \{i \in I : n(i) \geq n\}$ are in $\mathcal{U}$, but $\bigcap_n I_n$ is empty. It is obvious that any countably incomplete ultrafilter is free (it contains no singleton) and also that every free ultrafilter on $\mathbb{N}$ is countably incomplete.

4.1.2   Ultraproducts of Banach Spaces

Let $(X_i)_{i \in I}$ be a family of Banach spaces indexed by the set I and let $\mathcal{U}$ be an ultrafilter on I. The space of bounded families $\ell_\infty(I, X_i)$ endowed with the supremum norm is a Banach space, and $c_0^{\mathcal{U}}(X_i) = \{(x_i) \in \ell_\infty(I, X_i) : \lim_{\mathcal{U}} \|x_i\| = 0\}$ is a closed subspace of $\ell_\infty(I, X_i)$. The ultraproduct of the spaces $(X_i)_{i \in I}$ following $\mathcal{U}$ is defined as the quotient space

$$[X_i]_{\mathcal{U}} = \ell_\infty(I, X_i)/c_0^{\mathcal{U}}(X_i),$$

with the quotient norm. We denote by $[(x_i)]$ the element of $[X_i]_{\mathcal{U}}$ which has the family (x_i) as a representative. It is easy to see that $\|[(x_i)]\| = \lim_{\mathcal{U}} \|x_i\|$. In the case $X_i = X$ for all i, we denote the ultraproduct by $X_{\mathcal{U}}$, and call it the ultrapower of X following $\mathcal{U}$. If $T_i : X_i \to Y_i$ is a uniformly bounded family of operators, the ultraproduct operator $[T_i]_{\mathcal{U}} : [X_i]_{\mathcal{U}} \to [Y_i]_{\mathcal{U}}$ is given by $[T_i]_{\mathcal{U}}[(x_i)] = [T_i(x_i)]$. Quite clearly, $\|[T_i]_{\mathcal{U}}\| = \lim_{\mathcal{U}} \|T_i\|$.

4.1.3 The Set-Theoretic Ultraproduct

In some places of this monograph, we will require also the set-theoretic ultraproduct. Let us recall the definition and fix notations. Let $(S_i)_{i \in I}$ be a family of sets and $\mathcal{U}$ an ultrafilter on I. The set-theoretic ultraproduct $\langle S_i \rangle_{\mathcal{U}}$ is the product set $\prod_i S_i$ factored by the equivalence relation

$$(s_i) = (t_i) \iff \{i \in I : s_i = t_i\} \in \mathcal{U}.$$

The class of (s_i) in $\langle S_i \rangle_{\mathcal{U}}$ is denoted $\langle (s_i) \rangle$. If we are given functions $f_i : S_i \to K$, where K is some compact space, we can define another function $f : \langle S_i \rangle_{\mathcal{U}} \to K$ by

$$f(\langle (s_i) \rangle) = \lim_{\mathcal{U}(i)} f_i(s_i).$$

When X_i is a family of Banach spaces, there is an obvious connection between $[X_i]_{\mathcal{U}}$ and $\langle X_i \rangle_{\mathcal{U}}$. Indeed, the former space can be obtained from the latter, first taking the elements for which the function

$$\langle (x_i) \rangle \longmapsto \lim_{\mathcal{U}(i)} \|x_i\|$$

is finite (we may consider the original norms on the spaces X_i as taking values on the extended ray $[0, \infty]$), and then taking quotient by the kernel of the function.

4.2 Injectivity Properties of Ultraproducts of Type $\mathscr{L}_\infty$

The classes of $\mathscr{L}_{p,\lambda+}$ spaces are stable under ultraproducts [45, Proposition 1.22]. In the opposite direction, a Banach space is an $\mathscr{L}_{p,\lambda+}$ space if and only if some (or every) ultrapower is. In particular, a Banach space is an $\mathscr{L}_\infty$ space or a

Lindenstrauss space if and only if its ultrapowers are; see, e.g., [127]. However, one can make a Lindenstrauss space out of not-even-$\mathscr{L}_\infty$-spaces: indeed, if $p(i) \to \infty$ along $\mathcal{U}$, then the ultraproduct $[L_{p(i)}]_\mathcal{U}$ is a Lindenstrauss space (and, in fact, an abstract M-space; see [51, Lemma 3.2]). The "local" situation had been considered earlier in [135].

4.2.1 Separable Subspaces and Separable Injectivity

We present now the key result to study the structure of separable subspaces of ultraproducts of type $\mathscr{L}_\infty$. To avoid repetitions, throughout this Chapter $\mathcal{U}$ will denote a countably incomplete ultrafilter on a set I, we will consider families of Banach spaces (X_i) indexed by I, and we denote by $[X_i]_\mathcal{U}$ the corresponding ultraproduct.

Lemma 4.2 *Suppose* $[X_i]_\mathcal{U}$ *is an* $\mathscr{L}_{\infty,\lambda+}$-*space. Then each separable subspace of* $[X_i]_\mathcal{U}$ *is contained in a subspace of the form* $[F_i]_\mathcal{U}$, *where* $F_i \subset X_i$ *is finite dimensional and* $\lim_{\mathcal{U}(i)} d(F_i, \ell_\infty^{d(i)}) \leq \lambda$ *with* $d(i) = \dim(F_i)$.

Proof Let us assume S is an infinite-dimensional separable subspace of $[X_i]_\mathcal{U}$. Let (s^n) be a linearly independent sequence spanning a dense subset in S and, for each n, let (s_i^n) be a fixed representative of s^n in $\ell_\infty(I, X_i)$. Let $S^n = \text{span}\{s^1, \dots, s^n\}$. Since $[X_i]_\mathcal{U}$ is an $\mathscr{L}_{\infty,\lambda+}$-space there is, for each n, a finite dimensional $F^n \subset [X_i]_\mathcal{U}$ containing S^n with $d(F^n, \ell_\infty^{d(n)}) \leq \lambda + 1/n$, where $d(n) = \dim(F^n)$.

For fixed n, let (f^m) be a basis for F^n containing $s^1, \dots, s^n$. Choose representatives (f_i^m) such that $f_i^m = s_i^\ell$ if $f^m = s^\ell$. Moreover, let F_i^n be the subspace of X_i spanned by f_i^m for $1 \leq m \leq \dim F^n$. Let (I_n) a decreasing sequence of sets of $\mathcal{U}$ whose intersection is empty and, for each integer n, put

$$J_n' = \{i \in I : d(F_i^n, \ell_\infty^{d(n)}) \leq \lambda + 2/n\} \cap I_n.$$

Finally set $J_m = \bigcap_{n \leq m} J_n'$. The sets J_m are all in $\mathcal{U}$ although $\bigcap_m J_m = \varnothing$. Next we define a function $k : I \to \mathbb{N}$ as $k(i) = \sup\{n : i \in J_n\}$. For each $i \in I$, take $F_i = F_i^{k(i)}$. This is a finite-dimensional subspace of X_i whose Banach-Mazur distance to the corresponding ℓ_∞^k is at most $\lambda + 2/k(i)$. It is clear that $[F_i]_\mathcal{U}$ contains S and also that $k(i) \to \infty$ along $\mathcal{U}$, which completes the proof.   □

Lemma 4.3 *For every function* $k : I \to \mathbb{N}$, *the space* $[\ell_\infty^{k(i)}]_\mathcal{U}$ *is universally* 1-*separably injective.*

Proof Let Γ be the disjoint union of the sets $\{1, 2, \dots, k(i)\}$ viewed as a discrete set. We observe that $c_0^\mathcal{U}(\ell_\infty^{k(i)})$ is an ideal in $\ell_\infty(\ell_\infty^{k(i)}) = \ell_\infty(\Gamma) = C(\beta\Gamma)$ and apply Corollary 2.19.   □

Theorem 4.4 *If $[X_i]_{\mathcal{U}}$ is an $\mathscr{L}_{\infty,\lambda+}$-space, then $[X_i]_{\mathcal{U}}$ is universally λ-separably injective.*

Proof It is clear that a Banach space is λ-universally separably injective if and only if every separable subspace is contained in some larger λ-universally separably injective subspace. A Banach space is λ-universally separably injective provided it is linearly isomorphic to a 1-universally separably injective one through an isomorphism u satisfying $\|u\|\|u^{-1}\| \leq \lambda$. Now take a look to the two preceding Lemmata. Notice that, with the notations of Lemma 4.2, one can easily construct an isomorphism $u : [F_i]_{\mathcal{U}} \to [\ell_\infty^{k(i)}]_{\mathcal{U}}$ such that $\|u\|\|u^{-1}\| \leq \lambda$. Indeed, $d(F_i, \ell_\infty^{k(i)})$ is attained at some u_i for which we can assume $\|u_i\| = 1$. Taking $u = [u_i]_{\mathcal{U}}$ suffices.

$\square$

Recalling that Lindenstrauss spaces are precisely the $\mathscr{L}_{\infty,1+}$ spaces we obtain the following result.

Corollary 4.5 *If $[X_i]_{\mathcal{U}}$ is a Lindenstrauss space, then it is universally 1-separably injective.*

A closely related result will be proved in Chap. 5. The reader should compare the proof of preceding result to that of Theorem 5.15 which is reminiscent from the use of ultraproducts in the proof of the "compactness theorem" (see [98, Theorem 2.10]): if A_i is a sequence of "axioms" and for each n we have a "model" S_n which satisfies $A_1, A_2, \ldots A_n$, then, for any free ultrafilter on the integers $\mathcal{U}$, the set-theoretic ultraproduct $\langle S_n \rangle_{\mathcal{U}}$ satisfies every A_i.

4.2.2   Ultraproducts Are Never Injective

Despite the preceding results, infinite dimensional ultraproducts via countably incomplete ultrafilters are never injective. This was proved by Henson and Moore in [133, Theorem 2.6] using the language of nonstandard analysis. Here we present a generalization of Sims's "translation" for ultraproducts appearing in [234, Sect. 8].

Theorem 4.6 *Ultraproducts via countably incomplete ultrafilters are never injective, unless they are finite dimensional.*

Proof Recalling that injective Banach spaces are $\mathscr{L}_\infty$-spaces, assume that $[X_i]_{\mathcal{U}}$ is a $\mathscr{L}_\infty$-space. According to Lemma 4.2, if $[X_i]_{\mathcal{U}}$ is infinite dimensional, it contains some infinite dimensional complemented subspace isomorphic to $[\ell_\infty^{k(i)}]_{\mathcal{U}}$. Thus, it suffices to see that the later is not an injective space.

Let $\langle \{1, \ldots, k(i)\}\rangle_{\mathcal{U}}$ denote the set-theoretic ultraproduct of the sets $\{1, \ldots, k(i)\}$. We have

$$c_0(\langle\{1,\ldots,k(i)\}\rangle_{\mathcal{U}}) \subset [\ell_\infty^{k(i)}]_{\mathcal{U}} \subset \ell_\infty(\langle\{1,\ldots,k(i)\}\rangle_{\mathcal{U}}). \tag{4.1}$$

This should be understood as follows: each $[(f_i)] \in [\ell_\infty^{k(i)}]\mathcal{U}$ defines a function on $\langle\{1,\ldots,k(i)\}\rangle\mathcal{U}$ by the formula

$$f\langle(x_i)\rangle\mathcal{U} = \lim_{\mathcal{U}(i)} f_i(x_i).$$

In this way, $[\ell_\infty^{k(i)}]\mathcal{U}$ embeds isometrically as a subspace of $\ell_\infty(\langle\{1,\ldots,k(i)\}\rangle\mathcal{U})$ containing $c_0(\langle\{1,\ldots,k(i)\}\rangle\mathcal{U})$. Write $\Gamma = \langle\{1,\ldots,k(i)\}\rangle\mathcal{U}$ and $U = [\ell_\infty^{k(i)}]\mathcal{U}$, so that (4.1) becomes $c_0(\Gamma) \subset U \subset \ell_\infty(\Gamma)$. We will prove that the inclusion of $c_0(\Gamma)$ into U cannot be extended to $\ell_\infty^c(\Gamma)$, the space of all countably supported bounded families on Γ.

An internal subset of Γ is one of the form $\langle A_i\rangle\mathcal{U}$, where $A_i \subset \{1,\ldots,k(i)\}$ for each $i \in I$. The cardinality of any infinite internal sets is at least the continuum: just use an almost disjoint family. This is the basis of the ensuing argument: as U is spanned by the characteristic functions of the internal sets, if $f \in U$ is not in $c_0(\Gamma)$, then there is $\delta > 0$ and an infinite internal $A \subset \Gamma$ such that $|f| \geq \delta$ on A.

Suppose $L : \ell_\infty^c(\Gamma) \to U$ is an operator extending the inclusion of $c_0(\Gamma)$ into U. Given a countable $S \subset \Gamma$, let us consider $\ell_\infty(S)$ as the subspace of $\ell_\infty^c(\Gamma)$ consisting of all functions vanishing outside S and let us write L_S for the endomorphism of $\ell_\infty(S)$ given by $L_S(f) = 1_S L(f)$, where 1_S is the characteristic function of S. Notice that L_S cannot map $\ell_\infty(S)$ to $c_0(S)$ since c_0 is not complemented in ℓ_∞. Thus, given an infinite countable $S \subset \Gamma$, there is a norm one $f \in \ell_\infty(S)$ (the characteristic function of a countable subset of S, if you prefer), a number $\delta > 0$ and an infinite internal $A \subset \Gamma$ such that $|L(f)| \geq \delta$ on A, with $|A \cap S| = \aleph_0$. Let $\beta(S)$ denote the supremum of the numbers δ arising in this way. Also, if T is any subset of Γ, put $\beta[T] = \sup\{\beta(S) : S \subset T, |S| = \aleph_0\}$.

Let S_1 be a countable set such that $\beta(S_1) > \frac{1}{2}\beta[\Gamma]$ and let us take $f_1 \in \ell_\infty(S_1)$ such that $|L(f_1)| > \frac{1}{2}\beta(S_1)$ on an infinite internal set A^1 with $|A^1 \cap S_1| = \aleph_0$. Let S_2 be a countable subset of $A^1\backslash S_1$ (notice $|A^1\backslash S_1| \geq \mathfrak{c}$) such that $\beta(S_2) > \frac{1}{2}\beta[A^1\backslash S_1]$ and take a normalized $f_2 \in \ell_\infty(S_2)$ such that $|L(f_2)| \geq \frac{1}{2}\beta(S_2)$ on an infinite internal set $A^2 \subset A^1$ with $|A^2 \cap S_2| = \aleph_0$. Let S_3 be an infinite countable subset of $A^2\backslash(S_1 \cup S_2)$ such that

$$\beta(S_3) > \frac{1}{2}\beta[A^2\backslash(S_1 \cup S_2)]$$

and take a normalized $f_3 \in \ell_\infty(S_3)$ such that $|Lf_3| > \frac{1}{2}\beta(S_3)$ on certain internal $A^3 \subset A^2$ such that $|A^3 \cap S_3| = \aleph_0$ and so on.

Continuing in this way we get sequences (S_n), (f_n) and (A^n), where

- Each A^n is an infinite internal subset of Γ.
- $A^0 = \Gamma$ and $A^{n+1} \subset A^n$ for all n.
- S_{n+1} is a countable subset of $A^n\backslash\bigcup_{m=1}^n S_m$, and $\beta(S_{n+1}) > \frac{1}{2}\beta[A^n\backslash\bigcup_{m=1}^n S_m]$.
- f_n is a normalized function in $\ell_\infty(S_n)$.
- $|L(f_n)| > \frac{1}{2}\beta(S_n)$ on A^n.
- For each n one has $|A^n \cap S_n| = \aleph_0$.

Our immediate aim is to see that $\beta(S_n)$ converges to zero. Fix n and consider any $a \in A^{n+1}$ to define

$$h_n = \sum_{m=1}^{n} \text{sign}(Lf_m(a))f_m.$$

Clearly, $\|h_n\| = 1$ since the f_m's have disjoint supports. On the other hand,

$$\|L\| \geq \|Lh_n\| \geq Lh_n(a) = \sum_{m=1}^{n} |Lf_m(a)| \geq \frac{1}{2}\sum_{m=1}^{n} \beta(S_m),$$

so $(\beta(S_n))$ is even summable.

For each $n \in \mathbb{N}$, choose a point $a_n \in S_n$ and consider the set $S = \{a_n : n \in \mathbb{N}\}$. We achieve the final contradiction by showing that L_S maps $\ell_\infty(S)$ to $c_0(S)$, thus completing the proof. Indeed, pick $f \in \ell_\infty(S)$ and let us compute $\text{dist}(1_S L(f), c_0(S))$. For each $n \in \mathbb{N}$, set $R_n = \{a_m : m \geq n\}$. We have $f = 1_{R_n}f + (1_S - 1_{R_n})f$ and since $S\backslash R_n$ is finite, $Lf = L1_{R_n}f + L((1_S - 1_{R_n})f) = L1_{R_n}f + (1_S - 1_{R_n})f$. Moreover, the function $1_{R_n}f$ has countable support contained in $A^n\backslash \bigcup_{m=1}^{n} S_m$. So,

$$\text{dist}(1_S Lf, c_0(S)) = \text{dist}(1_S L(1_{R_n}f), c_0(S))$$
$$\leq \text{dist}(1_{R_n} L1_{R_n}f, c_0(R_n)) + \text{dist}(1_{S\backslash R_n}L(1_{R_n}f), c_0(S\backslash R_n))$$
$$= \text{dist}(1_{R_n} L(1_{R_n}f), c_0(R_n))$$
$$\leq \|1_{R_n}f\|\beta(R_n)$$
$$\leq \|f\|\beta\left[A^n\backslash \bigcup_{m=1}^{n} S_m\right]$$
$$\leq 2\|f\|\beta(S_{n+1}).$$

And since $\beta(S_{n+1}) \to 0$ we are done. $\square$

We can present a simple proof of Theorem 4.6 for "countable" ultraproducts based on Rosenthal's result quoted in Corollary 1.15—an injective Banach space containing $c_0(\Gamma)$ contains an isomorphic copy of $\ell_\infty(\Gamma)$ as well. Suppose I countable. Then $[\ell_\infty^{k(i)}]_{\mathcal{U}}$ is a quotient of $\ell_\infty = \ell_\infty(I, \ell_\infty^{k(i)})$, and so its density character is (at most) the continuum. On the other hand, if $[\ell_\infty^{k(i)}]_{\mathcal{U}}$ is infinite dimensional, then $\lim_{\mathcal{U}(i)} k(i) = \infty$, and using an almost disjoint family we see that the cardinality of $\Gamma = \langle\{1,\ldots,k(i)\}\rangle_{\mathcal{U}}$ equals the continuum. Thus, if $[\ell_\infty^{k(i)}]_{\mathcal{U}}$ were injective, as it contains $c_0(\Gamma)$, it should contain a copy of $\ell_\infty(\Gamma)$, which is not possible, because the later space has density character $2^{\mathfrak{c}}$.

4.2.3   Copies of c_0 in Ultraproducts

Recall that a Banach space X is said to be a Grothendieck space if every operator from X into a separable space is weakly compact. It follows from Theorem 4.4 and Proposition 2.8 that ultraproducts which are $\mathscr{L}_\infty$-spaces are always Grothendieck. An alternative and simpler proof can be obtained from Lemma 4.2, taking into account the following facts:

- $\ell_\infty(\Gamma)$ is a Grothendieck space for every Γ.
- Quotients of Grothendieck spaces are again Grothendieck spaces, in particular $[\ell_\infty^{k(i)}]_\mathcal{U}$ is a Grothendieck space for every ultrafilter $\mathcal{U}$.
- A Banach space is Grothendieck if (and only if) every separable subspace is contained in a Grothendieck subspace.

Therefore, ultraproducts which are $\mathscr{L}_\infty$-spaces cannot contain infinite dimensional separable complemented subspaces, in particular, c_0. We present a stronger result, which moreover improves Corollary 3.14 of Henson and Moore in [134].

Proposition 4.7 *No ultraproduct of Banach spaces over a countably incomplete ultrafilter contains a complemented subspace isomorphic to c_0.*

Proof Let $[X_i]_\mathcal{U}$ denote the ultraproduct of a family of Banach spaces $(X_i)_{i \in I}$ with respect to a countably incomplete ultrafilter $\mathcal{U}$. Assume $[X_i]_\mathcal{U}$ has a subspace isomorphic to c_0, complemented or not, and let $\imath : c_0 \to [X_i]_\mathcal{U}$ be the corresponding embedding.

Let $f^n = \imath(e_n)$, where (e_n) denotes the traditional basis of c_0, and let (f_i^n) be a representative of f^n in $\ell_\infty(I, X_i)$, with $\|(f_i^n)\|_\infty = \|f^n\|$. Then we have

$$\|\imath^{-1}\|^{-1}\|(t_n)\|_\infty \leq \Big\| \sum_n t_n f^n \Big\| \leq \|\imath\| \, \|(t_n)\|_\infty,$$

for all (t_n) in c_0. Fix $0 < c < \|\imath^{-1}\|^{-1}$ and $\|\imath\| < C$ and, for $k \in \mathbb{N}$ define

$$J_k = \left\{ i \in I : c\|(t_n)\|_\infty \leq \Big\| \sum_{n=1}^{k} t_n f_i^n \Big\|_{X_i} \leq C\|(t_n)\|_\infty \text{ for all } (t_n) \in \ell_\infty^k \right\}.$$

It is easily seen that J_k belongs to $\mathcal{U}$ for all k. Moreover, $J_1 = I$ and $J_{k+1} \subset J_k$ for all $k \in \mathbb{N}$. Now, for each $i \in I$, define $k : I \to \mathbb{N} \cup \{\infty\}$ taking $k(i) = \sup\{n : i \in J_n\}$.

Let us consider the ultraproduct $[c_0^{k(i)}]_\mathcal{U}$, where $c_0^k = \ell_\infty^k$ when k is finite and $c_0^k = c_0$ for $k = \infty$. We define operators $j_i : c_0^{k(i)} \to X_i$ taking $j_i(e_n) = f_i^n$ for $1 \leq n \leq k(i)$ for finite $k(i)$ and for all n if $k(i) = \infty$. These are uniformly bounded and so they define an operator $j : [c_0^{k(i)}]_\mathcal{U} \to [X_i]_\mathcal{U}$. Also, we define $\kappa : c_0 \to [c_0^{k(i)}]_\mathcal{U}$ taking $\kappa(x) = [(\kappa_i(x))]$, where κ_i is the obvious projection of c_0 onto $c_0^{k(i)}$. We claim that $j\kappa = \imath$. Indeed, for $n \in \mathbb{N}$, we have $\kappa_i(e_n) = e_n$ (at least) for all $i \in J_n$ and since $J_n \in \mathcal{U}$ we have $j \circ \kappa(e_n) = \imath(e_n)$ for all $n \in \mathbb{N}$. Now, if $p : [X_i]_\mathcal{U} \to c_0$ is a projection for ι, that is, $p\iota$ is the identity on c_0, then pj is a projection for $\kappa : c_0 \to [c_0^{k(i)}]_{\mathcal{U}}$, which cannot be since the latter is a Grothendieck space. $\square$

4.3 Basic Examples

4.3.1 Ultraproducts of Spaces of Continuous Functions

Due to their special character, we will review in this Section the basic properties of ultrapowers of C-spaces. If $(X_i)_{i \in I}$ is a family of Banach algebras, then $\ell_\infty(I, X_i)$ is also a Banach algebra, with the coordinatewise product. If $\mathcal{U}$ is an ultrafilter on I, $c_0^{\mathcal{U}}(X_i)$ is an ideal in $\ell_\infty(I, X_i)$ and $[X_i]_{\mathcal{U}}$ becomes a Banach algebra with product

$$[(x_i)] \cdot [(y_i)] = [(x_i \cdot y_i)].$$

In view of Albiac-Kalton characterization quoted in Sect. 2.2.1, if $(K_i)_{i \in I}$ is a family of compact spaces, the ultraproduct $[C(K_i)]_{\mathcal{U}}$ is canonically isomorphic to a $C(K)$ space, for some compact space K. This compact is called the (topological) ultracoproduct of $(K_i)_{i \in I}$, and it is denoted $(K_i)^{\mathcal{U}}$. Actually, $(K_i)^{\mathcal{U}}$ is the maximal ideal space of $[C(K_i)]_{\mathcal{U}}$ equipped with the relative weak* topology. Let us take a look at the topological spaces $(K_i)^{\mathcal{U}}$ and gather some elementary properties of these spaces that will be later needed.

According to Corollary 4.5 the next result is a consequence of the implication $(1) \Rightarrow (4)$ in Theorem 2.14. We give here a direct "functional" proof based on the fact that, in view of (4) in Theorem 2.14, normality can be seen as an "approximate variant" of being an F-space: indeed, to be an F-space means that for every $f \in C(K)$ there is $u \in C(K)$ so that $f = u|f|$; while normality is equivalent (modulo Urysohn's lemma) to "given $f \in C(K)$ and $\varepsilon > 0$ there is $u \in C(K)$ such that $\||f - u|f|\|| < \varepsilon$".

Proposition 4.8 *If $\mathcal{U}$ is countably incomplete, then $(K_i)^{\mathcal{U}}$ is an F-space.*

Proof Put $[C(K_i)]_{\mathcal{U}} = C(K)$ and pick $f \in C(K)$ which we write as $f = [(f_i)]$. We are looking for a decomposition $f = u|f|$, with $u \in C(K)$.

Let I_n a decreasing family of members of $\mathcal{U}$ with empty intersection. We may assume $I_1 = I$. Define functions as follows: for $i \in I_n \backslash I_{n+1}$, set $A_i = \{x \in K_i : |f_i(x)| \geq 1/n\}$ and let $u_i(x) = \operatorname{sign} f_i(x)$ for $x \in A_i$. By normality u_i can be extended to K_i keeping $\|u_i\| \leq 1$. If u_i denotes now the extension we have $\|f_i - u_i|f_i|\|_{K_i} \leq 2/n$. It is then clear that $f = u|f|$, where $u = [(u_i)]$. $\square$

Regarding the spaces K_i as pure sets, we may form the set-theoretic ultraproduct $\langle K_i \rangle_{\mathcal{U}}$. Each point $\langle (x_i) \rangle$ in $\langle K_i \rangle_{\mathcal{U}}$ defines a multiplicative functional on $[C(K_i)]_{\mathcal{U}}$ by the formula

$$[(f_i)] \longmapsto \lim_{\mathcal{U}(i)} f_i(x_i).$$

The limit depends only on $\langle (x_i) \rangle$ and so we have a map $\langle K_i \rangle_{\mathcal{U}} \to (K_i)^{\mathcal{U}}$ which is easily seen to be one-to-one. The following result shows that this map has a "very dense" range.

Lemma 4.9 *Every nonempty zero set in* $(K_i)^{\mathcal{U}}$ *meets* $\langle K_i \rangle_{\mathcal{U}}$.

Proof A zero set is one of the form $\{x \in (K_i)^{\mathcal{U}} : f(x) = 0\}$ for some continuous f that we may assume to be nonnegative. Take a representation $f = [(f_i)]$ and, for each $i \in I$, put $m_i = \inf\{f_i(x) : x \in K_i\}$ and choose $x_i \in K_i$ such that $m_i = f_i(x_i)$. Now, if f vanishes at some point of $(K_i)^{\mathcal{U}}$, then

$$0 = \lim_{\mathcal{U}(i)} m_i = \lim_{\mathcal{U}(i)} f_i(x_i) = f(\langle (x_i) \rangle),$$

as we required.   □

Moreover,

Lemma 4.10 *Every point of* $\langle K_i \rangle_{\mathcal{U}}$ *is a P-point in* $(K_i)^{\mathcal{U}}$.

Proof We shall see that if $f : (K_i)^{\mathcal{U}} \to \mathbb{R}$ vanishes at $\langle (x_i) \rangle$, then it vanishes on some neighborhood of $\langle (x_i) \rangle$ in $(K_i)^{\mathcal{U}}$. Let (f_i) be a representation of f such that $\lim_{\mathcal{U}(i)} f_i(x_i) = 0$, and let (I_n) be as in Definition 4.1. For $i \in I_n \setminus I_{n+1}$, set

$$A_i = \{y \in K_i : |f_i(y) - f_i(x_i)| \leq 1/n\}.$$

It is clear that the closure of $\langle A_i \rangle_{\mathcal{U}}$ in $(K_i)^{\mathcal{U}}$ is a neighborhood of $\langle (x_i) \rangle$ in $(K_i)^{\mathcal{U}}$ and also that f vanishes on $\langle A_i \rangle_{\mathcal{U}}$.   □

Recall that f is said to be an idempotent if $f^2 = f$. Idempotents of $C(K)$ are associated to clopen subsets of K in the sense that every idempotent has the form $f = 1_A$, where A is a clopen set.

Lemma 4.11 *Each idempotent of* $[C(K_i)]_{\mathcal{U}}$ *can be represented as* $[(f_i)]$, *where* f_i *is an idempotent of* $C(K_i)$.

Proof It is almost trivial to check that there is a function $\delta = \delta(\varepsilon)$ with $\delta(\varepsilon) \to 0$ as $\varepsilon \to 0$ such that, if t is a real number satisfying $|t^2 - t| \leq \varepsilon$, then $\min(|t|, |t - 1|) \leq \delta(\varepsilon)$: just think about solving the equations $t^2 - t \pm \varepsilon = 0$. It follows that if $f \in C(K)$ satisfies the inequality $\|f^2 - f\| \leq \varepsilon$, then there is a clopen subset $A \subset K$ such that $\|f - 1_A\| \leq \delta(\varepsilon)$.

Now, if $[(f_i)]$ is an idempotent of $[C(K_i)]_{\mathcal{U}}$, then $[(f_i)]^2 = [(f_i)]$ and $\varepsilon_i = \|f_i^2 - f_i\| \to 0$ along $\mathcal{U}$. For each $i \in I$ we may take a clopen $A_i \subset K$ such that $\|f_i - 1_{A_i}\| \leq \delta(\varepsilon_i)$ and since $\delta(\varepsilon_i) \to 0$ along $\mathcal{U}$ we have $[(f_i)] = [(1_{A_i})]$ in $[C(K_i)]_{\mathcal{U}}$.   □

The following observation of Bankston relates topological ultracoproducts of the Cantor set Δ to our old friend $\mathbb{N}^*$.

Proposition 4.12 (CH) *If* $\mathcal{U}$ *is a nontrivial ultrafilter on* $\mathbb{N}$ *then* $(\Delta)^{\mathcal{U}}$ *is homeomorphic to* $\mathbb{N}^*$. *Equivalently,* $C(\Delta)_{\mathcal{U}}$ *is isometric to* ℓ_∞ / c_0.

Proof Consider the following properties for a given compact Hausdorff space:

1. To be a totally disconnected *F*-space.
2. Not having isolated points.
3. Having topological weight $\mathfrak{c}$.
4. Every nonempty G_δ subset has nonempty interior.

An important result of Parovičenko [209] states that, under **CH**, any compact Hausdorff space having the preceding four properties has to be homeomorphic to $\mathbb{N}^*$; see [245, Chap. 3, pp. 80–83] for an exposition.

Let us check that $(\Delta)^{\mathcal{U}}$ fulfills those properties. That $(\Delta)^{\mathcal{U}}$ is an *F*-space was proved in Proposition 4.8. To verify that $(\Delta)^{\mathcal{U}}$ is totally disconnected just note that a compact space L is zero-dimensional if and only for every continuous $f : L \to [0, 1]$ there is an idempotent $g \in C(L)$ such that $\|f - g\|_\infty \leq \frac{1}{2}$. Clearly, this property passes from $C(L)$ to its ultrapowers. Property (2) follows from the observation that an isolated point of L corresponds to a minimal idempotent of $C(L)$. The topological weight of L is the density character of $C(L)$. It is clear that the density character of $C(\Delta)_{\mathcal{U}}$ is the continuum and this gives (3). To check (4) just observe that each nonempty G_δ contains a zero set which has to be a neighbourhood of any *P*-point it contains. Now, see Lemma 4.9.   □

The preceding result cannot be proved in **ZFC**, since $(\Delta)^{\mathcal{U}}$ has *P*-points, a fact that cannot be established for $\mathbb{N}^*$ in **ZFC** due to a result of Shelah [247]. Concerning Lemma 4.10 it is worth noticing that there are *P*-points in $(K_i)^{\mathcal{U}}\backslash[K_i]_{\mathcal{U}}$ as proved in [26, 2.3.16].

4.3.2 Other Classes of Lindenstrauss Spaces

Apart from the class of *C*-spaces, other interesting classes of Lindenstrauss spaces are:

- C_0-spaces: maximal ideals of *C*-spaces.
- *G*-spaces: Banach spaces of the form $X = \{f \in C(K) : f(x_i) = \lambda_i f(y_i)$ for all $i \in I\}$ for some compact space K and some family of triples (x_i, y_i, λ_i), where $x_i, y_i \in K$ and $\lambda_i \in \mathbb{R}$.
- *M*-spaces: *G*-spaces where $\lambda_i \geq 0$ for every $i \in I$; equivalently, the closed sublattices of the *C*-spaces.

It is perhaps worth noticing that all these classes admit quite elegant characterizations: C_0-spaces (*C*-spaces) are exactly those real Banach algebras X (with unit) satisfying the inequality $\|x\|^2 \leq \|x^2 + y^2\|$ for all $x, y \in X$, a classical result by Arens; see [1, Theorem 4.2.5]. Also, a Banach lattice X is representable as a concrete *M*-space if and only if one has $\|x + y\| = \max(\|x\|, \|y\|)$ whenever x and y are disjoint, that is $|x| \wedge |y| = 0$. Finally, *G*-spaces are exactly those Banach spaces that are contractively complemented in M-spaces. The preceding classes are therefore closed under ultraproducts and so is the class of affine spaces, that is, spaces of continuous affine functions on Choquet simplexes; see [127, Proposition 1].

4.3.3   Ultrapowers of the Gurariy Space

We focus now on spaces of almost universal disposition. Although some results in this Section might fit better in Sect. 4.6 we present them right now to emphasize that they can be proved without invoking any result from "model theory".

Recall from Chap. 3 that a Banach space U is of almost universal disposition (for finite-dimensional spaces) if, given an isometry $g : X \to Y$, where Y is finite-dimensional and X is a subspace of U and $\varepsilon > 0$, there is an ε-isometry $f : Y \to U$ such that $f(g(x)) = x$ for every $x \in X$. Replacing "finite-dimensional" by "separable" and allowing $\varepsilon = 0$ one obtains the notion of universal disposition for separable Banach spaces.

Proposition 4.13 *Ultraproducts of spaces of almost universal disposition (in particular, ultrapowers of the Gurariy space) with respect to countably incomplete ultrafilters are of universal disposition for separable Banach spaces.*

Proof Suppose X_i are of almost universal disposition and let $\mathcal{U}$ be a countably incomplete ultrafilter on I. Let X be a separable subspace of $[X_i]_\mathcal{U}$ and $g : X \to Y$ an isometry, where Y is any separable Banach space. We will prove that there is an isometry $f : Y \to [X_i]_\mathcal{U}$ such that $f(g(x)) = x$ for every $x \in X$.

Let (x^n) and (y^n) be normalized sequences whose linear span is dense in X and Y, respectively. We may assume (x^n) is linearly independent, for if not X has to be finite-dimensional and the proof is even simpler. Let X^n be the subspace spanned by $(x^1, \ldots, x^n)$ in X^n and Y^n the subspace spanned by $g[X^n]$ and $(y^1, \ldots, y^n)$ in Y. Also, let us fix representatives (x_i^n) so that $x^n = [(x_i^n)]$ for every n. We may assume $\|x_i^n\| = 1$ for every n and every i and also that for each fixed i the sequence (x_i^n) is linearly independent in X_i. For $i \in I$ and $n \in \mathbb{N}$, let us denote by X_i^n the subspace of X_i spanned by $(x_i^1, \ldots, x_i^n)$. We define a linear map $u_i^n : X_i^n \to X^n$ by letting $u_i^n(x_i^k) = x^k$ for $1 \le k \le n$ and linearly on the rest. Also, we define $g_i^n : X_i^n \to Y^n$ as the composition $g \circ u_i^n$, that is, $g_i^n(x_i^k) = y^k$ for $1 \le k \le n$. To proceed, we observe that the sets

$$I_\varepsilon^n = \{i \in I \text{ such that } u_i^n : X_i^n \to X^n \text{ is a strict } \varepsilon\text{-isometry}\}$$

are in $\mathcal{U}$ for every n and every $\varepsilon > 0$. In particular $I_{1/n}^n \in \mathcal{U}$ for all $n \in \mathbb{N}$. Take any function $m : I \to \mathbb{N}$ such that $m(i) \to \infty$ along $\mathcal{U}$ and define $n : I \to \mathbb{N}$ thus:

$$n(i) = \begin{cases} m(i) & \text{if } i \in I_{1/n}^n \text{ for all } n \\ \max\{n : i \in I_{1/n}^n\} & \text{otherwise} \end{cases}$$

Clearly, $n(i) \to \infty$ along $\mathcal{U}$. Indeed, for $k \in \mathbb{N}$ one has

$$n^{-1}[k, \infty) = \{i \in I : n(i) \geq k\} \supset \left(I^k_{1/k} \bigcap m^{-1}[k, \infty) \right)$$

and so $n^{-1}[k, \infty) \in \mathcal{U}$.

Let us form the ultraproducts of the operators $g_i^{n(i)} : X_i^{n(i)} \to Y^{n(i)}$ with respect to $\mathcal{U}$. We claim that:

- $X \subset [X_i^{n(i)}]_\mathcal{U}$ and $Y \subset [Y_i^{n(i)}]_\mathcal{U}$;
- $[g_i^{n(i)}]_\mathcal{U}$ is an isometry extending g.

Let us begin with the containment of X. It suffices to check that $x^n \in [X_i^{n(i)}]_\mathcal{U}$ for all n. But since the set $J = \{i \in I : n(i) \geq n\}$ belongs to $\mathcal{U}$ we may form the family

$$z_i = \begin{cases} x_i^n & \text{if } i \in J \\ 0 & \text{otherwise} \end{cases} \tag{4.2}$$

in which $z_i \in X_i^{n(i)}$ for all $i \in I$ and we have $[(z_i)] = [(x_i^n)] = x^n$. Similarly, if we regard Y as the diagonal subspace of $Y_\mathcal{U}$ and $[Y^{n(i)}]_\mathcal{U}$ as a subspace of $Y_\mathcal{U}$ in the obvious way, then the same argument gives that $[Y^{n(i)}]_\mathcal{U}$ contains Y. To prove that $[g_i^{n(i)}]_\mathcal{U}$ is an isometry just observe that, by the very definition of the sets $I^n_{1/n}$ the operator $g_i^{n(i)} : X_i^{n(i)} \to Y^{n(i)}$ is an $n(i)^{-1}$ isometry and that $n(i)^{-1}$ tends to 0 along $\mathcal{U}$. This immediately gives that the ultraproduct operator $[g_i^{n(i)}]_\mathcal{U}$ preserves the norm. To see that it extends g it is enough to check that $[g_i^{n(i)}]_\mathcal{U}(x^n) = [(y^n)]$ for every $n \in \mathbb{N}$. Take J and z_i as in (4.2). Then

$$[g_i^{n(i)}]_\mathcal{U}(x^n) = [g_i^{n(i)}]_\mathcal{U}[(z_i)] = [g_i^{n(i)}(z_i)] = [g_i^{n(i)}(x_i^n)] = [(y^n)],$$

as required.

Recalling that $g_i^{n(i)} : X_i^{n(i)} \to Y^{n(i)}$ is a strict $n(i)^{-1}$-isometry, Lemma 3.28 gives an $n(i)^{-1}$-isometry $f_i : Y^{n(i)} \to X_i$ such that $f_i(g_i(x)) = x$ for each $x \in X_i^{n(i)}$. It is then obvious that the ultraproduct operator

$$[f_i]_\mathcal{U} : [Y^{n(i)}]_\mathcal{U} \to [X_i]_\mathcal{U}$$

preserves the norm and besides the composition $[f_i]_\mathcal{U} \circ [g_i^{n(i)}]_\mathcal{U}$ is the inclusion of $[X_i^{n(i)}]_\mathcal{U}$ into $[X_i]_\mathcal{U}$. The restriction of $[f_i]_\mathcal{U}$ to Y does what we need.   □

Thus, a natural juxtaposition of the preceding Proposition and Theorem 3.34 yields

Corollary 4.14 *Ultrapowers of Banach spaces of almost universal disposition (in particular, of the Gurariy space) with respect to countably incomplete ultrafilters are not complemented in any M-space.*

Proposition 4.13 gives an alternative construction of spaces of universal disposition to that presented in Chap. 3. Taking Proposition 3.16 into account one gets:

Theorem 4.15 (CH) *The following Banach spaces are isometric:*

- *Any space of density character $\mathfrak{c}$ of universal disposition for separable spaces.*
- *The Kubiś space.*
- *Any ultrapower of the Gurariy space built over a free ultrafilter on the integers.*

It is unclear whether spaces of (almost) universal disposition for finite dimensional spaces are (almost) isotropic or not. However, it follows from Proposition 3.41 and its corollary that ultrapowers of the Gurariy space built over a countably incomplete ultrafilter on $\mathbb{N}$ are isotropic and even separably transitive. We also know from the combination of Theorem 4.4 (ultrapowers of $\mathscr{L}_\infty$ spaces are universally separably injective) and Proposition 2.52 (universally separably injective spaces are separably automorphic) that ultrapowers of the Gurariy space are separably automorphic. We can give a unified proof for both facts improving the second:

Proposition 4.16 *Let $\mathcal{U}$ be a countably incomplete ultrafilter. Suppose $t : X \to Y$ is a linear isomorphism, where X and Y are separable subspaces of $\mathscr{G}_\mathcal{U}$. Then there is an automorphism T of $\mathscr{G}_\mathcal{U}$ extending t, with*

$$\|T\| = \|t\| \quad \text{and} \quad \|T^{-1}\| = \|t^{-1}\|.$$

In particular, ultrapowers of the Gurariy space with respect to countably incomplete ultrafilters are separably transitive.

Proof We fix a (normalized) linearly independent sequence (x^n) spanning X and we denote by X^n the subspace spanned by $x^1, \ldots, x^n$ in X and we set $Y^n = t[X^n]$. Also, we denote by $t^n : X^n \to Y^n$ the restriction of t.

For each n we fix a normalized family $(x_i^n)_i$ representing x^n. Taking $y^n = t(x^n)$ we may choose families (y_i^n) representing y^n, with $\|t^{-1}\|^{-1} \leq \|y_i^n\| \leq \|t\|$ for all i and n. We may assume an do that for each fixed $i \in I$ the sequences $(x_i^n)_n$ and $(y_i^n)_n$ are linearly independent in E. Put $X_i^n = \text{span}\{x_i^1, \ldots, x_i^n\}$ and $Y_i^n = \text{span}\{y_i^1, \ldots, y_i^n\}$. We define operators $u_i^n : X_i^n \to X^n$ by letting $u_i^n(x_i^k) = x^k$ for $1 \leq k \leq n$. Similarly, we define $v_i^n : Y_i^n \to Y^n$ by letting $v_i^n(y_i^k) = y^k$ and extending by linearity on the rest. All these operators are isomorphisms. Moreover, for each fixed n, the four families of real numbers $\|u_i^n\|$, $\|(u_i^n)^{-1}\|$, $\|v_i^n\|$ and $\|(v_i^n)^{-1}\|$ converge to 1 along $\mathcal{U}$.

Define $t_i^n : X_i^n \to Y_i^n$ by the composition $t_i^n = (v_i^n)^{-1} \circ t^n \circ u_i^n$, that is, the following is a commutative diagram:

$$
\begin{array}{ccc}
X_i^n & \xrightarrow{\;t_i^n\;} & Y_i^n \\[2pt]
{\scriptstyle u_i^n}\Big\downarrow & & \Big\downarrow{\scriptstyle v_i^n} \\[2pt]
X^n & \xrightarrow{\;t^n\;} & Y^n
\end{array}
$$

Since $\|t^n\| \le \|t\|$ and $\|(t^n)^{-1}\| \le \|t^{-1}\|$ for all n, the sets

$$ I_\varepsilon^n = \{i \in I : \|t_i^n\| < (1+\varepsilon)\|t\| \text{ and } \|(t_i^n)^{-1}\| < (1+\varepsilon)\|t^{-1}\|\} $$

are all in $\mathcal{U}$. In particular, $I_{1/n}^n \in \mathcal{U}$ for all $n \in \mathbb{N}$. Take any function $m : I \to \mathbb{N}$ such that $m(i) \to \infty$ along $\mathcal{U}$ and define $n : I \to \mathbb{N}$ thus:

$$
n(i) = \begin{cases}
m(i) & \text{if } i \in I_{1/n}^n \text{ for all } n \\[4pt]
\max\{n : i \in I_{1/n}^n\} & \text{otherwise}
\end{cases}
$$

Clearly, $n(i) \to \infty$ along $\mathcal{U}$. Let us form the ultraproduct of the operators $t_i^{n(i)} : X_i^{n(i)} \to Y_i^{n(i)}$. We claim that:

- $X \subset [X_i^{n(i)}]\mathcal{U}$ and $Y \subset [Y_i^{n(i)}]\mathcal{U}$;
- $\tau = [t_i^{n(i)}]\mathcal{U}$ is an isomorphism extending t, with $\|\tau\| = \|t\|$ and $\|\tau^{-1}\| = \|t^{-1}\|$.

This is proved as we did in Proposition 4.13. One takes $x^n \in X$ and considers the family

$$
z_i = \begin{cases}
x_i^n & \text{if } n(i) \ge n \\
0 & \text{otherwise}
\end{cases}
\tag{4.3}
$$

in which $z_i \in X_i^{n(i)}$ for all $i \in I$ and we have $[(z_i)] = [(x_i^n)] = x^n$. This shows that $[X_i^{n(i)}]\mathcal{U}$ contains X and the same argument gives $Y \subset [Y_i^{n(i)}]\mathcal{U}$.

To compute the norm of $\tau = [t_i^{n(i)}]\mathcal{U}$ just observe that, by the very definition of the sets $I_{1/n}^n$ one has $\|t_i^{n(i)} : X_i^{n(i)} \to Y_i^{n(i)}\| < (1+n(i)^{-1})\|t\|$ and since $n(i)^{-1}$ tends to 0 along $\mathcal{U}$ we see that $\|\tau\| = \lim_{\mathcal{U}} \|t_i^{n(i)}\| \le \|t\|$. The same argument yields $\|\tau^{-1}\| \le \|t^{-1}\|$. We check that τ extends t, which automatically gives the estimates $\|\tau\| \ge \|t\|$ and $\|\tau^{-1}\| \ge \|t^{-1}\|$. As before it is enough to see that $\tau(x^n) = y^n$ for every $n \in \mathbb{N}$. Take z_i as in (4.3). Then

$$ \tau(x^n) = [t_i^{n(i)}]\mathcal{U}(x^n) = [t_i^{n(i)}]\mathcal{U}[(z_i)] = [t_i^{n(i)}(z_i)] = [t_i^{n(i)}(x_i^n)] = [(y_i^n)] = y^n. $$

To end, we apply Lemma 3.29 to each isomorphism $t_i^{n(i)}$. This gives a family of automorphisms T_i of E such that T_i extends $t_i^{n(i)}$, with $\|T_i\| \leq (1 + n(i)^{-1})\|t\|$ and $\|T_i^{-1}\| \leq (1 + n(i)^{-1})\|t^{-1}\|$. Thus, the ultraproduct operator $T = [T_i]_{\mathcal{U}}$ is the required extension.   □

4.4 Lifting Operators Taking Values in Ultraproducts

In this Section we study the exact sequence associated to an ultraproduct, namely

$$0 \longrightarrow c_0^{\mathcal{U}}(I, E_i) \longrightarrow \ell_\infty(I, E_i) \longrightarrow [E_i]_{\mathcal{U}} \longrightarrow 0,$$

where (E_i) is now an arbitrary family of Banach spaces. The first thing one must know about the preceding sequence is:

Lemma 4.17 $c_0^{\mathcal{U}}(I, E_i)$ is an M-ideal in $\ell_\infty(I, E_i)$.

Proof The notion of an M-ideal was introduced in Sect. 2.2.2. Obviously it is very difficult to manage the dual of $\ell_\infty(I, E_i)$ and so we need a different approach avoiding duality. It is proved in [121, Theorem 2.2] that J is an M-ideal in X if and only if it satisfies the following condition: given a finite family of closed balls $B(x^k, r_k)$ in X such that $B(x^k, r_k) \cap J \neq \varnothing$ for all k and

$$\bigcap_k B(x^k, r_k) \neq \varnothing,$$

one has

$$\bigcap_k B(x^k, r_k + \varepsilon) \cap J \neq \varnothing$$

for each $\varepsilon > 0$.

Let us check this condition for $c_0^{\mathcal{U}}(I, E_i)$. Let $B(x^k, r_k)$ be the corresponding balls and take $x = (x_i)$ in their intersection. Also, for each k, pick $y^k \in B(x^k, r_k) \cap c_0^{\mathcal{U}}(I, E_i)$. Now, given $\varepsilon > 0$, as $\|y_i^k\| \to 0$ along $\mathcal{U}$ we may find I_ε in $\mathcal{U}$ such that $\|y_i^k\| \leq \varepsilon$ for all k and all $i \in I_\varepsilon$. We define $y = (y_i)$ taking

$$y_i = \begin{cases} 0 & \text{for } i \in I_\varepsilon \\ x_i & \text{otherwise} \end{cases}$$

It is clear that $y \in \bigcap_k B(x^k, r_k + \varepsilon) \cap c_0^{\mathcal{U}}(I, E_i)$.   □

Corollary 4.18 *Let $T : X \to [E_i]_{\mathcal{U}}$ be an operator. If T factors through a separable Banach space with the BAP then T can be lifted to an operator $X \to \ell_\infty(I, E_i)$.*

Proof The hypothesis means that there is a separable space Y having the BAP and operators $R : X \to Y$ and $S : Y \to [E_i]_{\mathcal{U}}$ such that $T = S \circ R$. Now, by Theorem 2.20 (1), S lifts to $\ell_\infty(I, E_i)$. Composing R with the lifting of S gives a lifting of T to $\ell_\infty(I, E_i)$. □

Note that the lifting occurs whatever the spaces E_i are when X is separable and has the BAP. But even if X lacks the BAP, the lifting is still possible if the spaces E_i have the (joint) UAP. This follows from the fact that, if (E_i) has the joint UAP, then $[E_i]_{\mathcal{U}}$ has the BAP [126, Theorem 9.1], and that every separable subspace of a Banach space with the BAP is contained in a further separable subspace with the BAP [60, Theorem 9.7]. In particular, (the inclusion of) every separable subspace of $[E_i]_{\mathcal{U}}$ with the BAP lifts to $\ell_\infty(I, E_i)$. The analogous statement for finite-dimensional subspaces of ultrapowers is often used in proving the finite representability of ultrapowers in the base space, see [126, Proposition 6.1]:

Corollary 4.19 *Every separable complemented subspace of $[E_i]_{\mathcal{U}}$ having the BAP embeds as a complemented subspace of $\ell_\infty(I, E_i)$.*

This unifies some old results on complemented subspaces of $\ell_\infty(I, E_i)$: for instance, $\ell_\infty(\mathbb{N}, \ell_p^n)$ contains a complemented copy of L_p for $1 \le p < \infty$. See [73] for more information on this topic.

4.5 Duality, Twisted Sums, and the BAP

We discuss now some applications to twisted sums.

Theorem 4.20 *Let Z be a separable Banach space and E a Banach space such that $\mathrm{Ext}(Z, E) = 0$. Suppose that either Z has the BAP or E has the UAP. Then $\mathrm{Ext}(Z, E_{\mathcal{U}}) = 0$ for all ultrafilters $\mathcal{U}$.*

Proof We write the proof in the case where Z has the BAP and leave the case where E has the UAP to the reader. Let $q : \ell_1 \to Z$ be any quotient map. Consider the exact sequence

$$0 \longrightarrow \ker q \longrightarrow \ell_1 \overset{q}{\longrightarrow} Z \longrightarrow 0.$$

It is well-known that, given a Banach space X, the condition $\mathrm{Ext}(Z, X) = 0$ is equivalent to: "every operator $v : \ker q \to X$ has an extension $\tilde{v} : \ell_1 \to X$"; see Lemma A.20. If so, by the open mapping theorem, there is some $C > 0$ so that this can be can be done with $\|\tilde{v}\| \le C\|v\|$.

Thus, let $u : \ker q \to E_{\mathcal{U}}$ be an operator. We know from Lusky [189, 190] (there are also improved versions in [66, 95]) that $\ker q$ has the BAP when Z has the BAP and so u lifts to an operator $U : \ker q \to \ell_\infty(I, E)$ that we may write as $U = (u_i)$, with $u_i \in L(\ker q, E)$. The following diagram can be helpful.

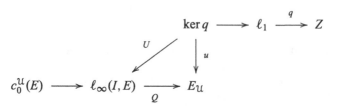

Since $\mathrm{Ext}(Z, E) = 0$ each u_i extends to an operator $\tilde{u}_i : \ell_1 \to E$, with $\|\tilde{u}_i\| \leq C\|u_i\|$. Thus $(\tilde{u}_i) : \ell_1 \longrightarrow \ell_\infty(I, E)$ is an operator and composing with the natural quotient map Q we obtain the required extension of u.   □

It is a standard fact that the dual spaces of even order of a Banach space Y are complemented subspaces of suitably ultrapowers of Y (just iterate Proposition A.8); one thus has.

Corollary 4.21 *Let Z be a separable Banach space and Y a Banach space. If Z has the BAP or Y has the UAP and $\mathrm{Ext}(Z, Y) = 0$ then, for all $n \in \mathbb{N}$, one has $\mathrm{Ext}(Z, Y^{(2n)}) = 0$.*

The hypotheses of Corollary 4.18 cannot be easily removed:

- Separability is necessary because if $\mathcal{U}$ is a free ultrafilter on $\mathbb{N}$ then $[\ell_2^n]_{\mathcal{U}}$ is a (nonseparable) Hilbert space that cannot be embedded into the subspace $\ell_\infty(\mathbb{N}, \ell_2^n)$ of ℓ_∞, whose weakly compact sets are separable.
- As for the BAP, we can adapt Lusky's example in [187]. Let E be a separable Banach space and let (E_n) be an increasing sequence of finite dimensional subspaces whose union is dense in E. If $\mathcal{U}$ is a free ultrafilter on $\mathbb{N}$, E embeds (isometrically) as a subspace of $[E_n]_U$: given $x \in E$, let us choose $x_n \in E_n$ in such a way that $\|x - x_n\| \to 0$ as $n \to \infty$ and define $\iota(x) = [(x_n)]$. Suppose ι lifts to an operator $L : E \to \ell_\infty(\mathbb{N}, E_n)$. Writing $L(x) = (L_n(x))$ we have that $L_n : E \to E_n$ is a uniformly bounded sequence of finite-rank operators such that $\lim_{\mathcal{U}} \|L_n(x) - x\| = 0$ for all $x \in E$, from where it follows that E has the BAP. Hence if E lacks the BAP, the operator ι cannot be lifted.

It would be interesting to know if the approximation properties are truly necessary in Theorem 4.20 and its Corollary 4.21. Separability is really necessary in Theorem 4.20: since $[\ell_\infty]_{\mathcal{U}}$ is not injective, for some Banach space Z one has $\mathrm{Ext}(Z, (\ell_\infty)_{\mathcal{U}}) \neq 0$, while $\mathrm{Ext}(Z, \ell_\infty) = 0$.

A somehow unexpected consequence of Corollary 4.21 is the following.

Corollary 4.22 *Let Z be a separable Banach space and Y a Banach space such that $\mathrm{Ext}(Z, Y) = 0$. If either Z has the BAP or Y has the UAP, then $\mathrm{Ext}(Y^*, Z^*) = 0$.*

(Observe that if Y has the UAP then also Y^* has the UAP). This is a formal consequence of Corollary 4.21 in view of the duality formula $\text{Ext}(A, B^*) = \text{Ext}(B, A^*)$ (that holds for all Banach spaces A and B) which can be explicitly seen in [64, 141], and is implicit in [54], and implies that $\text{Ext}(Y^*, Z^*) = \text{Ext}(Z, Y^{**})$.

It is an open problem whether $\text{Ext}(Z, Y) = 0$ always implies $\text{Ext}(Y^*, Z^*) = 0$. The obvious difficulty for a "straightforward" proof is the existence of exact sequences $0 \longrightarrow Z^* \longrightarrow X \longrightarrow Y^* \longrightarrow 0$ in which the space X is not a dual space or even complemented in its bidual (see [53]); therefore the sequence above cannot be a dual sequence. In other words, there are elements in $\text{Ext}(Y^*, Z^*)$ not induced by elements of $\text{Ext}(Z, Y)$.

4.6 Ultra-Isomorphic Spaces

As we mentioned before, the study of the isometric equivalence of ultrapowers goes back to the inception of the ultraproduct construction in Banach space theory and has produced many interesting results in the "model theory of Banach spaces". In this Section we will rather consider the isomorphic variation introduced by Henson and Moore [134, p.106] and so we address the question of when two given Banach spaces have isomorphic ultrapowers.

We begin with the following observation which needs a bit of model theory. Recall from the preliminaries that $\sim$ stands for "linearly isomorphic" while $\approx$ stands for "linearly isometric".

Lemma 4.23 *Let X and Y be Banach spaces. The following are equivalent:*

- *There is an ultrafilter $\mathcal{U}$ such that $X_{\mathcal{U}}$ and $Y_{\mathcal{U}}$ are isomorphic.*
- *There are ultrafilters $\mathcal{U}$ and $\mathcal{V}$ such that $X_{\mathcal{U}}$ and $Y_{\mathcal{V}}$ are isomorphic.*

Also, the following are equivalent

- *There is an ultrafilter $\mathcal{U}$ such that $X_{\mathcal{U}}$ and $Y_{\mathcal{U}}$ are isometric.*
- *There are ultrafilters $\mathcal{U}$ and $\mathcal{V}$ such that $X_{\mathcal{U}}$ and $Y_{\mathcal{V}}$ are isometric.*

Proof We make the proof of the (•) statements; the proof for the (○) statements is analogous.

The iteration of ultrapowers produces new ultrapowers. Indeed, suppose that $\mathcal{U}, \mathcal{V}$ are ultrafilters on I and J respectively. Let $\mathcal{W}$ denote the family of all subsets W of $K = I \times J$ for which the set $\{j \in J : \{i \in I : (i, j) \in W\} \in \mathcal{U}\}$ belongs to $\mathcal{V}$. Then $\mathcal{W}$ is an ultrafilter, often denoted by $\mathcal{U} \times \mathcal{V}$, and moreover, one has $Z_{\mathcal{W}} = (Z_{\mathcal{U}})_{\mathcal{V}}$ for all Banach spaces Z. On the other hand, the Banach space version of the Keisler-Shelah isomorphism theorem due to Stern [237, Theorem 2.1] establishes that given a Banach space X and two ultrafilters $\mathcal{U}, \mathcal{V}$ then there is another ultrafilter $\mathcal{W}$ such that $(X_{\mathcal{U}})_{\mathcal{W}} \approx (X_{\mathcal{V}})_{\mathcal{W}}$.

Now, if $X_\mathcal{U} \sim Y_\mathcal{V}$, taking an ultrafilter $\mathcal{W}$ such that $(Y_\mathcal{U})_\mathcal{W} \approx (Y_\mathcal{V})_\mathcal{W}$ we have

$$X_{\mathcal{U} \times \mathcal{W}} = (X_\mathcal{U})_\mathcal{W} \sim (Y_\mathcal{V})_\mathcal{W} \approx (Y_\mathcal{U})_\mathcal{W} = Y_{\mathcal{U} \times \mathcal{W}},$$

as we wanted to prove.   □

Definition 4.24 We will say that X and Y are *ultraisomorphic* or that they have the same ultratype when they satisfy the equivalent conditions (•) of Lemma 4.23.

We will say that X and Y are *ultraisometric* or that they have the same isometric ultratype, when they satisfy the equivalent conditions (○) of Lemma 4.23.

The problem of the classification of $\mathscr{L}_\infty$-spaces by isomorphic ultratypes appears posed in [134, p. 106] and [128, p. 315] and was considered in [130]: How many ultratypes of $\mathscr{L}_\infty$-spaces are there?

Recalling that ultrapowers of M-spaces are again M-spaces, we can reformulate Corollary 4.14 as follows.

Proposition 4.25 *Spaces of almost universal disposition are not ultraisomorphic to complemented subspace of M-spaces. In particular $\mathscr{G}$ and c_0 do have different ultratypes.*

In the opposite direction there is a result proved by Henson long time ago [130, Corollary 3.11] asserting that all infinite dimensional C-spaces are ultraisomorphic (to be true Henson worked with nonstandard hulls instead of ultrapowers). We state now a slight generalization for M-spaces. The proof we present is based on ideas of [237] that can moreover be easily modified to the effect of proving also Proposition 4.27. To simplify the exposition, we will write $X \lhd Y$ to mean that X is isomorphic to a complemented subspace of Y.

Proposition 4.26 *All infinite dimensional M-spaces have the same ultratype.*

Proof The key of the reasoning is the following nice result of Stern [237, Theorem 2.2]:

Let F be a separable subspace of the Banach space E. There exists a separable subspace L of E containing F and an ultrafilter $\mathcal{U}$ such that $L_\mathcal{U} \approx E_\mathcal{U}$. If E is a Banach lattice then L can be chosen to be a sublattice of E.

This implies that every M-space X has an ultrapower isometric to an ultrapower of some separable M-space Y. Since separable M-spaces are isomorphic to C-spaces (Benyamini [31]), what we have to show is that all C-spaces have the same ultratype.

Thus, let X be a separable C-space. By the very definition, a separable $\mathscr{L}_\infty$-space X embeds into an ultraproduct $[\ell_\infty^n]_\mathcal{U}$, where $\mathcal{U}$ is any free ultrafilter on the integers. Therefore X embeds as a subspace of $(c_0)_\mathcal{U}$. By Stern's result quoted above there is a separable C-space L of $(c_0)_\mathcal{U}$ which contains a copy of X and an ultrafilter $\mathcal{V}$ such that $L_\mathcal{V} \approx (c_0)_{\mathcal{U} \times \mathcal{V}}$. Since:

• X is isomorphic to its square (Bessaga-Pełczyński [37, Theorem 3]);
• L contains a complemented copy of X (Pełczyński [211, Theorem 1]) .

We arrive to:

$$X_\mathcal{V} \lhd L_\mathcal{V} \approx (c_0)_{\mathcal{U}\times\mathcal{V}} \lhd X_{\mathcal{U}\times\mathcal{V}}.$$

Now we can apply Keisler-Shelah-Stern's theorem to get an ultrafilter $\mathcal{W}$ such that $(X_\mathcal{V})_\mathcal{W} \approx (X_{\mathcal{U}\times\mathcal{V}})_\mathcal{W}$. Letting $\mathcal{J} = (\mathcal{U} \times \mathcal{V}) \times \mathcal{W}$ we have

$$X_\mathcal{J} \approx (X_\mathcal{V})_\mathcal{W} \lhd ((c_0)_{\mathcal{U}\times\mathcal{V}})_\mathcal{W} = (c_0)_\mathcal{J}.$$

On the other hand, $c_0 \lhd X$ and thus $(c_0)_\mathcal{J} \lhd X_\mathcal{J}$. Since both spaces X and c_0 are isomorphic to their squares the same is true for their ultrapowers and Pełczyński's decomposition method yields $X_\mathcal{J} \approx (c_0)_\mathcal{J}$.   □

This result can be "extended" to:

Proposition 4.27 *A Banach space that is isomorphic to its square and complemented in an M-space is ultraisomorphic to* ℓ_∞.

Proof Suppose X is isomorphic to its square and complemented in an M-space E. As E has the same ultratype as ℓ_∞ there is an ultrafilter $\mathcal{U}$ such that $E_\mathcal{U} \sim (\ell_\infty)_\mathcal{U}$ and so $X_\mathcal{U} \lhd (\ell_\infty)_\mathcal{U}$. But X is an infinite dimensional $\mathcal{L}_\infty$-space and so ℓ_∞ embeds as a subspace of $X_\mathcal{U}$. Hence $\ell_\infty \lhd X_\mathcal{U} \lhd (\ell_\infty)_\mathcal{U}$. Let $\mathcal{V}$ be an ultrafilter such that $(\ell_\infty)_\mathcal{V} \approx (\ell_\infty)_{\mathcal{U}\times\mathcal{V}}$. One has

$$(\ell_\infty)_{\mathcal{U}\times\mathcal{V}} \approx (\ell_\infty)_\mathcal{V} \lhd X_{\mathcal{U}\times\mathcal{V}} \lhd (\ell_\infty)_{\mathcal{U}\times\mathcal{V}}$$

and since X and ℓ_∞ and their ultrapowers are all isomorphic to their squares we can apply Pełczyński's decomposition method again and we are done.   □

We now prove that the Banach spaces constructed in Chap. 3 all have the same isometric ultratype. As a preparation, we (partially) complete Proposition 4.13:

Lemma 4.28 *A Banach space is of almost universal disposition if and only if so are its ultrapowers with respect to countably incomplete ultrafilters.*

Proof The "only if" part is obvious. To prove the converse suppose $X_\mathcal{U}$ is of almost universal disposition (for finite dimensional spaces). Let $g : E \to F$ be an isometry, where F is finite dimensional and E is a subspace of X.

Fix $\varepsilon > 0$ and consider X (hence E) as a subspace of $X_\mathcal{U}$ through the diagonal embedding. Then there is an $\varepsilon/2$-isometry $f : F \to X_\mathcal{U}$ such that $f(g(x)) = [(x)_i]$ for every $x \in E$. Writing $f = [(f_i)]_\mathcal{U}$ for suitable operators $f_i : F \to X$ we see that for most "indices" $i \in I$:

- $f_i : F \to X$ is an ε-isometry, and
- $\|f_i(g(x)) - x\|_X \le \varepsilon\|x\|_X$ for all $x \in E$.

By Lemma 3.27, X is of almost universal disposition.   □

Proposition 4.29 *A Banach space is of almost universal disposition if and only if it has the same isometric ultratype as $\mathscr{G}$. In particular, all spaces of almost universal disposition have the same ultratype.*

Proof If X is ultraisometric to $\mathscr{G}$ then some ultrapower of X is of almost universal disposition and so is X, by the preceding Lemma.

To check the converse, suppose X is of almost universal disposition. Then, according to Stern's result quoted at the beginning of the proof of Proposition 4.26, there is a separable subspace $L \subset X$ and an ultrafilter $\mathcal{U}$ such that $L_\mathcal{U} \approx X_\mathcal{U}$. Since $X_\mathcal{U}$ and $L_\mathcal{U}$ are of almost universal disposition the same happens to L and so L is almost isometric to $\mathscr{G}$ by Proposition 3.29 or Theorem 3.38. Hence $X_\mathcal{U} \approx L_\mathcal{U} \approx \mathscr{G}_\mathcal{U}$.   □

The following example shows that it would be a mistake to think that the different ultratypes of C-spaces and spaces of almost universal disposition is due to the fact that $\mathscr{G}$ is not complemented in any C-space:

Proposition 4.30 *There is a (nonseparable) Lindenstrauss space which is complemented in no C-space but has an ultrapower isomorphic to a C-space.*

Proof The example can be found in Sect. 2.2.6; it is Benyamini's construction of a nonseparable M-space which is complemented in no C-space. That space has an ultrapower isomorphic to a C-space, by Theorem 4.27.   □

Back to the comment at the end of Sect. 3.3.4, and taking into account that ultrapowers of C^*-algebras are again C^*-algebras, we see that every ultrapower of a separable (complex) Lindenstrauss space lives 1-complemented in an ultrapower of the CAR algebra A. It follows that every Lindenstrauss space $\mathscr{L}$, separable or not, has an ultrapower 1-complemented in a C^*-algebra. Indeed, applying Stern's result quoted in the proof of Proposition 4.26 one obtains a separable subspace $L \subset \mathscr{L}$ and an ultrafilter $\mathcal{U}$ such that $L_\mathcal{U} \approx \mathscr{L}_\mathcal{U}$. But L must be a Lindestrauss space, by [127, Theorem 2.1], so L is 1-complemented in A and therefore $L_\mathcal{U}$, hence $\mathscr{L}_\mathcal{U}$, is 1-complemented in $A_\mathcal{U}$.

4.7   Notes and Remarks

4.7.1   Measurable Cardinals

In this Chapter we have considered ultraproducts based on countably incomplete ultrafilters only. As the reader can imagine, an ultrafilter is said to be countably complete if it is closed under countable intersections. A set I supporting a free, countably complete ultrafilter is said to have measurable cardinal. Otherwise we call the cardinal of I nonmeasurable.

While questions around the existence of measurable cardinals spurred a considerable interest in set theory and logic, ultraproducts based on countably complete ultrafilters should not be very interesting to us. In fact, if $\mathcal{U}$ is countably complete and $|X|$ is nonmeasurable, then $X_{\mathcal{U}} = X$ in the sense that the diagonal embedding is onto. This is so because if $\mathcal{U}$ is countably complete, one has $\langle X_i \rangle_{\mathcal{U}} = [X_i]_{\mathcal{U}}$ for all families of Banach spaces in view of the remark following Definition 4.1 and the diagonal embedding of X into $\langle X \rangle_{\mathcal{U}}$ is onto according to [75, Corollary 4.2.8].

Assuming that all cardinals are nonmeasurable is consistent with ZFC, the usual setting of set theory, with the axiom of choice. The cardinals

$$\aleph_0, \aleph_1, \ldots, \aleph_\omega, \ldots, \aleph_{\omega_1}, \ldots, \aleph_{\omega_\omega}, \ldots \qquad \text{and} \qquad \mathfrak{c}, 2^{\mathfrak{c}}, 2^{2^{\mathfrak{c}}}, \ldots$$

are all nonmeasurable, and so are cardinals that can be obtained from nonmeasurable cardinals by the standard processes of cardinal arithmetic. It is conceivable that no measurable cardinal exists and actually much of the research on measurable cardinals has been done with the purpose of establish that they do not exist at all! In any case, measurable cardinals, if they exist, should be very large; see, for instance, [75, Sect. 4.2] or [159, Chap. 1, Sect. 2]. Lately the existence of a measurable cardinal is often treated as an additional axiom.

4.7.2 Topological Ultracoproducts

There is a description of the ultracoproduct construction in purely topological terms. Let (K_i) be a family of compact spaces indexed by I and let $\mathcal{U}$ be an ultrafilter on I. Let $S = \bigsqcup_i K_i$ be the topological direct sum of the family and βS its Stone-Čech compactification. Then $(K_i)^{\mathcal{U}}$ is obtained as the intersection of those closed sets $C \subset \beta S$ for which $\{i \in I : K_i \subset C\}$ belongs to $\mathcal{U}$. This definition is the usual one in topology, see [26, 27]. It is equivalent to that appearing in Sect. 4.3.1 in view of [119, Proposition 2].

It seems to be part of the "topological folklore" that F-spaces tend to be disconnected (cf. [245, Exercise 3D2, p. 93] or [26, 2.3.9. Remark]). However, according to Proposition 4.8 and Lemma 4.11 one has:

Corollary 4.31 *If $(K_i)_{i \in I}$ is a family of connected compact spaces, then the ultracoproduct $(K_i)^{\mathcal{U}}$ is a connected F-space.*

This is nearly obvious once one realizes that K is connected if and only if there are no idempotents of $C(K)$, apart from 0 and 1. Also, Proposition 4.8 easily leads to compact F-spaces having any prescribed (covering) dimension.

Proposition 4.32 *Let (K_i) be a family of compacta indexed by I and let $\mathcal{U}$ be a countably incomplete ultrafilter on I. Then $(K_i)^{\mathcal{U}}$ is an F-space and $\dim((K_i)^{\mathcal{U}}) = \lim_{\mathcal{U}(i)} \dim K_i$.*

This was first proved by Bankston in [26, Theorem 2.2.2]. Let us indicate an easier "functional" proof.

Proof A closed subalgebra A of $C(K)$ is said to be analytic if $f \in A$ provided $f^2 \in A$. If B is any subset of $C(K)$, then the analytic algebra of base B is the smallest analytic subalgebra of $C(K)$ that contains B. A well known result of Katetov [107, Theorem 16.35] states that dim $K \leq d$ if and only if every finite subfamily of $C(K)$ is contained in some analytic subalgebra having a base of cardinal d. The result follows.   □

Long time ago Bade and Curtis [24] proved that if B is a (complex) commutative Banach algebra whose spectrum $\mathfrak{M}$ is a totally disconnected F-space, then the Gelfand homomorphism is surjective; hence $B = C(\mathfrak{M})$. In [228, Corollary 2] Seever gave another proof of this fact, and asked if the hypothesis of total disconnectedness can be dropped. Proposition 4.8 shows that this is not the case. Indeed, let A be the disk algebra, whose spectrum is the closed disk $\mathbb{D}$. If $\mathcal{U}$ is a free ultrafilter on the integers, then it is easily seen that the spectrum of $B = A_\mathcal{U}$ equals that of $C(\mathbb{D})_\mathcal{U}$, which is the topological ultracoproduct $\mathbb{D}^\mathcal{U}$. This is an F-space, but the Gelfand map $A_\mathcal{U} \to C(\mathbb{D})_\mathcal{U}$ is not onto.

4.7.3   Digression About Stern's Lemma

The following statement appears, without proof, as Lemma 4.2 (ii) in Stern's paper [237]:

- If $\mathcal{U}$ is a countably incomplete ultrafilter and H is the corresponding ultrapower of c_0, then H contains a complemented subspace isometric to $c_0(H)$.

Here $c_0(H)$ is the space of sequences converging to zero in H, with the sup norm. This statement, however, turns out to be false since $c_0(H)$ contains a complemented subspace isometric to c_0 and we have seen in Proposition 4.7 that H cannot. Unfortunately, Stern's Lemma infected the proofs of a number of results in the nonstandard theory and ultraproduct theory of Banach spaces. We can mention:

a) If E is isomorphic to a complemented subspace of a C-space, then E has an ultrapower isomorphic to a C-space (Stern [237, Theorem 4.5(ii)] and also Henson-Moore [134, Theorem 6.6 (c)]).
b) If E is isomorphic to a complemented subspace of an M-space, then E has an ultrapower isomorphic to a C-space (Heinrich-Henson [128, Theorem 12(c)]).
c) If E is an M-space then E has an ultrapower isomorphic to an ultrapower of ℓ_∞ (Henson-Moore [134, Theorem 6.7]).
d) Ultrapowers of the Gurariy space with respect to countably incomplete ultrafilters are not complemented in any C-space (Henson-Moore [134, Theorem 6.8]).

Regarding these statements we have proved (c) and (d) in Proposition 4.26 and Proposition 4.25 respectively, while we have rescued (b) [hence (a)] in Proposition 4.27 under the additional hypothesis that E is isomorphic to its square.

The following problem was considered by Henson and Moore in [134, Problem 21]: Does every (infinite-dimensional, separable) Banach space X have an ultrapower isomorphic to its square? An affirmative answer would imply that the hypothesis of being isomorphic to its square is superfluous in Proposition 4.27.

And what if X is an $\mathscr{L}_\infty$-space? It is perhaps worth noticing that Semadeni proved in [229] that the space of continuous functions on the first uncountable ordinal is not isomorphic to its square. Under CH, the same happens to $C(\mathscr{K})$, the space of continuous functions on Kunen's compact. According to Proposition 4.26 those spaces have ultrapowers which are isomorphic to their own squares. Shelah's space $\mathscr{S}_\diamond$ quoted in Sect. 3.4.1 is not isomorphic to its square either. However it has an ultrapower isomorphic to its own square, too. This follows from Proposition 4.29, taking into account that $\mathscr{G} \sim \mathscr{G} \oplus \mathscr{G}$ as we show now: To see that Gurariy space is isomorphic to its square we apply Proposition 3.30 to the Lindenstrauss space $c_0(\mathscr{G})$ to get that it is complemented in $\mathscr{G}$. If H is a complement of $c_0(\mathscr{G})$ in $\mathscr{G}$ we have

$$\mathscr{G} \sim c_0(\mathscr{G}) \oplus H = \mathscr{G} \oplus c_0(\mathscr{G}) \oplus H \sim \mathscr{G} \oplus \mathscr{G}.$$

4.7.4 Ultra-Roots

Heinrich undertook in [127] the classification of Lindenstrauss spaces up to ultra-isometry. Amongst the many interesting results he proved one finds that the class of C-spaces is closed under "isometric ultra-roots": this just means that if a Banach space X has an ultrapower isometric to a C-space then X is itself isometric to a C-space. A similar result holds for G-spaces; see [127, Theorems 2.7 and 2.10] and spaces of almost universal disposition for finite dimensional spaces; see Lemma 4.28.

At the end of [127] Heinrich asked whether the classes of C_0-spaces and M-spaces enjoy the same property. In a subsequent paper [134] it is claimed that there is a Banach space that fails to be isometric to a Banach lattice but is ultraisometric to c_0. Since c_0 is both a C_0-space and an M-space this would imply a negative solution for both questions. Unfortunately, a close inspection to the example reveals that it is indeed a C_0-space since it is a subalgebra of ℓ_∞. Indeed, the closed linear span of the characteristic functions of the sets of an almost disjoint family of subsets of $\mathbb{N}$ plus c_0 is always a subalgebra of ℓ_∞. Thus, the following should be considered as an open problem: Are the classes of C_0-spaces and M-spaces closed under "isometric ultra-roots"?

The following problem appears both in [128] (see Problem 2 on p. 316) and [134] (see Problems 5 and 7 on pp. 103 and 104): Does Gurariy space have an ultrapower isometric (or isomorphic) to an ultraproduct of finite dimensional spaces? Of course the hypothesized finite dimensional spaces could not be at uniform distance from the corresponding ℓ_∞^n spaces.

4.7.5   Sources

This Chapter is largely based on the works [20] and [21]. The results of Sects. 4.2.1, 4.4 and 4.5 were taken from [20]. Theorem 4.6, Proposition 4.7 and the subsequent discussion in Sect. 4.6 were taken from [21].

In Sect. 4.3.1 we have followed Heinrich [126] to "define" the ultracoproduct and then Bankston [27] to study its main properties. Proposition 4.12 appears in [26, Proposition 2.4. 1]. Proposition 4.8 solves a problem raised by Bankston [26, 2.3.8. Question], who proved a similar result for totally disconnected spaces [26, 2.3.7. Theorem].

Proposition 4.13 appears in [19] with a different proof. The proof given here is more akin to the one given in [58]. Corollary 4.14 (hence Proposition 4.25) is basically due to Henson and Moore who solved in [134, Theorem 6.8] a problem posed by Stern [237, Problem 4.2]. The second part of Proposition 4.16 is an improvement of an observation by Henson [131, Proposition 3.2]. Proposition 4.26 appeared in [134, Theorem 6.7] with a proof infected by Stern's Lemma.

Chapter 5
ℵ-Injectivity

Many of the results presented in this monograph about (universal) separable injectivity can be formulated in terms of the extension of operators with separable range. It is natural to attempt to obtain analogous results under more relaxed conditions in the size of the range of the operators. In this Chapter we consider the notions of (universal) ℵ-injectivity obtained by allowing domain or ranges of operators to have larger density characters. As we shall see, some results easily generalize to the higher cardinal context, some present many difficulties, and some are simply impossible. And, of course, cardinal assumptions are necessary.

Here they are examples of each type: The homological characterizations for separable injectivity (Proposition 2.5) are straightforwardly transplanted to the higher cardinal ground. The ℵ-injective character of large ultrapowers can be obtained but there are considerable technical difficulties. The $\ell_\infty(\aleph)$-upper saturation of universally separably injective spaces is simply impossible, although a mildly satisfactory version (Theorem 5.10) can be obtained. Moreover, other topics, such as the higher cardinal injectivity properties of $C(\mathbb{N}^*)$ are still a mystery: we do not know even if it is $\aleph_2$-injective! And other basic questions such as if for every good ultrafilter $\mathcal{U}$ on a set of size $\aleph$ the ultrapower $X_\mathcal{U}$ of an $\mathscr{L}_\infty$-space X must be ℵ-injective remain open.

We begin this Chapter by describing the main properties of (universally) ℵ-injective Banach spaces and some basic examples. Next we consider specific properties and examples of (universally) $(1, \aleph)$-injective spaces, mainly $C(K)$ spaces and ultraproducts with respect to special ultrafilters. In particular, $(1, \aleph)$-injective C-spaces are characterized as those in which the underlying compact is an $F_\aleph$-space, which is a natural extension of the characterizations of 1-separably injective $C(K)$-spaces as those in which K is an F-space (Sect. 2.2.1) and of injective $C(K)$-spaces as those in which K is extremely disconnected (Proposition 1.19). Extremely disconnected compacta are precisely the projective elements in the category of compacta and continuous maps, a classical result by Gleason (see [108] or [245, Theorem 10.51]), which suggest to study the interplay between projectiveness

© Springer International Publishing Switzerland 2016
A. Avilés et al., *Separably Injective Banach Spaces*, Lecture Notes in Mathematics 2132, DOI 10.1007/978-3-319-14741-3_5

properties of the compact K and the $\aleph$-injective character of the corresponding space $C(K)$.

Definition 5.1

- A Banach space E is said to be $\aleph$-*injective* if for every Banach space X with dens $X < \aleph$ and each subspace $Y \subset X$ every operator $t : Y \to E$ can be extended to an operator $T : X \to E$.
- The space E is said to be *universally $\aleph$-injective* if for every space X and each subspace $Y \subset X$ with dens $Y < \aleph$, every operator $t : Y \to E$ can be extended to an operator $T : X \to E$.

When for every operator t there exists some extension T such that $\|T\| \le \lambda \|t\|$ we say that E is (universally) $(\lambda, \aleph)$-*injective*.

The case $\aleph = \aleph_1$ corresponds to (universal) separable injectivity studied so far. Thus, the resulting name for separable injectivity turns out to be $\aleph_1$-injectivity (not $\aleph_0$-injectivity), which is perhaps surprising. Nevertheless, we have followed the uses of set theory where properties labeled by a cardinal $\aleph$ always indicate that something happens for sets whose cardinality is strictly lesser than $\aleph$.

$\aleph_0$-injectivity is a bit singular: All Banach spaces are universally $\aleph_0$-injective since operators from finite dimensional spaces extend elsewhere. This occurs because $\aleph_0$ has countable cofinality and thus $\aleph_0$-injectivity does not imply the existence of a uniform constant λ bounding the norms of the extension operators (see Lemma 5.2 below). Instead, the spaces which are $(\lambda, \aleph_0)$-injective for some λ are quite clearly the λ-locally injective spaces.

5.1   Main Properties

We begin establishing a few facts in which the theory of $\aleph$-injectivity runs parallel to that of separable injectivity. Our first basic result needs cardinal assumptions.

Lemma 5.2 *If $\aleph$ is a cardinal with uncountable cofinality then every (universally) $\aleph$-injective Banach space is (universally) $(\lambda, \aleph)$-injective for some $\lambda \ge 1$.*

Proof Assume that there exists a Banach space E that is (universally) $\aleph$-injective but not (universally) $(\lambda, \aleph)$-injective for no $\lambda \ge 1$. Then we can pick spaces X_n (respectively Y_n) with density strictly smaller than $\aleph$, isometric embeddings $Y_n \to X_n$, and norm-one operators $t_n : Y_n \to E$ all whose extensions $X_n \to E$ have norm greater than n.

Consider the natural isometric embedding $\ell_1(\mathbb{N}, Y_n) \longrightarrow \ell_1(\mathbb{N}, X_n)$ and the norm-one operator $t : \ell_1(\mathbb{N}, Y_n) \longrightarrow E$ given by $t((x_n)) = \sum t_n x_n$. Since the density of $\ell_1(\mathbb{N}, X_n)$ [respectively $\ell_1(\mathbb{N}, Y_n)$] is still smaller than $\aleph$, and an extension $T : \ell_1(\mathbb{N}, X_n) \to E$ yields extensions for all t_n with norm at most $\|T\|$, such T cannot exist, in contradiction with the (universally) $\aleph$-injective character of E. $\qquad\square$

The proof of Lemma 5.2 does not work for cardinals with countable cofinality, and indeed the result is false as the trivial case of $\aleph_0$-injectivity shows. Proposition 2.5 however admits a straightforward generalization, with identical proof:

Proposition 5.3 *Let E be a Banach space and let $\aleph$ be an uncountable cardinal. The following are equivalent:*

1. *E is $\aleph$-injective.*
2. *For every index set Γ with $|\Gamma| < \aleph$, every operator from a subspace of $\ell_1(\Gamma)$ into E extends to $\ell_1(\Gamma)$.*
3. *For every Banach space X and each subspace Y such that $\mathrm{dens}(X/Y) < \aleph$, every operator $t : Y \to E$ extends to X.*
4. *If X is a Banach space containing E and $\mathrm{dens}(X/E) < \aleph$, then E is complemented in X.*
5. *$\mathrm{Ext}(Z, E) = 0$ for every Banach space Z with density character lesser than $\aleph$.*

Analogous characterizations can be given for universal separable injectivity as in Proposition 2.6.

Proposition 5.4 *Let E be a Banach space and let $\aleph$ be an uncountable cardinal. The following are equivalent:*

1. *E is universally $\aleph$-injective.*
2. *Every operator $t : S \to E$ from a Banach space S with density character lesser than $\aleph$ can be extended to an operator $T : \ell_\infty(\aleph) \to E$ through any embedding $S \to \ell_\infty(\aleph)$.*
3. *For every Banach space X and each subspace Y, every operator $t : Y \to E$ whose range has density character lesser than $\aleph$ extends to X.*

Proof The equivalence of (1) and (2) is clear: since $\ell_\infty(\aleph)$ is injective, once an operator can be extended from S to $\ell_\infty(\aleph)$ it can be extended anywhere. That (1) implies (3) only requires to draw a push-out diagram:

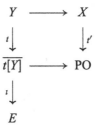

where $\imath$ denotes the canonical inclusion. Since the inclusion $\imath$ can be extended to an operator $I : \mathrm{PO} \to E$, the composition It' yields an extension of t. $\quad\square$

Regarding homological characterizations, we cannot prove that universal $\aleph$-injectivity is equivalent to $\mathrm{Ext}(\ell_\infty(\aleph)/K, E) = 0$ for all K with dens $K < \aleph$. Of course, since every space K with dens $K < \aleph$ is a subspace of $\ell_\infty(\aleph)$, given an operator $t : K \to E$ we can form the push-out diagram

$$
\begin{array}{ccccccccc}
0 & \longrightarrow & K & \longrightarrow & \ell_\infty(\aleph) & \longrightarrow & \ell_\infty(\aleph)/K & \longrightarrow & 0 \\
 & & \downarrow & & \downarrow & & \| & & \\
0 & \longrightarrow & E & \longrightarrow & \mathrm{PO} & \longrightarrow & \ell_\infty(\aleph)/K & \longrightarrow & 0
\end{array}
$$

to obtain that if E is complemented in every superspace X so that $X/E \sim \ell_\infty(\aleph)/K$ for some K with dens $K < \aleph$ then E is universally $\aleph$-injective. A homological characterization of $(2^\aleph)^+$-injectivity is however possible:

Proposition 5.5 *A Banach space E is $(2^\aleph)^+$-injective if and only if for every subspace K of $\ell_\infty(\aleph)$ one has* $\mathrm{Ext}(\ell_\infty(\aleph)/K, E) = 0$

Proof Every quotient of $\ell_\infty(\aleph)$ has density character at most $2^\aleph$; so the necessity is clear by Proposition 5.3 (5). To prove the sufficiency, let Z be a Banach space with density character at most $2^\aleph$ and let

$$
0 \longrightarrow E \longrightarrow X \longrightarrow Z \longrightarrow 0 \qquad (5.1)
$$

be an exact sequence. Let $\varpi : \ell_1(2^\aleph) \longrightarrow Z$ be a quotient map whose kernel is denoted by N. By the lifting property of $\ell_1(2^\aleph)$, one has a commutative diagram

$$
\begin{array}{ccccccccc}
0 & \longrightarrow & N & \longrightarrow & \ell_1(2^\aleph) & \overset{\varpi}{\longrightarrow} & Z & \longrightarrow & 0 \\
 & & \ell\downarrow & & L\downarrow & & \| & & \\
0 & \longrightarrow & E & \longrightarrow & X & \longrightarrow & Z & \longrightarrow & 0
\end{array} \qquad (5.2)
$$

where L is a lifting of ϖ and $\ell = L|_N$, which is necessarily a push out diagram; see Appendix A.4.5. Now, one has:

CLAIM 3 $\ell_1(2^\aleph)$ is a subspace of $\ell_\infty(\aleph)$.

Proof of the Claim Let B be the (closed) unit ball of $\ell_\infty(2^\aleph)$, equipped with the weak* topology induced by $\ell_1(2^\aleph)$. Clearly, B is homeomorphic to the product $[-1, 1]^{2^\aleph}$, which is a continuous image of $\{0, 1\}^{2^\aleph}$. This space has density $\aleph$ since the clopen sets of $2^\aleph$ form a dense set of $\{0, 1\}^{2^\aleph}$: Indeed, take a basic open set U; i.e., take points $p_1, \dots, p_n$ and $q_1, \dots, q_m$ from $2^\aleph$ and form the basic open set

$$
U = \{x \in \{0, 1\}^{2^\aleph} : \; x_{p_i} = 1 \quad \text{and} \quad x_{q_i} = 0\}.
$$

Find a clopen C of $2^\aleph$ such that $p_1, \dots, p_n$ are in C, but $q_1, \dots, q_n$ do not belong to C. Then the characteristic function of this set C belongs to U. Since $2^\aleph$ has $\aleph$ many clopen sets, we conclude that (B, w^*) has density $\aleph$. It follows that $\ell_1(2^\aleph)$ has density character $\aleph$ in the weak* topology induced by $c_0(2^\aleph)$; and therefore it can be embedded into $\ell_\infty(\aleph)$. END OF THE PROOF OF THE CLAIM.

Thus, there is some embedding $\kappa : \ell_1(2^{\aleph}) \to \ell_\infty(\aleph)$ which provides the pullback diagram

$$
\begin{array}{ccccccccc}
0 & \longrightarrow & N & \xrightarrow{\kappa|_N} & \ell_\infty(\aleph) & \longrightarrow & \ell_\infty(\aleph)/\kappa[N] & \longrightarrow & 0 \\
& & \Big\| & & \kappa\Big\uparrow & & \overline{\kappa}\Big\uparrow & & \\
0 & \longrightarrow & N & \longrightarrow & \ell_1(2^{\aleph}) & \xrightarrow{\varpi} & Z & \longrightarrow & 0
\end{array}
\qquad (5.3)
$$

Assembling Diagrams (5.2) and (5.3) we obtain the following commutative pullback/push-out diagram

$$
\begin{array}{ccccccccc}
0 & \longrightarrow & N & \xrightarrow{\kappa|_N} & \ell_\infty(\aleph) & \longrightarrow & \ell_\infty(\aleph)/\kappa[N] & \longrightarrow & 0 \\
& & \Big\| & & \kappa\Big\uparrow & & \overline{\kappa}\Big\uparrow & & \\
0 & \longrightarrow & N & \longrightarrow & \ell_1(2^{\aleph}) & \xrightarrow{\varpi} & Z & \longrightarrow & 0 \\
& & \ell\Big\downarrow & & L\Big\downarrow & & \Big\| & & \\
0 & \longrightarrow & E & \longrightarrow & X & \longrightarrow & Z & \longrightarrow & 0
\end{array}
$$

But making pull-back and making push-out are operations that commute (see Appendix A.4.6), so the lower sequence in the commutative push-out/pull-back diagram

$$
\begin{array}{ccccccccc}
0 & \longrightarrow & N & \xrightarrow{\kappa|_N} & \ell_\infty(\aleph) & \longrightarrow & \ell_\infty(\aleph)/\kappa[N] & \longrightarrow & 0 \\
& & \ell\Big\downarrow & & \Big\downarrow & & \Big\| & & \\
0 & \longrightarrow & E & \longrightarrow & \mathrm{PO}(\ell, \kappa|_N) & \longrightarrow & \ell_\infty(\aleph)/\kappa[N] & \longrightarrow & 0 \\
& & \Big\| & & \Big\uparrow & & \overline{\kappa}\Big\uparrow & & \\
0 & \longrightarrow & E & \longrightarrow & \mathrm{PB} & \longrightarrow & Z & \longrightarrow & 0
\end{array}
$$

is equivalent to (5.1).

The middle sequence splits since $\mathrm{Ext}(\ell_\infty(\aleph)/\kappa[N], E) = 0$, and then so does the lower sequence, hence (5.1). Thus, E is $(2^{\aleph})^+$-injective by Proposition 5.3 (5). □

This immediately yields:

Corollary 5.6

- [CH] *A Banach space E is $\mathfrak{c}^+$-injective if and only if it satisfies $\mathrm{Ext}(\ell_\infty/X, E) = 0$ for every subspace X of ℓ_∞.*
- [GCH] *A Banach space E is $\aleph^{++}$-injective if and only if for every subspace X of $\ell_\infty(\aleph)$ one has $\mathrm{Ext}(\ell_\infty(\aleph)/X, E) = 0$.*

Some stability properties of (universal) separable injectivity have identical statements and proofs (compare to Proposition 2.11):

Proposition 5.7 *Let* ℵ *be an uncountable cardinal.*

1. *The class of* ℵ-*injective spaces has the 3-space property. In particular, products of two* ℵ-*injective spaces are* ℵ-*injective.*
2. *The quotient of an* ℵ-*injective space by an* ℵ-*injective space subspace is again* ℵ-*injective.*
3. *The product of two universally* ℵ-*injective spaces is universally* ℵ-*injective.*
4. *The quotient of a universally* ℵ-*injective space by an* ℵ-*injective subspace is universally* ℵ-*injective.*
5. *Complemented subspaces of (universally)* ℵ-*injective spaces are (universally)* ℵ-*injective. In particular, 1-complemented subspaces of (universally)* $(\lambda, \aleph)$-*injective spaces are (universally)* $(\lambda, \aleph)$-*injective.*

The translation of other properties, however, presents serious difficulties. For example, Theorem 2.26 has no analogue for universal ℵ-injectivity. Indeed, the obvious extension of this result fails because there exist injective Banach spaces with arbitrarily large density character, like the spaces $L_\infty(\mu)$ with μ the product measure on $[0, 1]^\aleph$, that do not contain subspaces isomorphic to $\ell_\infty(\aleph_1)$. This is so since a family of mutually disjoint sets of positive measure on a finite measure space must be countable. A partial extension can be obtained by introducing the following concept.

Definition 5.8 Let ℵ be an infinite cardinal. We say that a subspace Y of a Banach space X is $c_0(\aleph)$-*supplemented* if there exists another subspace Z of X isomorphic to $c_0(\aleph)$ such that $Y \cap Z = 0$ and the sum $Y + Z$ is closed. In this case we also say that Z is a $c_0(\aleph)$-supplement of Y.

Lemma 5.9 *Every subspace of* $\ell_\infty(\aleph)$ *with density* ℵ *or less is* $c_0(\aleph)$-*supplemented.*

Proof Let I have cardinality ℵ and let $\{I_j : j \in J\}$ be a family of disjoint subsets of I with $|I_j| = \aleph$ for every j and $|J| = \aleph$. Let Y be a subspace of $\ell_\infty(I)$ having density ℵ (or less). Since dens$\left(\ell_\infty(I_j)\right) > \aleph$, for each $j \in J$ we can find $x_j \in \ell_\infty(I_j)$ with $\|x_j\| = 1$ and dist$(x_j, Y) > 1/2$. In this way we obtain a family $\{x_j : j \in J\}$ in $\ell_\infty(I)$ isometrically equivalent to the basis of $c_0(I)$.

Let $\pi : \ell_\infty(I) \to \ell_\infty(I)/Y$ denote the quotient map. Since

$$\inf\{\|\pi(x_j)\| : j \in J\} \geq 1/2 > 0,$$

by Rosenthal's [223, Theorem 3.4] there exists $J_1 \subset J$ with $|J_1| = |J|$ such that the restriction of π to the closed subspace generated by $\{x_j : j \in J_1\}$ is an isomorphism. That subspace is a $c_0(\aleph)$-supplement of Y.   □

The partial extension of Theorem 2.26 we were looking for is:

Theorem 5.10 *Let X be a universally $\aleph^+$-injective Banach space and let Y be a $c_0(\aleph)$-supplemented subspace of X with $\mathrm{dens}(Y) \leq \aleph$. Then Y is contained in a subspace of X isomorphic to $\ell_\infty(\aleph)$.*

Proof Let Y_0 be a subspace of $\ell_\infty(\aleph)$ isomorphic to Y and let $t : Y_0 \to Y$ be an isomorphism with $\|t^{-1}\| = 1$. By Lemma 5.9, the subspace Y_0 is $c_0(\aleph)$-supplemented in $\ell_\infty(\aleph)$.

We can find projections P on X and Q on $\ell_\infty(\aleph)$ such that $Y \subset \ker P$, $Y_0 \subset \ker Q$, and both ranges $\mathrm{ran}\, P$ and $\mathrm{ran}\, Q$ are isomorphic to $\ell_\infty(\aleph)$. Indeed, let $\pi :$ $X \to X/Y$ be the quotient map. Using the universal $\aleph^+$-injectivity of X and the $c_0(\aleph)$-supplements we may find an operator $I : \ell_\infty(\aleph) \to X$ such that πI is an isomorphism on a copy of $c_0(\aleph)$. By Theorem 1.15 πI is an isomorphism on a copy of $\ell_\infty(\aleph)$, too. Therefore, there exists a subspace M of X isomorphic to $\ell_\infty(\aleph)$ where the restriction of π is an isomorphism and we have $X/Y = \pi[M] \oplus N$, where N is a complement of $\pi[M]$ in X/Y. Hence $X = M \oplus \pi^{-1}[N]$, and it is enough to take as P the projection with range M and kernel $\pi^{-1}[N]$. Since $\ker P$ and $\ker Q$ are universally $\aleph$-injective spaces, we can take operators $U : X \to \ker Q$ and $V : \ell_\infty(\aleph) \to \ker P$ such that $V|_{Y_0} = t$ and $U|_Y = t^{-1}$. Note that $\|U\| \geq 1$. From here the proof is entirely similar to that of Theorem 2.26: just replace ℓ_∞ there by $\ell_\infty(\aleph)$ here. $\square$

Regarding infinite products, it is obvious that if $(E_i)_{i \in I}$ is a family of (universally) $(\lambda, \aleph)$-injective Banach spaces, then $\ell_\infty(I, E_i)$ is (universally) $(\lambda, \aleph)$-injective. And, of course, $c_0(I, E_i)$ is not: If I is infinite then $c_0(I)$ is not $\aleph_2$-injective just because its complemented subspace c_0 is not, since there exist nontrivial sequences

$$0 \longrightarrow c_0 \longrightarrow X \longrightarrow c_0(\aleph_1) \longrightarrow 0;$$

see Sect. 2.2.4.

The simplest examples of (universally) $\aleph$-injective spaces arise restricting the size of the support of bounded functions. Indeed, let us consider the spaces

$$\ell_\infty^{\aleph}(I) = \{x \in \ell_\infty(I) : |\{i : x(i) \neq 0\}| < \aleph\}.$$

When $\aleph$ is a cardinal with uncountable cofinality, they coincide with the spaces

$$\ell_\infty^{<\aleph}(I) = \bigcap_{\varepsilon>0} \{x \in \ell_\infty(I) : |\{i : |x(i)| > \varepsilon\}| < \aleph\}$$

introduced by Pełczyński and Sudakov [214] and studied in Chap. 1, where it is proven that $\ell_\infty^{<\aleph}(I)$ is not injective as soon as $\aleph_0 < \aleph \leq |I|$; see Corollary 1.27. When $\aleph$ is an uncountable *regular* cardinal, which means that the union of less than $\aleph$ sets of less than $\aleph$ elements each has less than $\aleph$ elements, the space $\ell_\infty^{\aleph}(I)$ is $(1, \aleph)$-injective. If, moreover, $|I| = \aleph$ then it cannot be $(2^{\aleph})^+$-injective by the Pełczyński-Sudakov result since $\ell_\infty^{\aleph}(\aleph) \subset \ell_\infty(\aleph)$, which has density character $2^{\aleph}$.

More interesting examples of ℵ-injective spaces will appear in the next Section: suitable ultraproducts of $\mathscr{L}_\infty$-spaces and $C(K)$-spaces for K a $F_ℵ$-space.

5.2   (1, ℵ)-Injective Spaces

Let us begin with the following observation of Neville.

Proposition 5.11 *A Banach space which is* $(1, ℵ^+)$-*injective and has density character* ℵ *is necessarily 1-injective.*

Proof By a typical application of Zorn lemma, it suffices to see that if Y is a one-codimensional subspace of X every norm-one operator $t : Y \to E$ extends to X without increasing the norm. Once again look at the diagram

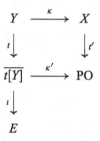

where ι and κ are the corresponding inclusions maps. Since $\mathrm{PO}\,/\overline{t[Y]}$ is isomorphic to X/Y we have $\mathrm{dens}(\mathrm{PO}) = \mathrm{dens}\,\overline{t[Y]} \leq \mathrm{dens}(E)$ and there is $I : \mathrm{PO} \to E$ such that $\iota = I \circ \kappa'$, with $\|I\| = 1$. Letting $T = I \circ t'$ one obtains the required extension. □

Now we give a characterization of $(1, ℵ)$-injectivity by intersection properties of balls, which is the promised generalization of Proposition 2.30.

Proposition 5.12 *A Banach space E is* $(1, ℵ)$-*injective if and only if every family of less than* ℵ *mutually intersecting balls of E has nonempty intersection.*

Proof

SUFFICIENCY.   Take an operator $t : Y \to E$, where Y is a closed subspace of a Banach space X with $\mathrm{dens}\,X < ℵ$. We may and do assume $\|t\| = 1$. Let $z \in X\backslash Y$ and let Y_0 be a dense subset of Y forming a linear space over the rational numbers with $|Y_0| < ℵ$ and, for each $y \in Y_0$, consider the ball $B(ty, \|y - z\|)$ in E. Any two of these balls intersect, since for $y_1, y_2 \in Y_0$ we have

$$\|ty_2 - ty_1\| \leq \|t\|\|y_2 - y_1\| \leq \|y_2 - z\| + \|y_1 - z\|.$$

Thus the hypothesis implies that there exists

$$f \in \bigcap_{y \in Y_0} B(ty, \|y - z\|) = \bigcap_{y \in Y} B(ty, \|y - z\|).$$

It is clear that the map $T : Y + \langle z \rangle \to E$ defined by $T(y + cz) = ty + cf$ is an extension of t with $\|T\| = 1$. The rest is clear using Zorn's lemma.

NECESSITY. Let E be $(1, ℵ)$-injective and suppose $B(e_i, r_i)$ is a family of less than ℵ mutually intersecting balls in E. Let Y be the closed subspace of E spanned by the centers, so that dens $Y < ℵ$, and let $j : Y \to \ell_\infty(\Gamma)$ be any isometric embedding. Notice that even if the family of balls $B_Y(e_i, r_i) = B(e_i, r_i) \cap Y$ are not mutually intersecting in Y, any two balls of the family $B(j(e_i), r_i)$ meet in $\ell_\infty(\Gamma)$ because the distance between the centers does not exceed the sum of the radii. Therefore the intersection

$$\bigcap_i B(j(e_i), r_i)$$

contains some point, say $x \in \ell_\infty(\Gamma)$. Let X be the subspace spanned by x and $j(Y)$ in $\ell_\infty(\Gamma)$, so that $\dim X/Y \leq 1$. The hypothesis on the space E allows one to extend the inclusion of Y into E to X through the embedding $j : Y \to X$ without increasing the norms. It is pretty clear that the image of x in E under any such extension lives in every $B(e_i, r_i)$. □

Proposition 2.29 (namely that 1-separable injectivity and universal 1-separable injectivity are equivalent properties under CH) admits a higher cardinal counterpart. Let ℵ be a cardinal. Since a set of cardinal $ℵ^+$ can be written as the union of an increasing chain of sets of cardinal ℵ, the same method of Lemma 2.28 applies to show that when E is $(1, ℵ)$-injective and Y a subspace of X, with dens$(X) \leq ℵ^+$ then every operator $t : Y \to E$ can be extended to an operator $T : X \to E$ with the same norm. Next recall that a Banach space with density character κ embeds in $\ell_\infty(\kappa)$ which, under GCH, has density character κ^+. Therefore, if $\kappa < ℵ$ also $\kappa^+ \leq ℵ$ and one gets:

Proposition 5.13 (GCH) *Every $(1, ℵ)$-injective Banach space is $(1, ℵ)$-universally injective.*

We now pass to obtain the new announced examples of ℵ-injective spaces. The crucial notion here will be that of *good* ultrafilter, an invention by Keisler [27, 75]. Recall that fin(S) denotes the set of finite subsets of a given set S. Given an ultrafilter $\mathcal{U}$, we say that a map $f : \text{fin}(S) \to \mathcal{U}$ is *monotone* if $f(A) \supset f(B)$ whenever $A \subset B$; moreover, we say that the map f is *multiplicative* if $f(A \cup B) = f(A) \cap f(B)$.

Definition 5.14 An ultrafilter $\mathcal{U}$ on I is said to be ℵ-*good* if, for every set S with $|S| < ℵ$, and every monotone map $f : \text{fin}(S) \to \mathcal{U}$, there is a multiplicative map $g : \text{fin}(S) \to \mathcal{U}$ such that $g(A) \subset f(A)$ for all A.

Every set of cardinality $\aleph$ supports countably incomplete, $\aleph^+$-good ultrafilters; see [75, Theorem 6.1.4] or [77, Theorem 10.4]. No more can be expected, since an $\aleph^{++}$-good ultrafilter on a set of cardinality $\aleph$ has to be fixed (by "saturation" and [75, Proposition 4.2.2]) and so $\aleph^+$-good ultrafilters based on sets of cardinality $\aleph$ will be simply called "good ultrafilters". Every countably incomplete ultrafilter is $\aleph_1$-good (cf. [75, Exercise 6.1.2])

Theorem 5.15 *Let $\mathcal{U}$ be a countably incomplete, $\aleph$-good ultrafilter on I and let $(X_i)_{i \in I}$ be a family of Banach spaces. If the ultraproduct $[X_i]_{\mathcal{U}}$ is a Lindenstrauss space, then it is $(1, \aleph)$-injective.*

Proof Here we need the saturation property of the set-theoretic ultraproducts via good ultrafilters:

Let $(S_i)_{i \in I}$ be a family of sets and let $\mathcal{U}$ be an ultrafilter on I. A subset A of $\langle S_i \rangle_{\mathcal{U}}$ is called *internal* if there are sets $A_i \subset S_i$ such that $A = \langle A_i \rangle_{\mathcal{U}}$. It can be proved that if $\mathcal{U}$ is countably incomplete and $\aleph$-good, then every family of less than $\aleph$ internal subsets of $\langle S_i \rangle_{\mathcal{U}}$ having the finite intersection property has nonempty intersection; see [75, Theorem 4.2.5] or [77, 13.9].

Let $(B^\alpha)_{\alpha \in \Gamma}$ be a family of mutually intersecting balls in $[X_i]_{\mathcal{U}}$, with $|\Gamma| < \aleph$. Let us write $B^\alpha = B(x^\alpha, r_\alpha)$ and let $x^\alpha = (x_i^\alpha)$ be. Clearly, $\langle B(x_i^\alpha, r_\alpha + 1/m) \rangle_{\mathcal{U}}$ is a lifting of $B(x^\alpha, r_\alpha + 1/m)$ in the set-theoretic ultraproduct $\langle X_i \rangle_{\mathcal{U}}$. As $[X_i]_{\mathcal{U}}$ is a Lindenstrauss space, the original family (B^α) has the finite intersection property. This implies the same for the family of internal sets

$$(\langle B(x_i^\alpha, r_\alpha + 1/m) \rangle_{\mathcal{U}})_{(\alpha, m) \in \Gamma \times \mathbb{N}}.$$

Indeed, if F is a finite subset of $\Gamma \times \mathbb{N}$, we may assume it is of the form $E \times \{1, \dots, k\}$ for some finite $E \subset \Gamma$. Then there exists $z \in \bigcap_{\alpha \in E} B^\alpha$. Thus, if (z_i) is a representative of z, the sets $\{i \in I : \|x_i^\alpha - z_i\| \leq 1/k\}$ belong to $\mathcal{U}$ for every $\alpha \in E$ and $\langle (z_i) \rangle_{\mathcal{U}} \in \bigcap_{(\alpha, m) \in F} \langle B(x_i^\alpha, r_\alpha + 1/m) \rangle_{\mathcal{U}}$.

Since $|\Gamma \times \mathbb{N}| < \aleph$ and $\mathcal{U}$ is $\aleph$-good, there is $x \in \langle X_i \rangle_{\mathcal{U}}$ in the nonempty intersection

$$\bigcap_{(\alpha, m) \in \Gamma \times \mathbb{N}} \langle B(x_i^\alpha, r_\alpha + 1/m) \rangle_{\mathcal{U}}.$$

It is clear that if (x_i) is any representation of x, then

$$[(x_i)] \in \bigcap_{\alpha, m} B(x^\alpha, r_\alpha + 1/m) = \bigcap_{\alpha \in \Gamma} B^\alpha,$$

which completes the proof.   □

We turn now our attention to spaces of continuous functions. Theorem 2.14 provided several characterizations of 1-separably injective C-spaces. The following more general result summarizes all we know about the interplay between the topological properties of K, $(1, \aleph)$-injectivity of $C(K)$, and its lattice structure.

Recall a cozero set of K is one of the form $\{t \in K : f(t) \neq 0\}$ for some $f \in C(K)$.

Theorem 5.16 *For a compact space K and a cardinal number $\aleph$, the following statements are equivalent:*

1. *$C(K)$ is $(1, \aleph)$-injective.*
2. *Given subsets L and U of $C(K)$ with $|L|, |U| < \aleph$ such that $f \leq g$ for every $f \in L$ and $g \in U$, there exists $h \in C(K)$ separating them, that is, such that $f \leq h \leq g$ for all $f \in L$ and $g \in U$.*
3. *Every family of mutually intersecting balls in $C(K)$ of cardinal lesser than $\aleph$ has nonempty intersection.*
4. *Every couple of disjoint open sets G and H of K which are the union of less than $\aleph$ many closed sets have disjoint closures.*
5. *Every couple of disjoint open sets G and H of K which are the union of less than $\aleph$ many cozero sets have disjoint closures.*

Proof We first prove the implications $(1) \Rightarrow (2) \Rightarrow (3) \Rightarrow (1)$, in that order.

Let L and U be as in (2). We consider $C(K)$ as a subalgebra of $\ell_\infty(K)$. Let $\eta \in \ell_\infty(K)$ such that $f \leq \eta \leq g$ for all $f \in L$ and $g \in U$. Let A be the least unital closed subalgebra of $\ell_\infty(K)$ containing L, U and η, and let $B = A \cap C(K)$. Clearly, $\mathrm{dens}\, A < \aleph$. By (1), the inclusion of B into $C(K)$ extends to a norm-one operator $I : A \to C(K)$. Let M be the maximal ideal space of A and N that of B. By general representation theorems we have $A = C(N)$ and $B = C(M)$, and we get a commutative diagram

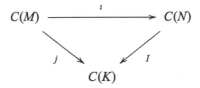

By Lemma 2.38, the operator I is positive, hence $I(\eta)$ separates L from U.

We check now $(2) \Rightarrow (3)$. Let $(B_i)_{i \in I}$ be a family of mutually intersecting balls, where $|I| < \aleph$. Writing $B_i = B(f_i, r_i)$, we have $\|f_i - f_j\| \leq r_i + r_j$ for all $i, j \in I$, that is,

$$f_i - r_i \leq f_j + r_j \qquad (i, j \in I).$$

By (2) there is $h \in C(K)$ such that

$$f_i - r_i \leq h \leq f_j + r_j \qquad (i, j \in I).$$

In particular $f_i - r_i \leq h \leq f_i + r_i$, that is, $h \in \bigcap_i B_i$.

The implication (3) $\Rightarrow$ (1) is contained in Proposition 5.12.

We pass to prove the string (2) $\Rightarrow$ (4) $\Rightarrow$ (5) $\Rightarrow$ (2).

Assume that (2) holds and let G and H be open sets as in (4), so that $G = \bigcup_{\alpha \in I} C_\alpha$ and $H = \bigcup_{\alpha \in I} D_\alpha$, where C_α and D_α are closed subsets of K and $|I| < \aleph$. For every $\alpha < \aleph$, let $f_\alpha \in C(K)$, $0 \le f_\alpha \le 1$, such that $f_\alpha|_{K \setminus G} = 0$ and $f_\alpha|_{C_\alpha} = 1$, and let $g_\alpha \in C(K)$, $0 \le g_\alpha \le 1$, such that $g_\alpha|_{K \setminus H} = 1$ and $g_\alpha|_{D_\alpha} = 0$. The sets $L = \{f_\alpha : \alpha \in I\}$ and $U = \{g_\alpha : \alpha \in I\}$ satisfy the assumptions of condition (2). The function $h \in C(K)$ that separates L and U has the property that $h|_G = 1$ and $h|_H = 0$, hence $\overline{G} \cap \overline{H} = \varnothing$. That (4) implies (5) is a consequence of the fact that each cozero set is the union of countably many closed sets, namely for $f \in C(K)$,

$$\{x \in K : f(x) \ne 0\} = \bigcup_{n \in \mathbb{N}} \{x \in K : |f(x)| \ge 1/n\}.$$

Assume now that (5) holds. As a first step towards (2), we prove it modulo a given positive ε.

CLAIM Given U and L like in (2) and given $\varepsilon > 0$, there exists $h \in C(K)$ such that $f - \varepsilon \le h \le g + \varepsilon$ for every $f \in L$ and every $g \in U$.

Proof of the claim By homogeneity, it is enough to consider the case $\varepsilon = 1$.

Let $N \in \mathbb{N}$ be such that $-N < f_0 \le g_0 < N$ for some $f_0 \in L$ and $g_0 \in U$. Let $I = \{n \in \mathbb{N} : -N < n < N\}$. For every $n \in I$, let

$$G_n = \{x \in K : f(x) > n \text{ for some } f \in L\} = \bigcup_{f \in L} f^{-1}(n, +\infty),$$

$$H_n = \{x \in K : g(x) < n \text{ for some } g \in U\} = \bigcup_{g \in U} g^{-1}(-\infty, n).$$

For each n, G_n and H_n are disjoint open sets which are the union of less than $\aleph$ cozero sets, because $|L|, |U| < \aleph$. Hence $\overline{G_n} \cap \overline{H_n} = \varnothing$, therefore there exists $h_n \in C(K)$, $-1 \le h_n \le 1$ such that $h_n|_{G_n} = 1$ and $h_n|_{H_n} = -1$. We shall check that $h = \frac{1}{2} \sum_{n \in I} h_n \in C(K)$ is the desired function. For $f \in L$ and $x \in K$,

$$h(x) = \frac{1}{2} \sum_{n \in I} h_n = \frac{1}{2} \left(\sum_{n \in I, n < f(x)} (1) + \sum_{n \in I, n \ge f(x)} h_n(x) \right)$$

$$\ge \frac{|\{n \in I, n < f(x)\}| - |\{n \in I, n \ge f(x)\}|}{2}$$

$$\ge f(x) - 1.$$

Similarly, one gets that $h(x) \le g(x) + 1$ for all $g \in U$ and $x \in K$. END OF THE PROOF OF THE CLAIM.

Now, if U and L are sets like in (2) we construct inductively a sequence of new sets $U_n, L_n \subset C(K)$ and functions $h_n \in C(K)$ as follows:

- $L_0 = L, U_0 = U$.
- For every $f \in L_n$ and every $g \in U_n$ one has $f - 2^{-n} \le h_n \le g + 2^{-n}$.
- $L_{n+1} = L_n \cup \{h_n - 2^{-n}\}$ and $U_{n+1} = U_n \cup \{h_n + 2^{-n}\}$.

This can be performed because of the preceding claim. Notice that the sequence $(h_n)_{n \in \mathbb{N}}$ is uniformly convergent because for $m < n$, $h_m - 2^{-m} \in L_n$, $h_m + 2^{-m} \in U_n$, hence $h_m - 2^{-m} - 2^{-n} \le h_n \le h_m + 2^{-m} + 2^{-n}$ and so $\|h_n - h_m\| \le 2^{-m+1}$. Thus, we can consider $h = \lim_n h_n$. Then $h \in C(K)$ and satisfies $f \le h \le g$ for $f \in L$ and $g \in U$. $\qquad\square$

We conclude this section tuning the construction of a 1-universally separably injective space not isomorphic to any $C(K)$ space presented in Chaps. 3 and 4. The presence of an extreme point in Part (2) is reminiscent from the early papers on injectivity.

Proposition 5.17 *For every cardinal ℵ there exists a Banach space of density $2^{\aleph}$ such that*

1. *It is $(1, \aleph^+)$-injective but it is not isomorphic to a complemented subspace of any M-space.*
2. *After suitable renorming, it is still $(1, \aleph^+)$-injective and its unit ball has extreme points.*

Proof

1. If $\mathcal{U}$ is a countably incomplete good ultrafilter on a set of cardinality ℵ, then $\mathscr{G}_{\mathcal{U}}$ is an $(1, \aleph^+)$-injective Banach space of density $2^{\aleph}$ by Theorem 5.15. The assertion (1) follows from Corollary 4.14.
2. The space $\mathscr{G}$ is isomorphic to the space $A(P)$ of continuous affine functions on the Poulsen simplex as proved by Lusky [187]. See also [97, 186]. Hence $\mathscr{G}_{\mathcal{U}}$ is isomorphic to $A(P)_{\mathcal{U}}$, in turn isometric to the space of continuous affine functions on certain simplex S, by [127, Proposition 2.1]. Thus, the unit ball of $A(S) = A(P)_{\mathcal{U}}$ has extreme points: 1_S is one. However, $A(S)$, being isomorphic to $\mathscr{G}_{\mathcal{U}}$ cannot be complemented in an M-space. As before, the density character of $A(S)$ equals $2^{\aleph}$ and $A(S)$ is $(1, \aleph^+)$-injective. $\qquad\square$

The preceding examples are as bad as the generalized continuum hypothesis allows. Indeed, if a Banach space is $(1, \aleph^+)$-injective and has density character ℵ, then it is 1-injective and therefore isometric to a $C(K)$-space, by Proposition 5.11.

5.3 $\aleph$-Injectivity Properties of $C(\mathbb{N}^*)$

We have already shown in Sect. 2.5 that $C(\mathbb{N}^*)$ is $(1, \aleph_1)$-injective. Maintaining the parameter 1, no cardinal improvement is possible:

Proposition 5.18 $C(\mathbb{N}^*)$ *is not* $(1, \aleph_2)$*-injective.*

Proof A classical construction in set theory known as the Hausdorff gap [123] asserts the existence of two ω_1-sequences of clopen sets in $\mathbb{N}^*$, say (a_i) and (b_i) where $i \in \omega_1$, such that (a_i) is increasing, (b_i) is decreasing, $a_i \subset b_j$ for all i, j, and with the additional property that there exists no clopen set c such that $a_i \subset c \subset b_j$ for all $i, j \in \omega_1$. Considering the characteristic functions of those clopen sets, condition (2) of Theorem 5.16 is violated for $\aleph = \aleph_2$. Keep in mind that zero-dimensional compacta are in fact strongly zero-dimensional, that is, disjoint zero sets can be put into disjoint clopen sets. $\qquad\square$

It follows from Proposition 2.43 that $C(\mathbb{N}^*)$ is not $\mathfrak{c}^+$-injective. So there is some room for improvement in the isomorphic case. We refer to Chap. 6 for details.

5.4 Projectiveness Properties of Compact Spaces

The compact spaces arising in Theorem 5.16 constitute a well known class (see [26, 27, 238]) that we consider now.

Definition 5.19 A compact space K is said to be an $F_\aleph$*-space* if every couple of disjoint open subsets of K which are the union of less than $\aleph$ many closed sets have disjoint closures.

Thus, the $F_{\aleph_1}$-spaces are simply the F-spaces.

Corollary 5.20 *Every topological ultracoproduct of compact spaces via an $\aleph$-good ultrafilter is an $F_\aleph$-space.*

As we mentioned in Chap. 1, a Banach space is 1-injective if and only if it is isometrically isomorphic to $C(K)$ for some extremely disconnected compact space K. We observe that such compacta are precisely the projective elements in the category of formed by compact spaces and continuous maps (see Gleason [108] or [245, Theorem 10.51]). This means that if $\sigma : L \to M$ is a continuous surjection between compact spaces and K is an extremely disconnected compact space, then any continuous map $\varphi : K \to M$ lifts to L, in the sense that there is a continuous map $\tilde{\varphi} : K \to L$ such that $\varphi = \sigma \circ \tilde{\varphi}$. Of course this can be rephrased by saying that $C(K)$ is injective in the category of commutative C^*-algebras.

One may wonder if some natural relativization of this result holds, meaning whether the fact that the space $C(K)$ is injective with respect to a subcategory of Banach spaces is reflected dually by K being projective with respect to some subcategory of compact spaces.

If $\mathfrak{C}$ is some class of continuous surjections between compact spaces, we say that a compact space K is *projective with respect to* $\mathfrak{C}$ if for every continuous surjection $\pi : L \longrightarrow M$ that belongs to $\mathfrak{C}$ and every continuous map $f : K \longrightarrow M$ there exists a continuous function $g : K \longrightarrow M$ such that $\pi g = f$. The first guess would be that $C(K)$ being injective with respect to Banach spaces of density character lesser than $\aleph$ should be equivalent to the compact space K being projective with respect to compact spaces of weight lesser than $\aleph$. There is however a serious obstruction for this: if π is any surjection from the Cantor set Δ onto the unit interval and K is any connected F-space, then the only liftable maps $f : K \to [0, 1]$ are the constant ones (and there are many connected F-spaces; among them those produced by Corollary 4.31). There are two ways of avoiding this problem. The first one is to assume K to be totally disconnected (Theorem 5.22). The second one is to reduce the subcategory we are dealing with and to consider only compact convex sets and affine maps between them (Theorem 5.29).

Lemma 5.21 *Let $\aleph$ be a cardinal number, and let K be a compact space. The following assertions are equivalent:*

1. *Every open cover of every subspace of K has a subcover of cardinality lesser than $\aleph$.*
2. *Every open subset of K is the union of less than $\aleph$ many closed subsets of K.*

Proof Suppose (1) holds and let U be an open subset of K. Simply consider an open cover of U by open sets V with $\overline{V} \subset U$. Conversely, assume (2) and let $S \subset K$ and $\{U_i : i \in I\}$ a cover of S by open subsets of K. Consider $U = \bigcup_{i \in I} U_i$. By (2), U is the union of less than $\aleph$ many compact sets, so it is enough to take a finite subcover of each. $\square$

We denote by $HL_\aleph$ the class of compact spaces satisfying the conditions of the preceding lemma. Observe that the class $HL_\aleph$ is stable under continuous images, and that it contains all compact spaces of weight lesser than $\aleph$. Moreover a compact space belongs to $HL_{\aleph_1}$ if and only if it is hereditarily Lindelöf, or equivalently, if it is perfectly normal.

An example of a hereditarily Lindelöf space of uncountable weight is the double arrow space: the lexicographical product of ordered sets $[0, 1] \times \{0, 1\}$ endowed with the order topology.

Theorem 5.22 *For a compact space K the following assertions are equivalent:*

1. *K is a zero-dimensional $F_\aleph$-space.*
2. *K is projective with respect to surjections $\pi : L \longrightarrow M$ such that $w(L) < \aleph$, where $w(L)$ denotes the weight of L.*
3. *K is projective with respect to surjections $\pi : L \longrightarrow M$ such that $w(M) < \aleph$ and $w(L) \leq \aleph$.*
4. *K is projective with respect to surjections $\pi : L \longrightarrow M$ with $L \in HL_\aleph$.*

Proof First we prove that (2) implies (1).

In order to show that K is a zero-dimensional $F_\aleph$-space we shall show that for any disjoint open subsets A and B, which are the union of $\kappa < \aleph$ many closed subsets of K there exists disjoint clopen sets C and D such that $A \subset C$ and $B \subset D$. Suppose $A = \bigcup_{\alpha < \kappa} C_\alpha$ and $B = \bigcup_{\alpha < \kappa} D_\alpha$ where each C_α and each D_α are closed sets. For every $\alpha < \kappa$ let $f_\alpha : K \longrightarrow [-1, 1]$ be a continuous function such that

- $f_\alpha|_{C_\alpha} = -1$,
- $f_\alpha|_A \leq 0$,
- $f_\alpha|_{K \setminus (A \cup B)} = 0$,
- $f_\alpha|_B \geq 0$, and
- $f_\alpha|_{D_\alpha} = 1$.

Consider the map $f : K \longrightarrow [-1, 1]^\kappa$ given by $f(x) = (f_\alpha(x))_{\alpha < \kappa}$. Denote $L = [0, 1]^\kappa \times \{-1, 1\}$, let $\pi : L \longrightarrow [-1, 1]^\kappa$ be the map given by $\pi(x, t) = (t \cdot x_\alpha)_{\alpha < \kappa}$, and let $M = \pi(L)$. Notice that the image of f is contained in M, hence we can apply the projectivity property so show that there exists $g : K \longrightarrow L$ with $\pi g = f$. But then $g(A) \subset [0, 1]^\kappa \times \{-1\}$ and $g(B) \subset [0, 1]^\kappa \times \{1\}$, hence there are disjoint clopen sets which separate A and B.

Next we assume that K is a zero-dimensional $F_\aleph$ space and we shall prove (3) and (4). We suppose that we are given an onto map $\pi : L \longrightarrow M$ like either in (3) or (4), and $f : K \longrightarrow M$, and we will find $g : K \longrightarrow L$ with $\pi g = f$.

CASE 1.   We suppose that $M \in HL_\aleph$, $L \subset M \times \{0, 1\}$ and $\pi : L \longrightarrow M$ is the first-coordinate projection.

Consider the sets

$$A = K \setminus f^{-1}[\pi(L \cap M \times \{1\})],$$

$$B = K \setminus f^{-1}[\pi(L \cap M \times \{0\})].$$

These are two disjoint open subsets of K which are moreover the union of less than $\aleph$ many closed sets, because $M \in HL_\aleph$. Therefore, since K is a totally disconnected $F_\aleph$ space, there exists a clopen set $C \subset K$ such that $A \subset C$ and $B \cap C = \emptyset$. The desired function $g : K \longrightarrow L$ can be defined now as $g(x) = (x, 0)$ if $x \in C$ and $g(x) = (x, 1)$ if $x \notin C$.

CASE 2.   We suppose that $L \in HL_\aleph$, $L \subset M \times [0, 1]$ and $\pi : L \longrightarrow M$ is the first-coordinate projection.

Let $q : 2^\omega \longrightarrow [0, 1]$ be a continuous surjection from the Cantor set onto the unit interval. Let $L' = \{(x, t) \in M \times 2^\omega : (x, q(t)) \in L\}$ and $\pi' : L' \longrightarrow M$ the first coordinate projection. We shall find a continuous map $g' : K \longrightarrow L'$ such that $\pi' g' = f$. From g' we easily obtain the desired function g by composing with q in the second coordinate. For every $n < m \leq \omega$ let $p_n^m : M \times 2^m \longrightarrow M \times 2^n$ be the natural projection which forgets about coordinates $i \geq n$ in 2^m. Let $L_n = p_n^\omega(L')$. Each $L_n \subset L \times 2^n$ is a member of $HL_\aleph$. Hence, by repeated application of the Case 1 proved above, we can construct inductively continuous maps $g_n : K \longrightarrow L_n$ such that $g_0 = f$ and $\pi_n^{n+1} g_{n+1} = g_n$. These functions must be of the form

$$g_n(x) = (f(x), \gamma_0(x), \ldots, \gamma_{n-1}(x))$$

for some continuous functions $\gamma_i : K \longrightarrow 2$, $i < \omega$. The function $g' : K \longrightarrow L'$, where $L' \subset M \times 2^\omega$, is defined as $g'(x) = (f(x), \gamma_0(x), \gamma_1(x), \ldots)$.

GENERAL CASE. We view L as a closed subset of a cube $L \subset 2^\Gamma$, where Γ is some cardinal. If we are dealing with condition (3), then $\Gamma = \aleph$.

Let $G = \{(x, \pi(x)) : x \in L\} \subset 2^\Gamma \times M$ be the graph of π, and let $\pi_1 : G \longrightarrow L$ and $\pi_2 : G \longrightarrow M$ be the two coordinate functions. We shall find a continuous function $h : K \longrightarrow G$ such that $\pi_2 h = f$. From this we immediately get the desired lifting as $g = \pi_1 h$.

For every $\alpha < \beta \leq \Gamma$ let $p_\alpha^\beta : 2^\beta \times M \longrightarrow 2^\alpha \times M$ be the natural projection and let $G_\alpha = p_\alpha^\Gamma(G)$. If we assume condition (3) then all spaces G_α have weight lesser than $\aleph$, while if we assume (4), then all these spaces belong to $HL_\aleph$ because G is homeomorphic to L and this class is stable under taking continuous images. Thus we can construct by transfinite induction continuous functions $h_\alpha : K \longrightarrow G_\alpha$ such that $\pi_2 h_\alpha = f$ and such that they are coherent: $p_\alpha^\beta g_\beta = g_\alpha$ for $\alpha < \beta$.

In the successor step of the induction, in order to obtain $g_{\alpha+1}$ from g_α we apply Case 2 above. In the limit step, the function g_β is uniquely determined by the functions g_α with $\alpha < \beta$, similarly as we did in Case 2. $\square$

Definition 5.23 We say that a compact space is *metrically projective* if it is projective with respect to all continuous surjections between metrizable compacta.

Let us see some consequences of Theorem 5.22.

Corollary 5.24

- *The compact space $\mathbb{N}^*$ is metrically projective.*
- *Let $(K_i)_{i \in I}$ be a family of totally disconnected compacta and let $\mathcal{U}$ be a countably incomplete ultrafilter on I. Then $(K_i)^{\mathcal{U}}$ is metrically projective.*

Corollary 5.25 *Totally disconnected F-spaces are projective with respect to hereditarily Lindelöf compact spaces.*

In the following Corollary, we denote by $RO(X)$ the set of all regular open subsets of X, that is, those open sets which coincide with the interior of a closed set.

Corollary 5.26 *Let K be a totally disconnected $F_\aleph$-space. Then K is projective with respect to surjections $\pi : L \longrightarrow M$ in which $w(M) < \aleph$ and $|RO(M)| \leq \aleph$.*

Proof Let $p : G \longrightarrow M$ be the Gleason cover of M. We refer to [245] for an explanation of this concept. We just recall the facts that we need about it: the space G is an extremely disconnected space (that is, projective with respect to the full category of compact spaces), $w(G) = |RO(M)|$, and $p : G \longrightarrow M$ is an onto continuous map. Since $w(G) \leq \aleph$ and $w(M) < \aleph$, by Theorem 5.22 there exists

$h : K \longrightarrow G$ such that $ph = f$. Since G is projective, there exists $u : G \longrightarrow L$ such that $\pi u = p$. Thus we can take $g = uh$.   □

Corollary 5.27 *Suppose that κ is a cardinal for which $\kappa^+ = 2^\kappa$, and let K be a totally disconnected compact F_{κ^+}-space. Then K is projective with respect to all surjections $\pi : L \longrightarrow M$ such that $w(M) \leq \kappa$.*

Proof Apply the preceding Corollary for $\aleph = \kappa^+$, and notice that one always has $|RO(M)| \leq 2^{w(M)}$ because every open set is the union of a family of open sets from a basis. We conclude that K is projective with respect to all surjections $\pi : L \longrightarrow M$ such that $w(M) \leq \kappa$ and $w(L) \leq 2^\kappa$. But this includes the case when $L = \beta M$, the Čech-Stone compactification of M with the discrete topology, and $\pi_M : \beta M \longrightarrow M$ is the canonical surjection. Since βM is a projective compact space (it is $F_\aleph$ for any $\aleph$), it easily follows that K is projective with respect to any surjection $\pi : L \longrightarrow M$ such that $w(M) \leq \kappa$.   □

The assumption $\kappa^+ = 2^\kappa$ is necessary:

Proposition 5.28 *It is consistent that there exists a zero-dimensional compact F-space which is not projective with respect to surjections $\pi : L \longrightarrow M$ with $w(M) = \aleph_0$.*

Proof Under the assumption that $\mathfrak{c} = \aleph_2$ and that $\mathbb{P}(\mathbb{N})/\mathrm{fin}$ contains a chain of order type ω_2, Dow and Hart [87, Theorem 5.10] construct a zero-dimensional compact F-space K which does not map onto $\beta\mathbb{N}$. Let $M = \alpha\mathbb{N}$ be the one-point compactification of the natural numbers, $L = \beta\mathbb{N}$ and let $\pi : \beta\mathbb{N} \longrightarrow M$ be the map defined by $\pi(n) = n$ for $n \in \mathbb{N}$, and $\pi(x) = \infty$ if $x \in \beta\mathbb{N} \setminus \mathbb{N}$.

Let $f : K \longrightarrow M$ be a continuous surjection. We claim that any continuous map $g : K \longrightarrow L$ with $\pi g = f$ must be onto, hence there is no such g. The reason is that for every $n \in \mathbb{N}$, if x_n is such that $f(x_n) = n$, then $\pi g(x_n) = n$, hence $g(x_n) = n$. Therefore $\mathbb{N} \subset g(K)$ and $\mathbb{N}$ is dense in L.   □

In the next result, by a *compact convex set* we mean a compact convex set lying inside some locally convex space. Actually, every such set L is affinely homeomorphic to a closed convex subset of a cube $[0, 1]^\Gamma$, where the size of Γ can be as small as the weight of L.

Theorem 5.29 *Suppose $\aleph \geq \aleph_1$. For a compact space K the following assertions are equivalent:*

1. *K is an $F_\aleph$-space.*
2. *For every continuous affine surjection $\pi : L \longrightarrow M$ between compact convex sets with $w(L) < \aleph$, and every continuous function $f : K \longrightarrow M$, there exists a continuous function $g : K \longrightarrow L$ such that $\pi g = f$.*
3. *Idem with $w(M) < \aleph$ and $w(L) \leq \aleph$.*
4. *Idem with $L \in HL_\aleph$.*

Proof We suppose that (2) holds, and we shall show that the second condition of Theorem 5.16 holds for any cardinal $\Gamma < \aleph$.

Let $f_\alpha, g_\alpha : K \longrightarrow [0,1]$, with $\alpha < \Gamma$, be two families of continuous functions such that $f_\alpha \leq g_\beta$ for every $\alpha, \beta < \Gamma$. Consider the sets

$$M = \left\{ ((t_\alpha)_{\alpha < \Gamma}, (s_\alpha)_{\alpha < \Gamma}) \in [0,1]^\Gamma \times [0,1]^\Gamma : \sup_{\alpha < \Gamma} t_\alpha \leq \inf_{\alpha < \Gamma} s_\alpha \right\},$$

$$L = \left\{ ((t_\alpha)_{\alpha < \Gamma}, r, (s_\alpha)_{\alpha < \Gamma}) \in [0,1]^\Gamma \times [0,1] \times [0,1]^\Gamma : \sup_{\alpha < \Gamma} t_\alpha \leq r \leq \inf_{\alpha < \Gamma} s_\alpha \right\}.$$

Let $\pi : L \longrightarrow M$ be the natural surjection which forgets the intermediate coordinate r, and let $f : K \longrightarrow M$ be given by

$$f(x) = (f_\alpha(x)_{\alpha < \Gamma}, g_\alpha(x)_{\alpha < \Gamma}).$$

The statement of (2) implies the existence of a function $g : K \longrightarrow L$ such that $\pi(g(x)) = f(x)$. Now if we take the composition of g with the projection on the central coordinate r of L, we obtain a continuous function $h : K \longrightarrow [0,1]$ such that $f_\alpha \leq h \leq g_\alpha$ for every $\alpha < \Gamma$. This proves that K is an $F_\aleph$-space.

Now we suppose that K is an $F_\aleph$-space, $f : K \longrightarrow M$ is a continuous affine surjection and $\pi : L \longrightarrow M$. We assume that M is a closed convex subset of a cube, $M \subset [0,1]^\Gamma$, and we denote by $\pi_\alpha : M \longrightarrow [0,1]$ the projection on the αth coordinate. The first step is to find the desired function f under the following assumption (it is the analogue of considering a Banach superspace of codimension 1):

STEP 1.   We assume $M \in HL_\aleph$ and there exists a continuous affine function $\phi : L \longrightarrow [0,1]$ such that the map $(\pi, \phi) : L \longrightarrow M \times [0,1]$ given by $(\pi, \phi)(x) = (\pi(x), \phi(x))$ is one-to-one.

In this case, we shall view L as a closed convex subset of $M \times [0,1]$, so that π and ϕ are just the projections on the first and second coordinate. To find the desired function $g : K \longrightarrow L$ is equivalent to find a continuous function $\gamma : K \longrightarrow [0,1]$ such that $(f(x), \gamma(x)) \in L$ for every $x \in K$. Let $\{q_n : n < \omega\}$ be a countable dense subset of $[0,1]$. We shall define by induction continuous functions $\gamma_n^-, \gamma_n^+ : K \longrightarrow [0,1]$ such that $\gamma_n^- \leq \gamma_m^+$ for every n, m, and then γ will be chosen such that $\gamma_n^- \leq \gamma \leq \gamma_m^+$ for every n, m.
For each n, define

$$U_n^- = \{y \in M : (y, t) \notin L \text{ for every } t \in [q_n, 1]\} \setminus \pi(L \cap (M \times [q_n, 1])),$$

$$U_n^+ = \{y \in M : (y, t) \notin L \text{ for every } t \in [0, q_n]\} \setminus \pi(L \cap (M \times [0, q_n])),$$

which are disjoint open subsets of M. Then $f^{-1}(U_n^-)$ and $f^{-1}(U_n^+)$ are disjoint open subsets of K and, since $M \in HL_\aleph$, each of them is a union of less than $\aleph$ many closed sets. Since K is an $F_\aleph$-space, there exist continuous functions δ_n^- and δ_n^+ over K such that

$$0 \leq \delta_n^- \leq q_n, \quad \delta_n^-|_{f^{-1}(U_n^-)} \equiv 0, \quad \delta_n^-|_{f^{-1}(U_n^+)} \equiv q_n,$$

$$q_n \leq \delta_n^+ \leq 1, \quad \delta_n^+|_{f^{-1}(U_n^-)} \equiv q_n, \quad \delta_n^+|_{f^{-1}(U_n^+)} \equiv 1.$$

A priori, it may be false that $\delta_n^- \leq \delta_m^+$ for every n, m, so in order to make sure of this we define inductively:

$$\gamma_n^- = \min\{\delta_n^-, \gamma_m^+ : m < n\}, \quad \gamma_n^+ = \max\{\delta_n^+, \gamma_m^- : m < n\}.$$

Using the fact that if $q_i < q_j$, then $f^{-1}(U_i^-) \subset f^{-1}(U_j^-)$ and $f^{-1}(U_i^+) \supset f^{-1}(U_j^+)$, it is easy to see that these new functions still keep the key properties that

$$0 \leq \gamma_n^- \leq q_n, \quad \gamma_n^-|_{f^{-1}(U_n^-)} \equiv 0, \quad \gamma_n^-|_{f^{-1}(U_n^+)} \equiv q_n,$$

$$q_n \leq \gamma_n^+ \leq 1, \quad \gamma_n^+|_{f^{-1}(U_n^-)} \equiv q_n, \quad \gamma_n^+|_{f^{-1}(U_n^+)} \equiv 1.$$

Since K is in particular an F-space, there is a continuous function $\gamma : K \longrightarrow [0, 1]$ such that $\gamma_n^- \leq \gamma \leq \gamma_n^+$ for all n. We have to show that $(f(x), \gamma(x)) \in L$ for every $x \in K$. Given $x \in K$, let

$$I = \{t \in [0, 1] : (f(x), t) \in L\} = \phi(\pi^{-1}[f(x)]).$$

Since ϕ and π are affine, I is a closed interval $[a, b]$. In order to check that $\gamma(x) \in I$, we show that $q_n \leq \gamma(x) \leq q_m$ whenever $q_n < a$ and $q_m > b$. For example, if $q_n < a$, then this means that $f(x) \in U_n^+$, $x \in f^{-1}(U_n^+)$, so $q_n = \gamma_n^-(x) \leq \gamma(x)$. Analogously, if $q_m > b$, then $x \in f^{-1}(U_m^-)$, and $\gamma(x) \leq \gamma_m^+(x) = q_m$. This finishes the proof under the assumption made in Step 1.

GENERAL CASE. We view now L as a compact convex subset of the Hilbert cube $[0, 1]^\Gamma$ [with $\Gamma = \aleph$ when we are under the assumptions of case (3)] and we call $\chi_\alpha : L \longrightarrow [0, 1]$ the coordinate functions, $\alpha < \Gamma$. For every α, we consider the map $h_\alpha : L \longrightarrow M \times [0, 1]^\alpha$ given by $h_\alpha(z) = (\pi(z), \chi_\beta(z)_{\beta < \alpha})$, and we call $L_\alpha = h_\alpha(L) \subset M \times [0, 1]^\alpha$ the image of this continuous function. For $\alpha < \beta$, we also call $p_\alpha^\beta : L_\beta \longrightarrow L_\alpha$ the continuous surjection which forgets about coordinates t_i with $i \geq \alpha$. We will construct by transfinite induction a sequence of coherent liftings $g_\alpha : K \longrightarrow L_\alpha$, $\alpha < \Gamma$, that is, functions satisfying $g_0 = f$ and $p_\alpha^\beta g_\beta = g_\alpha$ whenever $\alpha < \beta$. Observe that this is actually equivalent to finding continuous functions $\gamma_\alpha : K \longrightarrow [0, 1]$ such that $g_\alpha(x) = (f(x), \gamma_\beta(x)_{\beta < \alpha}) \in L_\alpha$ for every $x \in K$ and $\alpha \leq \Gamma$.

In the inductive process $g_{\alpha+1}$ is obtained from g_α by applying Step 1, while in the limit ordinals one has to take $g_\beta(x) = (f(x), \gamma_\alpha(x)_{\alpha < \beta})$. Notice that Step 1 can be applied because $L_\alpha \in HL_\aleph$: if we are in case (3), we take $\Gamma = \aleph$, so $w(L_\alpha) < \aleph$, while in case (4) L_α is a continuous image of L and $L \in HL_\aleph$. Let $g_\Gamma : K \longrightarrow M \times L$ be the final output of this inductive construction. We have that $p_0^\Gamma g_\Gamma = f$. Let $g : K \longrightarrow L$ be obtained by projecting g_Γ on the second coordinate, so that we can write $g^\Gamma(x) = (f(x), g(x))$. The fact that $g^\Gamma(x) \in L_\Gamma$ implies that $\pi(g(x)) = f(x)$, so g is the map that we were looking for.   □

The fact that $C(K)$ is $(1, \aleph)$-injective when K is an $F_\aleph$-space is a consequence of Theorem 5.29. Suppose that Y is a subspace of a Banach space X with $\mathrm{dens}(X) < \aleph$ and let $t : Y \longrightarrow C(K)$ be a norm-one operator. We can apply Theorem 5.29(2) to $\pi : B_{X^*} \longrightarrow B_{Y^*}$ and the mapping $f : K \longrightarrow B_{Y^*}$ given by $f(x) = t^*(\delta_x)$. We obtain a weak*-continuous function $g : K \longrightarrow B_{X^*}$ such that $\pi g = f$. Then, the formula $T(x)(k) = \|x\| g(k)(x/\|x\|)$, $x \in X$, $k \in K$, defines a norm-one extension of t.

5.5 Notes and Remarks

5.5.1 On Sobczyk's Theorem

A neat difference between $\aleph$-injectivity and separable injectivity is that Sobczyk's theorem has no simple counterpart for higher cardinals.

In this sense, perhaps the rôle of c_0 could be played by Hasanov's "filter version" of c_0 (see [122]). Recall that a filter $\mathcal{F}$ on a set I is called an $\aleph$-*filter* if whenever $\mathcal{A}$ is a family of less than $\aleph$ elements of $\mathcal{F}$ then $\bigcap \mathcal{A}$ is again in $\mathcal{F}$. Hasanov's space $c_0^{\mathcal{F}}(I)$ is the closure in $\ell_\infty(I)$ of the subspace $\{x \in \ell_\infty(I) : \lim_{\mathcal{F}} x = 0\}$. Hasanov shows in [122] that if $\mathcal{F}$ is an $\aleph$-filter, then $c_0^{\mathcal{F}}(I))$ is at most 2-complemented in any superspace E such that $\mathrm{dens}(E/c_0^{\mathcal{F}}(I)) \leq \aleph$. Thus, $c_0^{\mathcal{F}}(I)$ is $(2, \aleph^+)$-injective.

5.5.2 $\aleph$-Injectivity and Universal Disposition

As in Chap. 3, we can adjust the input data in the device of Sect. 3.1.1 to obtain spaces of universal disposition with respect to the class of Banach spaces having density character smaller than a given $\aleph$, which will immediately provide new examples of $(1, \aleph)$-injective spaces which are complemented in no C-space. Precisely, fix a Banach space X and set:

- The class $\mathfrak{M}_\aleph$ of Banach spaces having density character lesser than $\aleph$ and a set of Banach spaces $\tilde{\mathfrak{M}}_\aleph$ containing an isometric representative of each Banach space in $\mathfrak{M}_\aleph$.
- The family of isometries acting between the elements of $\tilde{\mathfrak{M}}_\aleph$.
- The family of all isometries $M \to X$, with $M \in \tilde{\mathfrak{M}}_\aleph$.

Now, acting as in Sect. 3.1.2, iterate the construction until an ordinal κ having cofinality greater than any cardinal below $\aleph$ (one may take $\kappa = \aleph^+$ in any case). The resulting space

$$\mathcal{M}_{\aleph}^{\kappa}(X)$$

is a space of universal disposition for $\mathfrak{M}_{\aleph}$ and, consequently is $(1, \aleph)$-injective (as in Theorem 3.5). Changing the family of into isometries for "isometries into $\ell_{\infty}(\aleph)$" the resulting space becomes universally $(1, \aleph)$-injective (as in Proposition 3.6).

5.5.3 Automorphisms of ℵ-Injective Spaces

From Theorem 5.10 we can derive that universally $\aleph^+$-injective spaces are automorphic for $c_0(\aleph)$-supplemented subspaces with density character $\aleph$ or less.

Theorem 5.30 *Let X be a universally $\aleph^+$-injective Banach space, and let Y_1 and Y_2 be isomorphic $c_0(\aleph)$-supplemented subspaces of X with $\mathrm{dens}(Y_i) \leq \aleph$ for $i = 1, 2$. Then every isomorphism from Y_1 onto Y_2 extends to an automorphism of X.*

Proof Note that we can modify the proof of Theorem 5.10 in such a way that the subspace Z isomorphic $\ell_{\infty}(\aleph)$ that contains Y has a complement isomorphic to X.

Indeed, if we write $\mathrm{ran}(P)$ as the direct sum of two copies of $\ell_{\infty}(\aleph)$ and take W so that its image is contained in one of the summands, then the complement Z' of Z in X contains a subspace isomorphic to $\ell_{\infty}(\aleph)$; hence

$$Z' \sim Z'' \oplus \ell_{\infty}(\aleph) \sim Z'' \oplus \ell_{\infty}(\aleph) \oplus \ell_{\infty}(\aleph) \sim Z' \oplus \ell_{\infty}(\aleph) \sim Z' \oplus Z \sim X.$$

So, for each $i = 1, 2$, we can assume that Y_i is contained in a subspace Z_i isomorphic to $\ell_{\infty}(\aleph)$ such that the complement of Z_i in X is isomorphic to X. Therefore, given an isomorphism $T : Y_1 \to Y_2$, since the quotients Z_1/X_1 and Z_2/X_2 are not reflexive, we can first extend T to an isomorphism from Z_1 onto Z_2 (see [182, Theorem 2.f.12]), and then extend it to an automorphism of X.   □

5.5.4 Sources

This Chapter is based on the work [22]. Some facts are more or less straightforward generalizations of the analogous results for the separably injective case and some others are rounded off forms of the corresponding results. Proposition 5.12 can be attributed to Lindenstrauss, and Theorem 5.16 to Aronszajn-Panitchpakdi [15], Henriksen [129] and Seever [228].

Good ultrafilters were introduced by Keisler who stablished in [162] that every set of cardinality α supports countably incomplete α^+-good ultrafilters assuming $\alpha^+ = 2^\alpha$. Later on Kunen removed that assumption.

Proposition 5.11 appears as Corollary 1 on page 210 in Neville [203] (there are two different results with the same label in the paper).

The equivalence of (1), (2) and (3) in Theorem 5.22 is due to Neville and Lloyd [204], and some particular facts concerning the relation of (1) and (4) were found by Przymusiński [218]. Corollary 5.27 is due to Neville and Lloyd [204].

Chapter 6
Open Problems

In this chapter we present and discuss in some detail problems that we encountered in the course of our work. Some of them have already been mentioned in previous chapters, others have appeared under different disguises and a few are new. The contents of the sections may freely overlap.

6.1 Characterizations of (Universally) Separably Injective Spaces

Many questions remain unanswered regarding the characterization and basic properties of separably and universally separably injective spaces. Analogously to what happens for injective spaces, it is reasonable to ask:

Problem 1 Is every universally separably injective space isomorphic to a universally 1-separably injective space? Must a λ-separably injective space, $\lambda < 2$, be isomorphic to a 1-separably injective space?

The second question has an affirmative answer for C-spaces (Proposition 2.34). Recall, however, that 2-separably injective spaces cannot, in general, be renormed to become λ-separably injective for $\lambda < 2$ (Proposition 2.32 and the fact that c_0 is 2-separably injective); let alone for $\lambda = 1$ (Proposition 2.31).

Since the first examples of (non-injective) universally separably injective spaces one encounters are $\ell_\infty^c(\Gamma)$ and ℓ_∞/c_0, it makes sense to ask for a pattern to construct explicit examples of operators into either $\ell_\infty^c(\Gamma)$ or ℓ_∞/c_0 that cannot be extended to some superspace. In the case of $\ell_\infty^c(\Gamma)$, the canonical embedding $c_0(\Gamma) \to \ell_\infty^c(\Gamma)$ is a reasonable candidate to be a non-extendable operator. Indeed, if some extension $T : \ell_\infty(\Gamma) \to \ell_\infty^c(\Gamma)$ would exist, by Rosenthal's result quoted

© Springer International Publishing Switzerland 2016
A. Avilés et al., *Separably Injective Banach Spaces*, Lecture Notes
in Mathematics 2132, DOI 10.1007/978-3-319-14741-3_6

in Theorem 1.14, this T should be an isomorphism on some copy of $\ell_\infty(\Gamma)$. Since

$$\text{dens } \ell_\infty^c(\Gamma) = |\Gamma|^{\aleph_0} \quad \text{and} \quad \text{dens } \ell_\infty(\Gamma) = 2^{|\Gamma|},$$

the embedding of $\ell_\infty(\Gamma)$ into $\ell_\infty^c(\Gamma)$ is impossible when $|\Gamma|^{\aleph_0} < 2^{|\Gamma|}$. This argument works for, say, $|\Gamma| = \mathfrak{c}$, while it fails—outside CH—for, say, $|\Gamma| = \aleph_1$ since it is consistent that $\aleph_1^{\aleph_0} = 2^{\aleph_1}$. A similar argument could work to show that no embedding $c_0(\Gamma) \to \ell_\infty/c_0$ extends to an operator $\ell_\infty(\Gamma) \to \ell_\infty/c_0$.

A different, although akin, topic is the topological characterization of separably injective and universally separably injective $C(K)$-spaces. The basic problem is:

Problem 2 Characterize the compact spaces K such that $C(K)$ is separably injective or universally separably injective.

Indeed, no known property of compacta seems to provide such characterizations. It is also an open problem to characterize Grothendieck C-spaces in terms of topological properties of the underlying compacta (see Sect. 6.8 for more information on Grothendieck spaces). We have already shown in Theorem 2.14 that K is an F-space if and only if $C(K)$ is 1-separably injective; hence, it is a Grothendieck space and, under CH (see Proposition 2.29), it is universally 1-separably injective. We do not know, however if, in ZFC, the fact that K is an F-space still implies that $C(K)$ is universally separably injective or even must contain ℓ_∞. When K is σ-Stonian every non-weakly compact operator $C(K) \to Y$ is an isomorphism on some copy of ℓ_∞ [223, Theorem 3.7] and thus $C(K)$ must necessarily contain ℓ_∞ (Dashiell [79] extends this result to different $C(K)$, including Baire classes—see last paragraph in Sect. 6.4.2). It is reasonable to conjecture that $C(K)$ is universally separably injective when K is σ-Stonian (in ZFC). We have even shown in Theorem 2.39 that there is a consistent example of 1-separably injective $C(K)$-space that is not universally 1-separably injective, but we do not know whether that example is universally separably injective or even if it contains ℓ_∞. Rosenthal asks in [223] whether there exists an F-space K such that $C(K)$ is injective but K is not σ-Stonian; and remarks that the answer is affirmative assuming the existence of a measurable cardinal.

It would be interesting to characterize 1-separable injectivity for the most popular classes of Lindenstrauss spaces, namely, M-spaces, G-spaces, and the like. In particular we ask for a characterization of those compact convex sets K for which $A(K)$, the space of continuous affine functions on K, is 1-separably injective. The following condition is sufficient (see Theorem 2.14): Given countable subsets L and U of $A(K)$ such that $f \leq g$ for every $f \in L$ and $g \in U$, there exists $h \in A(K)$ such that $f \leq h \leq g$ for all $f \in L$ and $g \in U$. Is the converse true? What if K is a simplex? (See [97] for the basics on simplex spaces). Related to this we have mentioned that if $(S_i)_{i \in I}$ is a family of simplices and $\mathcal{U}$ an ultrafilter on I, then $A(S_i)_{\mathcal{U}} = A(S)$ for some simplex S. Actually, S is unique by results of Rao [220]. It is also interesting to know how S is obtained from $(S_i)_{i \in I}$.

The study of $(1, \aleph)$-injectivity in terms of families of balls presented in Propositions 2.30 and 2.62 (for $\aleph = \aleph_1$) and Lemma 5.12 has no known analogue for universal injectivity. Precisely:

Problem 3   Can universal $(1, \aleph)$ injectivity be characterized in terms of intersection of families of balls?

6.2   The 3-Space Problem for Universal Separable Injectivity

The main problem we have been unable to solve is the 3-space problem for universal separable injectivity.

Problem 4   Is universal separable injectivity a 3-space property?

This problem has a surprising number of connections and ramifications, as we shall see. An affirmative answer would provide nice characterizations of that property and unexpected examples and counterexamples. As we already mentioned in Sect. 2.1.3, in [20, Proposition 3.7 (3)] it was claimed that universal separable injectivity is a 3-space property; but the proof contains a gap we have been unable to fill and a few statements in [20] and in [21] were infected. Let us clarify the situation about what is actually known:

Proposition 6.1   *The following assertions are equivalent:*

1. *Universal separable injectivity is a 3-space property.*
2. *Upper ℓ_∞-saturation is a 3-space property.*
3. *$\text{Ext}(\ell_\infty, U) = 0$ for every universally separably injective space U.*
4. *$\text{Ext}(\ell_\infty/S, U) = 0$ for every universally separably injective space U and every separable space S.*

Proof   It is clear that (1) and (2) are equivalent; see Theorem 2.26.

We prove that (1) implies (3) by showing that "¬(3) ⇒ ¬(1)". The idea is to prove that if a nontrivial exact sequence

$$0 \longrightarrow U \overset{i}{\longrightarrow} X \overset{p}{\longrightarrow} \ell_\infty \longrightarrow 0 \qquad (6.1)$$

with U universally separably injective exists, one arrives to another exact sequence

$$0 \longrightarrow \ell_\infty(\Gamma, U) \longrightarrow X' \longrightarrow \ell_\infty \longrightarrow 0$$

in which X' lacks Rosenthal's property (V) and thus it cannot be universally separably injective by Proposition 2.8.

Partington's distortion theorem for ℓ_∞ [210] establishes that any Banach space isomorphic to ℓ_∞ contains, for every $\varepsilon > 0$, an $(1+\varepsilon)$-isomorphic copy of ℓ_∞ (see also Dowling [90]). Let Γ be the set of all the 2-isomorphic copies of ℓ_∞ inside ℓ_∞. For each $E \in \Gamma$ let $\imath_E : E \to \ell_\infty$ be the inclusion map, p_E a projection of ℓ_∞ onto E of norm at most 2 and let $u_E : E \to \ell_\infty$ be a surjective 2-isomorphism.

Assume that (6.1) is a nontrivial exact sequence, with U universally separably injective. There is no loss of generality in assuming that $i : U \to X$ is the canonical inclusion map. We consider, for each $E \in \Gamma$, a copy of (6.1) and form the product of all these copies

$$0 \longrightarrow \ell_\infty(\Gamma, U) \longrightarrow \ell_\infty(\Gamma, X) \xrightarrow{\; p_\infty \;} \ell_\infty(\Gamma, \ell_\infty) \longrightarrow 0$$

Let us consider the operator $J : \ell_\infty \to \ell_\infty(\Gamma, \ell_\infty)$ given by $J(f)(E) = u_E p_E(f)$ and then form the pull-back sequence

$$
\begin{array}{ccccccccc}
0 & \longrightarrow & \ell_\infty(\Gamma, U) & \longrightarrow & \ell_\infty(\Gamma, X) & \xrightarrow{\; p_\infty \;} & \ell_\infty(\Gamma, \ell_\infty) & \longrightarrow & 0 \\
  &   & \big\| &   & \big\uparrow{\scriptstyle 'J} & & \big\uparrow{\scriptstyle J} & & \\
0 & \longrightarrow & \ell_\infty(\Gamma, U) & \longrightarrow & \mathrm{PB} & \xrightarrow{\; b_\infty \;} & \ell_\infty & \longrightarrow & 0
\end{array}
$$

Let us show that b_∞ cannot be an isomorphism on any copy of ℓ_∞ inside PB. Otherwise, it would have a right inverse on some $E \in \Gamma$ and thus the new pull-back sequence

$$
\begin{array}{ccccccccc}
0 & \longrightarrow & \ell_\infty(\Gamma, U) & \longrightarrow & \mathrm{PB} & \xrightarrow{\; 'p_\infty \;} & \ell_\infty & \longrightarrow & 0 \\
  &   & \big\| &   & \big\uparrow{\scriptstyle '\imath_E} & & \big\uparrow{\scriptstyle \imath_E} & & \\
0 & \longrightarrow & \ell_\infty(\Gamma, U) & \xrightarrow{\; \kappa \;} & \mathrm{PB}_E & \xrightarrow{\; ''b_\infty \;} & E & \longrightarrow & 0
\end{array}
$$

would split. Therefore, if $\pi_E^U : \ell_\infty(\Gamma, U) \longrightarrow U$ denotes the canonical projection onto the E-th copy of U, and similarly for π_E^X and $\pi_E^{\ell_\infty}$, the lower push-out sequence in the diagram

$$0 \longrightarrow \ell_\infty(\Gamma, U) \xrightarrow{\ \kappa\ } \mathrm{PB}_E \xrightarrow{\ ''p_\infty\ } E \longrightarrow 0$$

$$\pi_E^U \downarrow \qquad\qquad \downarrow (\pi_E^U)' \qquad \|$$

$$0 \longrightarrow \quad U \quad \xrightarrow{\ \kappa'\ } \mathrm{PO} \longrightarrow E \longrightarrow 0$$

also splits. We want to show that this lower sequence is isomorphically equivalent to
the starting sequence (6.1); more precisely, there is an isomorphism $\gamma : \mathrm{PO} \longrightarrow X$
rendering commutative the diagram

$$0 \longrightarrow U \xrightarrow{\ \kappa'\ } \mathrm{PO} \longrightarrow E \longrightarrow 0$$

$$\| \qquad\qquad \gamma \downarrow \qquad\qquad \downarrow u_E$$

$$0 \longrightarrow U \xrightarrow{\ i\ } X \xrightarrow{\ p\ } \ell_\infty \longrightarrow 0$$

In particular, one sequence splits if and only if the other does.

To obtain γ, observe the commutative square

$$\ell_\infty(\Gamma, U) \xrightarrow{\ \kappa\ } \mathrm{PB}_E$$

$$\pi_E^U \downarrow \qquad\qquad \downarrow \pi_E^X o' J o' \iota_E$$

$$U \quad \xrightarrow{\ i\ } \quad X$$

It is commutative since the restriction of $'J o' \iota_E$ to $\ell_\infty(\Gamma, U)$ is just the inclusion
into $\ell_\infty(\Gamma, X)$. Now, the universal property of the push-out construction yields an
operator γ making commutative the following diagram:

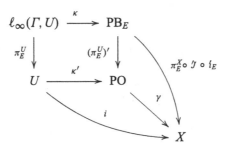

For the sake of clarity let us display all the data in the same drawing

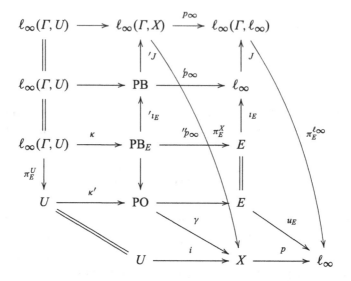

Let us see that the "horizontal flat" portion of this diagram is commutative. The left square is commutative by the very definition. As for the right square since $PB_E \to$ PO is onto it suffices to check that the composition

$$PB_E \longrightarrow PO \xrightarrow{\ \gamma\ } X \xrightarrow{\ p\ } \ell_\infty$$

is the same as

$$PB_E \xrightarrow{\ 'p_\infty\ } E \xrightarrow{\ u_E\ } \ell_\infty$$

which is obvious after realizing that $u_E = \pi_E^{\ell_\infty} \circ J \circ \iota_E$. And this is trivial since given $f \in E$ one has

$$\pi_E^{\ell_\infty}(J(f)) = J(f)(E) = u_E(p_E(f)) = u_E(f).$$

That (3) implies (4) is easy: Let S be separable and let U be universally separably injective. The homology sequence (see Appendix A.4.8) obtained from

$$0 \longrightarrow S \longrightarrow \ell_\infty \longrightarrow \ell_\infty/S \longrightarrow 0$$

is the exact sequence

$$\cdots \longrightarrow L(\ell_\infty, U) \longrightarrow L(S, U) \longrightarrow \mathrm{Ext}(\ell_\infty/S, U) \longrightarrow \mathrm{Ext}(\ell_\infty, U)$$

If $\text{Ext}(\ell_\infty, U) = 0$, the map $L(S, U) \longrightarrow \text{Ext}(\ell_\infty/S, U)$ must be surjective and thus every exact sequence $0 \longrightarrow U \longrightarrow X \longrightarrow \ell_\infty/S \longrightarrow 0$ fits in a push-out diagram

$$
\begin{array}{ccccccccc}
0 & \longrightarrow & S & \longrightarrow & \ell_\infty & \longrightarrow & \ell_\infty/S & \longrightarrow & 0 \\
& & \downarrow & & \downarrow & & \| & & \\
0 & \longrightarrow & U & \longrightarrow & X & \longrightarrow & \ell_\infty/S & \longrightarrow & 0.
\end{array}
$$

Since U is universally separably injective the operator $S \to U$ extends to an operator $\ell_\infty \to U$ and thus the lower sequence splits according to the splitting criterion for push-out sequences (Lemma A.20).

That (4) implies (3) is obvious, so both assertions are equivalent.

We show now that (3) implies (1): Let

$$
0 \longrightarrow Y \longrightarrow X \xrightarrow{\;q\;} Z \longrightarrow 0
$$

be an exact sequence in which both Y, Z are universally separably injective; let $j : S \to \ell_\infty$ be an into isomorphism with S separable, and let $t : S \to X$ be an operator. Since Z is universally separably injective, the operator qt admits an extension $T : \ell_\infty \to Z$. We can therefore form the pull-back diagram:

$$
\begin{array}{ccccccccc}
0 & \longrightarrow & Y & \longrightarrow & X & \xrightarrow{\;q\;} & Z & \longrightarrow & 0 \\
& & \| & & \uparrow & & \uparrow{\scriptstyle T} & & \\
0 & \longrightarrow & Y & \longrightarrow & \text{PB} & \longrightarrow & \ell_\infty & \longrightarrow & 0
\end{array}
$$

Since $\text{Ext}(\ell_\infty, Y) = 0$, there is an operator $\tau : \ell_\infty \to X$ so that $q\tau = T$. Since $q\tau j - qt = Tj - qt = 0$ the operator $t - \tau j$ takes actually values in Y. Let $\theta : \ell_\infty \to X$ be an extension of $t - \tau j$; namely, $\theta j = t - \tau j$. The operator $\theta + \tau : \ell_\infty \to X$ is the desired extension of t:

$$
(\theta + \tau)j = t - \tau j + \tau j = t.
$$

This completes the proof.   □

The preceding result provides a number of reformulations for Problem 4. A mildly convincing argument to support the idea that universal separable injectivity (i.e., ℓ_∞-upper-saturation; see Definition 2.25) is a 3-space property is:

Proposition 6.2 c_0-*upper-saturation is a 3-space property.*

Proof Let $0 \longrightarrow Y \longrightarrow X \xrightarrow{\;q\;} Z \longrightarrow 0$ be an exact sequence in which both Y and Z are c_0-upper-saturated and let S be a separable subspace of X. Pick Z_0 a subspace of Z isomorphic to c_0 and containing $q[S]$. It is a standard fact that there is a separable subspace $X_S \subset X$ containing S and such that $q[X_S] = Z_0$. Thus we have a commutative diagram

$$
\begin{array}{ccccccccc}
0 & \longrightarrow & Y & \longrightarrow & X & \longrightarrow & Z & \longrightarrow & 0 \\
  &   & \uparrow &   & \uparrow &   & \uparrow &   &   \\
0 & \longrightarrow & Y_S & \longrightarrow & X_S & \longrightarrow & Z_0 & \longrightarrow & 0
\end{array}
$$

with $Y_S = X_S \cap Y$. We want to see that there is a subspace of X isomorphic to c_0 containing X_S. Let Y_0 be an isomorphic copy of c_0 such that $Y_S \subset Y_0 \subset Y$. Making push-out with the inclusion $Y_S \to Y_0$ and taking Sobczyk's theorem into account yields the commutative diagram

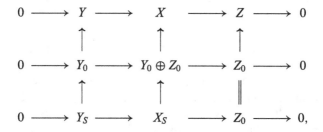

which shows that $S \subset Y_0 \oplus Z_0 \subset X$. Since $Y_0 \oplus Z_0 \sim c_0$ we are done. $\qquad\square$

On the other hand, a serious argument against could be that, analogously to what happens with Pełczyński's property (V), which is not a 3-space property (see [61]; see also [68]), one has:

Proposition 6.3 *Rosenthal's property (V) is not a 3-space property.*

Proof We start with a nontrivial exact sequence

$$
0 \longrightarrow \ell_2 \longrightarrow X \longrightarrow \ell_\infty \longrightarrow 0
$$

(see [54, Sect. 4.2]). Proceeding as in the proof that (1) implies (3) in Proposition 6.1 we construct an exact sequence

$$
0 \longrightarrow \ell_\infty(\Gamma, \ell_2) \longrightarrow X' \overset{q}{\longrightarrow} \ell_\infty \longrightarrow 0
$$

in which q cannot be an isomorphism on any copy of ℓ_∞. Thus, X' has not Rosenthal's property (V). The space $\ell_\infty(\Gamma, \ell_2)$ has Rosenthal's property (V) as a quotient of $\ell_\infty(\Gamma, \ell_\infty) = \ell_\infty(\mathbb{N} \times \Gamma)$, since Rosenthal's property (V) obviously passes to quotients. $\qquad\square$

The following consequence of an affirmative answer to the 3-space problem for universal separable injectivity was claimed in [20, Theorem 5.5]:

- A space U would be universally separably injective if and only if $\mathrm{Ext}(\ell_\infty/S, U) = 0$ for every separably space S.

The "only if" part is contained in Proposition 6.1. The other part does not depend on the solution to Problem 4:

Proposition 6.4 *If* $\mathrm{Ext}(\ell_\infty/S, X) = 0$ *for every separable space* S *then* X *is universally separably injective.*

Proof Let S be a separable Banach space, $t : S \to X$ an operator and $S \to \ell_\infty$ an embedding. Form the push-out diagram

$$
\begin{array}{ccccccccc}
0 & \longrightarrow & S & \longrightarrow & \ell_\infty & \longrightarrow & \ell_\infty/S & \longrightarrow & 0 \\
 & & {\scriptstyle t}\downarrow & & \downarrow & & \| & & \\
0 & \longrightarrow & X & \longrightarrow & \mathrm{PO} & \longrightarrow & \ell_\infty/S & \longrightarrow & 0.
\end{array}
$$

The lower sequence splits by the assumption $\mathrm{Ext}(\ell_\infty/S, X) = 0$ and so t extends to ℓ_∞, according to the splitting criterion for push-out sequences (Lemma A.20). □

In particular, as it was claimed in [20, Proposition 5.6], one would have that $\mathrm{Ext}(C(\mathbb{N}^*), C(\mathbb{N}^*)) = 0$ since $C(\mathbb{N}^*) = \ell_\infty/c_0$. This would rank ℓ_∞/c_0 into the exclusive list of spaces X for which $\mathrm{Ext}(X, X) = 0$, currently formed by

- c_0, by Sobczyk's theorem.
- Injective spaces, by the very definition.
- $L_1(\mu)$-spaces, by Lindenstrauss' lifting (Proposition A.18).

It is definitely not true however that $\mathrm{Ext}(U, V) = 0$ for all universally separably injective spaces U and V. For instance, consider the exact sequence $0 \to \ell_\infty^c(\Gamma) \to \ell_\infty(\Gamma) \to \ell_\infty(\Gamma)/\ell_\infty^c(\Gamma) \to 0$, where Γ is an uncountable set. Since the subspace is universally 1-separably injective (Example 2.4), the quotient is universally separably injective by Proposition 2.11(3). Actually it is even universally 1-separably injective, by Theorem 2.18. The sequence does not split because $\ell_\infty^c(\Gamma)$ is not injective (Proposition 1.28). Each universally separably injective non-injective space produces a similar counterexample. Moreover, it is easy to see that there exist universally separably injective spaces U such that $\mathrm{Ext}(U, U) \neq 0$: if V is a universally separably injective non-injective space then every exact sequence $0 \to V \to \ell_\infty(\Gamma) \to \ell_\infty(\Gamma)/V \to 0$ is not trivial, by Proposition 2.11. The space $W = \ell_\infty(\Gamma)/V$ is universally separably injective and, obviously, $\mathrm{Ext}(W, V) \neq 0$. The product space $U = V \oplus W$ is universally separably injective and $\mathrm{Ext}(U, U) \neq 0$.

The following problem seems quite interesting to us:

Problem 5 Characterize the $C(K)$ spaces so that $\mathrm{Ext}(C(K), C(K)) = 0$.

Probably a step in this direction would be to know whether the following generalization(s) of Problem 4 are possible:

Problem 6 Are there homological characterizations of $\aleph$-injectivity and universal $\aleph$-injectivity? In particular: Is it true that

- A Banach space E is $\aleph^+$-injective if and only if $\mathrm{Ext}(c_0(\aleph), E) = 0$?

- A Banach space E is universally $\aleph^+$-injective if and only if $\mathrm{Ext}(\ell_\infty(\aleph), E) = 0$?
- $\mathrm{Ext}(\ell_\infty(\Gamma), U) = 0$ for every universally separably injective space U?

Recall that the information we currently have is:

Theorem 6.5 *A Banach space E is*

- *Separably injective if and only if* $\mathrm{Ext}(Q, E) = 0$ *whenever Q is a quotient of* $C[0, 1]$.
- *Universally separably injective if* $\mathrm{Ext}(Q, E) = 0$ *whenever Q is a quotient of ℓ_∞ by a separable subspace (Proposition 6.4).*
- *[CH] $\aleph_2$-injective if and only if* $\mathrm{Ext}(Q, E) = 0$ *whenever Q is a quotient of ℓ_∞ (Corollary 5.6).*
- *[GCH] $\aleph^+$-injective if and only if* $\mathrm{Ext}(Q, E) = 0$ *whenever Q is a quotient of $\ell_\infty(\aleph)$ (Corollary 5.6).*

Proof Only the first point has not been explicitly done: if E is separably injective, then $\mathrm{Ext}(Q, E) = 0$ for every quotient of $C[0, 1]$; and conversely, if $\tau : S \to E$ is an operator from any separable Banach space S, pick an embedding $S \to C[0, 1]$ and form the push-out diagram:

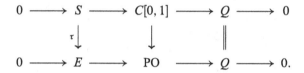

Since $\mathrm{Ext}(Q, E) = 0$, the lower sequence splits and τ can be extended to an operator $C[0, 1] \to E$, which shows that E is separably injective. $\qquad\square$

A different way of looking at these questions is the following: A result of Johnson, Rosenthal and Zippin [148], see also [182], asserts that every separable Banach space S fits into an exact sequence

$$0 \longrightarrow A \longrightarrow S \longrightarrow B \longrightarrow 0$$

in which both A and B have the BAP. Since a Banach space E is separably injective when $\mathrm{Ext}(S, E) = 0$ for all separable Banach spaces S, a 3-space argument yields that a Banach space E is separably injective when $\mathrm{Ext}(S, E) = 0$ for all separable Banach spaces S with the BAP. And since there exist a separable space $\mathscr{K}$ with the BAP complementably universal for all separable spaces with the BAP [150, 213], it follows that E is separably injective if and only if $\mathrm{Ext}(\mathscr{K}, E) = 0$. Therefore, there exist (separable) Banach spaces that "test" the separable injectivity. The question is then whether

- c_0, or its quotients, could be test spaces for separable injectivity.
- $c_0(\aleph)$, or its quotients, could be test spaces for $\aleph^+$-separable injectivity.

- ℓ_∞, or (some of) its quotients, could be test spaces for universal separable injectivity.
- $\ell_\infty(\aleph)$, or (some of) its quotients, could be test spaces for $\aleph^+$-universal injectivity.

And if quotients of c_0 are a puzzle, quotients of ℓ_∞ are a conundrum, as we discuss next.

6.3 Subspaces and Quotients of ℓ_∞

Many results and ideas in this monograph wheel around the question about to what extent universally separably injective spaces are "like" ℓ_∞. This suggests:

Problem 7 Do there exist universally separably injective subspaces of ℓ_∞ different from ℓ_∞?

Nonseparable separably injective subspaces of ℓ_∞ different from ℓ_∞ exist: indeed, Marciszewski and Pol show in [193] that there exist at least $2^{\mathfrak{c}}$ non-isomorphic C-spaces arising as twisted sums of c_0 and $c_0(\mathfrak{c})$. These are associated to different choices of almost disjoint families (see Sect. 2.2.4) which yields $2^{\mathfrak{c}}$ non-mutually isomorphic separably injective subspaces of ℓ_∞. None of them can be universally separably injective since the pull-back space in a diagram of the form

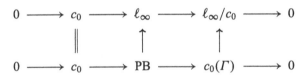

cannot be universally separably injective: otherwise the two sequences above are one pull-back of the other and then the diagonal principles yield

$$PB \oplus (\ell_\infty/c_0) \sim \ell_\infty \oplus c_0(\Gamma),$$

implying that $c_0(\Gamma)$ should also be universally separably injective, which is not.

The same question for quotients has an obvious answer: ℓ_∞/c_0. A further result in this direction follows from Dow and Vermeer [88]: if CH is assumed, every compact F-space of weight $\aleph_2$ (or less) embeds as a closed subset of an extremely disconnected compact space. Which implies that, under CH, every 1-separably injective C-space of density character $\aleph_2$ (or less) is an isometric quotient of a 1-injective Banach space.

As we have already remarked, there are subspaces G of ℓ_∞ that are not $\mathscr{L}_\infty$ spaces but such that $\ell_\infty/G \sim \ell_\infty$. No characterization is known of the subspaces X of ℓ_∞ for which ℓ_∞/X is (universally) separably injective. Quotients of ℓ_∞ must be Grothendieck spaces and quotients $\ell_\infty/\mathscr{L}_\infty$ of ℓ_∞ by a unspecified $\mathscr{L}_\infty$-space must be of type $\mathscr{L}_\infty$ since $\ell_\infty^{**} \sim \mathscr{L}_\infty^{**} \oplus (\ell_\infty/\mathscr{L}_\infty)^{**}$. We are specially interested in the following case:

Problem 8 Is $\ell_\infty/C[0,1]$ separably injective?

Recall that since ℓ_∞ is separably automorphic the space $\ell_\infty/C[0,1]$ is well defined. But we do not know if $\ell_\infty/C[0,1]$ is even isomorphic to a (complemented subspace of a) C-space. Additional information is contained in the following proposition.

Proposition 6.6 *Let $\mathscr{L}_\infty$ denote an arbitrary separable $\mathscr{L}_\infty$ space and let $\mathfrak{U}$ be a free ultrafilter on $\mathbb{N}$. Then the following statements are equivalent:*

1. *$\ell_\infty/\mathscr{L}_\infty$ is separably injective.*
2. *$(\mathscr{L}_\infty)\mathfrak{u}/\mathscr{L}_\infty$ is separably injective.*
3. *$[\ell_\infty^n]\mathfrak{u}/\mathscr{L}_\infty$ is separably injective.*

Proof We already know that both $[\ell_\infty^n]\mathfrak{u}$ and $(\mathscr{L}_\infty)\mathfrak{u}$ are separably automorphic so there is no need to particularize which embeddings $\mathscr{L}_\infty \to [\ell_\infty^n]\mathfrak{u}$ or $\mathscr{L}_\infty \to (\mathscr{L}_\infty)\mathfrak{u}$ are used. Look at the lower exact sequence in the complete push-out diagram

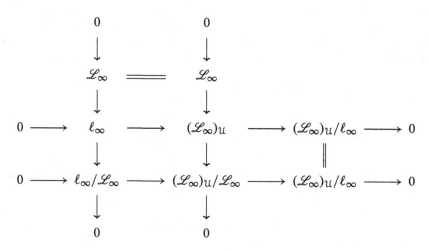

Since the middle horizontal sequence splits, the lower sequence also splits and

$$(\mathscr{L}_\infty)\mathfrak{u}/\mathscr{L}_\infty \sim (\ell_\infty/\mathscr{L}_\infty) \oplus ((\mathscr{L}_\infty)\mathfrak{u}/\ell_\infty).$$

Thus, since $(\mathscr{L}_\infty)\mathfrak{u}/\ell_\infty$ is separably injective, the space $(\mathscr{L}_\infty)\mathfrak{u}/\mathscr{L}_\infty$ is separably injective if and only if $\ell_\infty/\mathscr{L}_\infty$ is. We draw now the complete pull-back

diagram

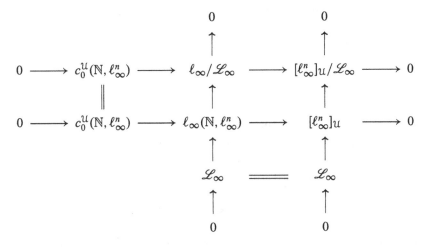

and recall that $c_0^{\mathcal{U}}(\mathbb{N}, \ell_\infty^n)$ is separably injective (Theorem 2.21 plus Lemma 4.17). Thus, if $[\ell_\infty^n]\mathcal{U}/\mathscr{L}_\infty$ is separably injective then $\ell_\infty/\mathscr{L}_\infty$ is separably injective by a 3-space argument applied to the upper sequence. If, however, $\ell_\infty/\mathscr{L}_\infty$ is separably injective then its quotient $[\ell_\infty^n]\mathcal{U}/\mathscr{L}_\infty$ is separably injective by Proposition 2.11.

□

Quotients of ℓ_∞ by separable subspaces are also intriguing. The following question was posted in Mathoverflow (http://mathoverflow.net/questions/148956/quo\discretionary-tients-of-ell-infty-by-separable-subspaces) by the authors:

Problem 9 Under which conditions on a separable subspace M of ℓ_∞ is the quotient ℓ_∞/M isomorphic to a subspace of ℓ_∞?

No complete answer to the question above is known, but the following partial results that appeared in that page are due to Bill Johnson, with slightly different proofs:

Proposition 6.7 Let X and Y be subspaces of ℓ_∞.

1. If X and Y are isomorphic, then ℓ_∞/X embeds into ℓ_∞ if and only if ℓ_∞/Y embeds into ℓ_∞.
2. Suppose that $Y \subset X$ and that ℓ_∞/X embeds into ℓ_∞. Then ℓ_∞/Y embeds into ℓ_∞ if and only if X/Y embeds into ℓ_∞.
3. If X is isomorphic to V^* for some separable space V, then ℓ_∞/X embeds into ℓ_∞.
4. If X is isomorphic to a subspace of a separable dual, then ℓ_∞/X embeds into ℓ_∞.
5. If X contains c_0 and X/c_0 embeds into ℓ_∞ then ℓ_∞/X does not embed into ℓ_∞.
6. If X embeds into $L_1[0, 1]$, then ℓ_∞/X embeds into ℓ_∞.

Proof

1. Let $\alpha : X \to Y$ be an isomorphism and consider the diagram

$$
\begin{array}{ccccccccc}
0 & \longrightarrow & X & \longrightarrow & \ell_\infty & \longrightarrow & \ell_\infty/X & \longrightarrow & 0 \\
  &   & \alpha\downarrow &   &   &   &   &   &   \\
0 & \longrightarrow & Y & \longrightarrow & \ell_\infty & \longrightarrow & \ell_\infty/Y & \longrightarrow & 0.
\end{array}
$$

By the injectivity of ℓ_∞, both α and α^{-1} can be extended to operators $\ell_\infty \to \ell_\infty$. Thus, each sequence is a push-out of the other and the diagonal principle (Proposition A.22) yields an isomorphism $(\ell_\infty/X) \oplus \ell_\infty \sim (\ell_\infty/Y) \oplus \ell_\infty$.

2. In this case one has an exact sequence

$$
0 \longrightarrow X/Y \longrightarrow \ell_\infty/Y \longrightarrow \ell_\infty/X \longrightarrow 0.
$$

Since "to be a subspace of ℓ_∞" is a 3-space property [61, Theorem 3.2.h], the result is clear.

3. Let $q : \ell_1 \to V$ be a quotient map. Then $V^* \sim (\ker q)^\perp$, so we can assume $X = (\ker q)^\perp$ by (1). Hence $\ell_\infty/X \sim (\ker q)^*$, which is a subspace of ℓ_∞.

   Part (4) is a direct consequence of (3), (2), and the fact that separable subspaces embed in ℓ_∞.

   Part (5) follows from (1), (2) and the fact that ℓ_∞/c_0 does not embed into ℓ_∞ since it contains an isometric copy of $c_0(\Gamma)$ with $|\Gamma| = 2^{\aleph_0}$.

6. The case $X = L_1[0, 1]$ follows from (2) and the fact that $L_1[0, 1]$ embeds into $C[0, 1]^*$ as a complemented subspace: Since $\ell_\infty/C[0, 1]^*$ and the complement of $L_1[0, 1]$ in $C[0, 1]^*$ embed into ℓ_∞, so does $\ell_\infty/L_1[0, 1]$. The general case can be proved as (4).

   $\square$

Assertion (3) in Proposition 6.7 can be completed with:

Proposition 6.8 *Let X be a separable Banach space such that $\mathrm{Ext}(X, \ell_2) \neq 0$. Then $\ell_\infty/j[X^*]$ is not an $\mathscr{L}_\infty$-space for any embedding $j : X^* \longrightarrow \ell_\infty$.*

Proof Let $q : \ell_1 \longrightarrow X$ be a quotient map. Since $\mathrm{Ext}(X, \ell_2) \neq 0$, there exists a nontrivial exact sequence $0 \longrightarrow \ell_2 \longrightarrow Y \longrightarrow X \longrightarrow 0$, and thus an operator $\ker q \longrightarrow \ell_2$ that cannot be extended to ℓ_1. Since every operator from an $\mathscr{L}_1$ into a Hilbert space is 2-summing, it follows that $\ker q$ cannot be an $\mathscr{L}_1$-space and thus $(\ker q)^* = \ell_\infty/q^*[X^*]$ cannot be an $\mathscr{L}_\infty$-space.

Now, let $j : X^* \longrightarrow \ell_\infty$ be any embedding. The two sequences in the diagram

$$
\begin{array}{ccccccccc}
0 & \longrightarrow & X^* & \xrightarrow{\ j\ } & \ell_\infty & \longrightarrow & \ell_\infty/j[X^*] & \longrightarrow & 0 \\
 & & \| & & & & & & \\
0 & \longrightarrow & X^* & \xrightarrow{\ q^*\ } & \ell_\infty & \longrightarrow & (\ker q)^* & \longrightarrow & 0
\end{array}
$$

are one pull-back of the other since both q^* and j can be extended to operators $\ell_\infty \to \ell_\infty$, so the diagonal principle (Proposition A.22) yields that $\ell_\infty \oplus (\ell_\infty/j[X^*])$ is isomorphic to $\ell_\infty \oplus (\ker q)^*$; and thus $\ell_\infty/j[X^*]$ cannot be an $\mathscr{L}_\infty$-space. □

In particular, it could be interesting to know if there exists an infinite dimensional separable subspace S of ℓ_∞ for which $\ell_\infty/S \sim \ell_\infty$.

6.4 Examples of Separably Injective Spaces

We have seen that there are many natural examples of (universally) separably injective spaces. A few more or less classical spaces could also enjoy separable injectivity properties:

6.4.1 Tensor Products

The fact that when E is separably injective so is $c_0(E) = c_0 \check{\otimes}_\varepsilon E$ suggests:

Problem 10 Must $F \check{\otimes}_\varepsilon E$ be separably injective if both E and F are?

We do not know the answer even if F is a $C(K)$-space, in which case $C(K) \check{\otimes}_\varepsilon E = C(K, E)$. A particularly interesting test case is that of the space

$$
\ell_\infty \check{\otimes}_\varepsilon \ell_\infty = C(\beta\mathbb{N}, C(\beta\mathbb{N})) = C(\beta\mathbb{N} \times \beta\mathbb{N}).
$$

The compact space $\beta\mathbb{N} \times \beta\mathbb{N}$ is not an F-space and it does not contain convergent sequences. However, the space $C(\beta\mathbb{N} \times \beta\mathbb{N}) = C(\beta\mathbb{N}, C(\beta\mathbb{N}))$ cannot be even a Grothendieck space as it can be inferred from the following result of Cembranos [72]: If K is an infinite compact and X is an infinite dimensional Banach space, then $C(K, X)$ contains a complemented subspace isomorphic to c_0. The validity of similar statements for $c_0(\aleph_1)$ under different axioms has been studied by several authors: Galego and Hagler [100] show that under CH there is a compact K so that $c_0(\aleph_1)$ embeds into $C(K \times K)$ but not in $C(K)$; Dow et al. [89, Example 2.16] show that in ZFC there exists $C(K)$ spaces with density $2^{\aleph_0}$ containing a copy of $c_0(\aleph_1)$ and such that $C(K \times K)$ does not contain $c_0(\aleph_1)$ complemented; Candido and Koszmider [59] show that it is consistent that if $C(K)$ has density character $\aleph_1$ and contains $c_0(\aleph_1)$ then $C(K \times K)$ contains $c_0(\aleph_1)$ complemented.

6.4.2 Baire Classes

The Baire classes of functions on $[0, 1]$ were studied by Bade in [23]. As we shall see, they share some of the properties of the universally separably injective spaces, and therefore can be considered natural candidates to provide new examples of separably injective spaces. We set $B_0 = C[0, 1]$, and denote by B_1 the class of all bounded functions which are pointwise limits of functions in B_0.

Definition 6.9 For each ordinal α with $1 \leq \alpha \leq \omega_1$, we define the class of Baire functions of order α, denoted B_α, as the class of all bounded functions which are pointwise limits of functions in $\bigcup_{\beta < \alpha} B_\beta$.

Bade shows in [23] that, for $1 \leq \alpha \leq \omega_1$, the space B_α is linearly isometric to $C(K_\alpha)$, where K_α is a totally disconnected compact space. He also shows that B_{ω_1} is the space of all bounded Borel measurable functions on $[0, 1]$. Dashiell shows in [78] (see also [80, Corollary 8]) the following results:

- For $\alpha < \beta \leq \omega_1$, B_α is not complemented in B_β.
- The spaces B_α are injective for no α.
- The dual spaces B_α^* are linearly isometric to ℓ_∞^*, for $1 \leq \alpha \leq \omega_1$.
- For $\alpha < \omega_1$, B_α is not isomorphic to B_{ω_1}.
- B_1 is not isomorphic to B_α if $\alpha > 1$.
- For $1 < \alpha < \beta < \omega_1$, the spaces B_α and B_β are not isometric. It is apparently unknown whether they are isomorphic.

Passing to separable injectivity affairs, Dashiell also shows that $B_{\omega_1} = C(K_{\omega_1})$, with K_{ω_1} σ-Stonean. Since each σ-Stonean compact is a totally disconnected F-space, B_{ω_1} is 1-separably injective. In Proposition 2.16 it was already shown that the space of all bounded Borel (resp. Lebesgue) measurable functions on the line is 1-separably injective (we do not know however if those spaces are universally 1-separably injective in ZFC). One however has:

Proposition 6.10 The space B_1 is not 1-separably injective.

Proof Let $Q = q_1, q_2, \ldots$ and $Q' = q_1', q_2', \ldots$ be two disjoint countable dense subsets of $[0, 1]$. We consider the following sequences of functions of the first Baire class:

$$f_n = 1_{q_n}, \quad g_n = 1_{[0,1]} - 1_{q_n'}.$$

Then $f_n \leq g_m$ for every n, m; hence, if B_1 were 1-separably injective, then there would exist $h \in B_1$ such that $f_n \leq h \leq g_m$ for every n, m. Consider $A =$

$h^{-1}(-\infty, 0.4]$ and $B = h^{-1}[0.6, +\infty)$. Then, since h is of the first Baire class, A and B must be G_δ sets, and since $Q \subset B$ and $Q' \subset A$, they are dense sets. But $A \cap B = \varnothing$ while the intersection of two dense G_δ sets is a dense G_δ set. $\qquad\square$

Thus, while B_1 is not 1-separably injective, $B_1 \cap \ell_\infty^c[0, 1]$ is 2-separably injective: see the discussion after Theorem 2.18. Moreover, for all $\alpha \geq 1$ the spaces B_α enjoy Rosenthal's property (V) and thus, in particular, they are Grothendieck spaces [79, Theorem 3.5].

Problem 11 For which $1 \leq \alpha < \omega_1$ is the space B_α (universally) separably injective?

However, the question that motivated our interest in Baire classes is the possibility of having the following type of "surrogate" separable injectivity: Does there exist a function $f : \omega_1 \to \omega_1$ such that given a separable Banach space X and a subspace $Y \subset X$ every operator $t : Y \longrightarrow B_\alpha$ admits an extension $T : X \longrightarrow B_{f(\alpha)}$?

6.4.3 $C(\mathbb{N}^*)$ and Its $\aleph$-Injectivity

We know that $C(\mathbb{N}^*)$ fails to be $\mathfrak{c}^+$-injective (Proposition 2.43) and that it fails to be $(1, \aleph_2)$-injective (Proposition 5.18). The next question is however open:

Problem 12 Is it consistent that $C(\mathbb{N}^*)$ is (universally) $\aleph_2$-injective?

Still, the answer is no for $\mathfrak{c} < 2^{\aleph_1}$: since $C(\mathbb{N}^*)$ contains $c_0(\aleph_1)$, it contains a $c_0(\aleph_1)$-supplemented copy; so, by Theorem 5.10, $C(\mathbb{N}^*)$ should contain $\ell_\infty(\aleph_1)$, which is impossible if $\mathfrak{c} < 2^{\aleph_1}$.

It is not difficult to see that a necessary condition for $C(\mathbb{N}^*)$ to be universally $\aleph_2$-injective is that every operator $c_0(\aleph_1) \to C(\mathbb{N}^*)$ can be extended to $C(\mathbb{N}^*)$. One thus has encountered the notion of space $c_0(\aleph_1)$-extensible (cf. [199]):

Definition 6.11 Let $\mathfrak{M}$ be a class of Banach spaces.

- A Banach space X is said to be $\mathfrak{M}$-extensible if every operator $A \to X$ with A a subspace of X in $\mathfrak{M}$ can be extended to an operator $X \to X$.
- A Banach space X is said to be $\mathfrak{M}$-automorphic if every isomorphism between two spaces $A, B \in \mathfrak{M}$ that are subspaces of X and for which X/A and X/B have the same density character can be extended to an automorphism of X.

For instance, the choice of $\mathfrak{M}$ as the class of all separable spaces leads to separably automorphic spaces; the choice $\mathfrak{M} = \{Y\}$ leads to Y-automorphic spaces. When $\mathfrak{M}$ are "all spaces" then we get the notions of extensible and automorphic space. For instance, ℓ_∞ is extensible but not automorphic while ℓ_2 is automorphic but not injective. It is proved in [199] that $\mathfrak{M}$-automorphic implies $\mathfrak{M}$-extensible.

We see in this way that the problem of injectivity-like properties of $C(\mathbb{N}^*)$ is connected at a deep level with its $c_0(\aleph_1)$-automorphic character: If $C(\mathbb{N}^*)$ is not $c_0(\aleph_1)$-extensible then it cannot be (universally) $\aleph_2$-injective. Moreover, if all copies of $c_0(\aleph_1)$ in $C(\mathbb{N}^*)$ are $c_0(\aleph_1)$-supplemented then by Theorem 5.30 the space $C(\mathbb{N}^*)$ is $c_0(\aleph_1)$-automorphic, hence $c_0(\aleph_1)$-extensible. Obviously, we do not know if $C(\mathbb{N}^*)$ is $c_0(\aleph_1)$-extensible:

Problem 13 Is $C(\mathbb{N}^*)$ a $c_0(\aleph_1)$-automorphic space?

The study of partially automorphic spaces goes beyond the scope of this monograph, and the interested reader is addressed to [17, 19, 63, 67, 199]. A few additional results will help to complete the picture about $C(\mathbb{N}^*)$ and will complement Sect. 2.6. We have already shown that $C(\mathbb{N}^*)$ is separably automorphic, a property somehow inherited from ℓ_∞. So, it would be nice to know "how much" automorphic the space is. In any case, it will be partially automorphic in a different sense from ℓ_∞, since ℓ_∞ is automorphic for subspaces X so that ℓ_∞/X contains ℓ_∞, while $C(\mathbb{N}^*)$ is not since $C(\mathbb{N}^*)$ is not $C(\mathbb{N}^*)$-automorphic. In contrast with Proposition 2.55 (1), we do not know if the quotient of a separably automorphic space by a separably injective space is separably automorphic.

It is obvious that in a c_0-automorphic space either all copies of c_0 are complemented or all of them are uncomplemented. It follows from Lemma 2.48 that C-spaces in which every copy of c_0 is complemented are c_0-automorphic. These spaces include C-spaces over Eberlein compacta by Proposition 2.57 or ordinal compacta. The C-spaces in which no copy of c_0 is complemented are Grothendieck, but their analysis is not so simple. In particular, some are c_0-automorphic (universally separably injective spaces, for instance) while others are not: if $\mathscr{H}$ denotes Haydon's Grothendieck C-space without copies of ℓ_∞, the space $\ell_\infty \oplus \mathscr{H}$ is not c_0-automorphic. Indeed, if $\sigma : \ell_\infty \oplus \mathscr{H} \longrightarrow \ell_\infty \oplus \mathscr{H}$ is an automorphism sending $c_0 \oplus 0$ to $0 \oplus c_0$ and $\pi_{\mathscr{H}} : \ell_\infty \oplus \mathscr{H} \longrightarrow \mathscr{H}$ denotes the projection onto the second coordinate, then $\pi_{\mathscr{H}}\sigma_{|\ell_\infty} : \ell_\infty \to \mathscr{H}$ cannot be weakly compact, hence must be an isomorphism on a copy of ℓ_∞, which is also impossible.

Copies of c_0 must also be complemented in any WCG Banach superspace, as it can be proved using the classical Amir-Lindenstrauss Theorem [6]. The natural question of whether the same happens for $c_0(\Gamma)$—must copies of $c_0(\Gamma)$ be complemented in every WCG superspace?—has a negative answer, as it was already mentioned in Sect. 2.2.4. More precisely, [13, 192], there exists under **GCH** an Eberlein compact $\mathscr{E}$ such that $C(\mathscr{E})$ contains an uncomplemented copy of $c_0(\aleph_\omega)$. The C-space $C(\mathscr{E}) \oplus c_0(\aleph_\omega)$ is therefore WCG and it is not $c_0(\aleph_\omega)$-automorphic. This shows that the situation is very different from that for c_0 and some restrictions in size are necessary. Indeed, if $\mathscr{E}$ is an Eberlein compact of weight $\aleph$ then, by Theorem 4.2 in [35], $C(\mathscr{E})$ contains a copy of $c_0(\aleph)$. For $\aleph = \aleph_n$, $n < \omega$, all those copies must be complemented in $C(\mathscr{E})$ due to the following result in [13]:

Theorem 6.12 *Let K be a Valdivia compact and let Γ be a set with cardinal $|\Gamma| = \aleph_n$ for some $n < \omega$. Then every λ-isomorphic copy of $c_0(\Gamma)$ inside $C(K)$ is $2^{n+1}\lambda$-complemented.*

However, we do not know if $C(\mathscr{E})$ is $c_0(\aleph_n)$-automorphic. A positive answer to the following question would imply that $C(\mathscr{E})$ is in fact H-automorphic for all subspaces $H \subset c_0(\aleph_n)$: Let $\mathcal{E}$ be an Eberlein compact and $\aleph < \aleph_\omega$. Does every complement of $c_0(\aleph)$ in $C(\mathcal{E})$ (having density character at least $\aleph$) contain a copy of $c_0(\aleph)$? Recall that it is even unknown if a complement of $c_0(\aleph_n)$ in $C(\mathcal{E})$ must be isomorphic to a C-space.

6.5 Ultraproblems

Since ultrapowers of $\mathcal{L}_\infty$ have emerged as unexpected universally separably injective spaces, questions involving ultraproducts are natural. The Henson-Moore classification problem of $\mathcal{L}_\infty$-spaces by isomorphic ultrapowers ([134, p. 106], [128, p. 315], [130]) is perhaps the deepest:

Problem 14 How many ultratypes of $\mathcal{L}_\infty$-spaces are there?

The results in Chap. 4 show only two different ultra-types of $\mathcal{L}_\infty$-spaces: that of C-spaces and that of spaces of almost universal disposition. It would be interesting to add some new classes here. If one is thinking about obtaining a third type, probably the best candidates are "the" quotient $C[0, 1]/\mathcal{G}$ (more natural should be thinking about the space $\ell_\infty/\mathcal{G}$, which is uniquely defined by the separably automorphic character of ℓ_∞) or the subspace $\ker \pi$ of a (rightly chosen) quotient map $\pi : C[0, 1] \to \mathcal{G}$ (see Sect. 6.6). Of course that we do not know whether $C[0, 1]/\mathcal{G}$ or $\ell_\infty/\mathcal{G}$ are C-spaces, whether they have the same ultratype of a C-space, or even if both have the same ultratype.

Even if we would pay for a third ultratype, some of the authors believe that:

Conjecture 1 There is a continuum of different ultratypes of $\mathcal{L}_\infty$-spaces.

Reasonable candidates to get such a continuum could then be hereditarily indecomposable $\mathcal{L}_\infty$-spaces [12, 240], exotic preduals of ℓ_1 as in [34, 104], some Bourgain and Pisier [47] or Bourgain and Delbaen [14, 46] spaces, or some of the $\mathcal{L}_\infty$-envelopes constructed in [69].

We pass now to problems involving ultraproducts and exact sequences. Recall from [140, 207] that if $0 \to Y \to X \to Z \to 0$ is an exact sequence and $\mathcal{U}$ is an ultrafilter then $0 \to Y_\mathcal{U} \to X_\mathcal{U} \to Z_\mathcal{U} \to 0$ is also exact (see [61, Lemma 2.2.g]). No criterion however is known to determine when the ultrapower sequence of a nontrivial exact sequence is again nontrivial.

Definition 6.13 We will say that an exact sequence ultra-splits if some of its ultrapower sequences split.

From the results in Sect. 3.3.4, and more specifically Corollary 4.14, one has:

Proposition 6.14 *Any exact sequence* $0 \longrightarrow Y \longrightarrow X \longrightarrow Z \longrightarrow 0$ *in which X is a C-space and either Y or Z are spaces of almost universal disposition does not ultra-split.*

Proposition 4.30 shows that the exact sequence

$$0 \longrightarrow \ell_\infty(c_0) \longrightarrow \ell_\infty(\mathbb{N}, \mathscr{B}_{\tau(n)}) \longrightarrow \ell_\infty(C(\mathbb{N}^*)) \longrightarrow 0$$

from Example 2.24 in Sect. 2.2.6 yields a nonseparable Lindenstrauss space which is complemented in no C-space, although it does have an ultrapower isomorphic to a C-space (cf. Proposition 4.27). We do not know, however, if the sequence above ultra-splits.

Ultra-splitting problems are connected to the 3-space problem for universal separable injectivity discussed in Sect. 6.2. Indeed, if universal separable injectivity were a 3-space property then one would have $\mathrm{Ext}(C(\mathbb{N}^*), C(\mathbb{N}^*)) = 0$, which implies, under CH, that all exact sequences of the form (here Δ is the Cantor set)

$$0 \longrightarrow C(\Delta) \longrightarrow X \longrightarrow C(\Delta) \longrightarrow 0 \qquad (6.2)$$

ultra-split no matter whether they are trivial or not. This was claimed in [21, Example 4.5(a)], but we do not know if it is true or not.

Indeed, assuming CH, one has $C(\Delta)_{\mathcal{U}} \approx C(\mathbb{N}^*)$ for all free ultrafilters on the integers $\mathcal{U}$ (Proposition 4.12). Hence the ultrapower sequence of (6.2) has the form

$$0 \longrightarrow C(\mathbb{N}^*) \longrightarrow X_{\mathcal{U}} \longrightarrow C(\mathbb{N}^*) \longrightarrow 0$$

and would split if $\mathrm{Ext}(C(\mathbb{N}^*), C(\mathbb{N}^*)) = 0$ were true. This could apply to the exact sequence

$$0 \longrightarrow C[0, 1] \longrightarrow \Omega \longrightarrow C[0, 1] \longrightarrow 0$$

constructed in [57, Corollary 2.4] which has the form (6.2) since $C[0, 1] \sim C(\Delta)$ by Milutin's theorem. Thus, if the assertion $\mathrm{Ext}(C(\mathbb{N}^*), C(\mathbb{N}^*)) = 0$ were true, and under CH, the space $\Omega_{\mathcal{U}} \sim C(\mathbb{N}^*) \oplus C(\mathbb{N}^*)$ would be isomorphic to a C-space, in spite of the fact [57, Corollary 2.4] that Ω is not even isomorphic to a quotient of a Lindenstrauss space.

Some of the authors believe that the following holds:

Conjecture 2 Every exact sequence $0 \longrightarrow \mathscr{L}_\infty \longrightarrow X \longrightarrow C \longrightarrow 0$ in which $\mathscr{L}_\infty$ denotes an arbitrary $\mathscr{L}_\infty$-space and C an arbitrary C-space ultra-splits.

We conclude this section with the explicit formulation of several open ends already mentioned though the text:

- Since both $\mathscr{G}$ and C-spaces are Lindenstrauss spaces, it makes sense to ask: Does every $\mathscr{L}_\infty$-space have an ultrapower isomorphic to a Lindenstrauss space?
- Does every (infinite-dimensional, separable) Banach space X have an ultrapower isomorphic to its square? What if X is an $\mathscr{L}_\infty$-space?
- Are the classes of C_0-spaces and M-spaces closed under "isometric ultra-roots"?

- Does the Gurariy space have an ultrapower isometric (or isomorphic) to an ultraproduct of finite dimensional spaces? Since spaces of universal disposition cannot be complemented in C-spaces, if $\mathscr{G}_\mathcal{U} = (G_i)_\mathcal{V}$ then the spaces G_i cannot be "uniformly injective".
- Are Lindenstrauss ultraproducts via $\aleph$-good ultrafilters universally $\aleph$-injective spaces in ZFC? In other words, can the conclusion of Theorem 5.15 be strengthened to obtain universal injectivity? Notice that this is sensitive to axioms. The answer is of course affirmative under GCH by Theorem 5.15 and Proposition 5.13.
- Another question regarding a possible generalization of Theorem 5.15 is whether the hypothesis "Lindenstrauss" can be weakened to just "$\mathscr{L}_\infty$-space": Namely, prove or disprove that every ultraproduct built over an $\aleph$-good ultrafilter is $\aleph$-injective as long as it is an $\mathscr{L}_\infty$-space.

6.6 Spaces of Universal Disposition

In this section, if no further specification is made, universal disposition means "with respect to finite dimensional spaces". Up to now, under CH, we have encountered two non-isomorphic spaces of universal disposition: The Grothendieck space $\mathscr{S}^{\omega_1}$ and the non-Grothendieck (since it contains c_0 complemented) space $\mathscr{F}^{\omega_1}$.

Problem 7 of Sect. 6.3 can be reformulated here for spaces of universal disposition:

Problem 15 Do there exist subspaces of ℓ_∞ of universal disposition?

Observe that a space of universal disposition for separable spaces cannot be a subspace of ℓ_∞ since it must contain copies of all spaces of density character $\aleph_1$, such as $\ell_2(\aleph_1)$. It could help to decide whether $\mathscr{F}^{\omega_1}$ is a subspace of ℓ_∞ to know if it contains a subspace isomorphic to $\ell_2(\aleph_1)$.

The same question for quotients of ℓ_∞ has an affirmative answer, at least under CH: Johnson and Zippin proved in [147] that every separable Lindenstrauss space is a quotient of $C(\Delta)$, where Δ is the Cantor set. So one has an exact sequence

$$0 \longrightarrow \ker \pi \longrightarrow C(\Delta) \overset{\pi}{\longrightarrow} \mathscr{G} \longrightarrow 0. \tag{6.3}$$

No ultrapower of this sequence splits by the results in Sect. 3.3.4; see also Proposition 6.14. Under CH, the ultrapower sequence with respect to any free ultrafilter $\mathcal{U}$ on $\mathbb{N}$ has the form

$$0 \longrightarrow (\ker \pi)_\mathcal{U} \longrightarrow \ell_\infty/c_0 \longrightarrow \mathscr{G}_\mathcal{U} \longrightarrow 0$$

and thus $\mathscr{G}_\mathcal{U}$ is a quotient of ℓ_∞. Observe that $\mathscr{F}^{\omega_1}$ cannot be a quotient of ℓ_∞ since c_0 is not.

Regarding the nature of $\ker \pi$, Pełczyński posed on the blackboard to us the question of whether it is possible to identify the kernel(s) of the sequence(s) (6.3), and in particular if some such kernel can be a C-space. It is not hard to check that $\ker \pi$ is an $\mathscr{L}_\infty$-space when π is an "isometric" quotient in the sense that it maps the open unit ball of $C(\Delta)$ onto that of $\mathscr{G}$. On the other hand, it is possible to get another quotient map $\varpi : C(\Delta) \longrightarrow \mathscr{G}$ whose kernel is not an $\mathscr{L}_\infty$-space: to this end, recall that Bourgain has shown that ℓ_1 contains an uncomplemented copy of itself. An obvious localization argument yields an exact sequence $0 \longrightarrow B_* \longrightarrow c_0 \longrightarrow c_0 \longrightarrow 0$ in which $\mathscr{B}_*$ cannot be an $\mathscr{L}_\infty$-space. See [45, Appendix 1]. "Multiplying" the sequence above by any exact sequence $0 \longrightarrow Y \longrightarrow C(\Delta) \longrightarrow \mathscr{G} \longrightarrow 0$ one gets the exact sequence $0 \longrightarrow Y \oplus B_* \longrightarrow C(\Delta) \oplus c_0 \longrightarrow \mathscr{G} \oplus c_0 \longrightarrow 0$. Since both $C(\Delta)$ and $\mathscr{G}$ have complemented subspaces isomorphic to c_0, this sequence can be written as $0 \longrightarrow Y \oplus \mathscr{B}_* \longrightarrow C(\Delta) \longrightarrow \mathscr{G} \longrightarrow 0$ in which the kernel $Y \oplus \mathscr{B}_*$ is not even an $\mathscr{L}_\infty$-space.

Again, a positive measure subset of authors believes that the following problem has an affirmative answer:

Problem 16 Is there a continuum of mutually non-isomorphic spaces of universal disposition having density character $\mathfrak{c}$?

The connection between universal disposition and transitivity is not yet clearly understood. In particular is not clear if every space of universal disposition for finite dimensional spaces must be $\mathfrak{F}$-transitive or whether every space of universal disposition for separable spaces must be separably transitive. Ultrapowers are also involved into these affairs: Since it is well known that ultrapowers of almost isotropic spaces are isotropic, one is tempted to believe that the proof for the following question is at hand

Problem 17 Do ultrapowers of almost $\mathfrak{F}$-transitive spaces must be $\mathfrak{F}$-transitive?

More yet: Is every ultrapower of a space of almost universal disposition separably transitive?

In a different direction it would be interesting to know if the class of almost isotropic spaces is "axiomatizable", equivalently if every Banach space whose ultrapowers are isotropic is itself almost isotropic; see [28] for a related discussion.

6.7 Asplund Spaces

A Banach space is called an Asplund space if every separable subspace has separable dual. One of the referees of this work formulated the problem of whether a classification of Asplund separably injective spaces is possible. More precisely, he asked:

Problem 18 Is it true that every Asplund separably injective space is c_0-upper-saturated? Does there exist an Asplund separably injective space that contains an infinite dimensional reflexive subspace?

Observe that Proposition 2.10 can be translated into:

Proposition 6.15 *A separably injective space is Asplund if and only if it does not contain $C[0, 1]$.*

Bourgain [43] (see also [105]) proved that any operator $T : C[0, 1] \longrightarrow X$ that fixes a subspace of finite cotype also fixes a subspace isomorphic to $C[0, 1]$. Thus, an Asplund and separably injective space X cannot contain finite cotype (in particular, superreflexive) subspaces: the corresponding embedding would extend to $C[0, 1]$ providing a copy of $C[0, 1]$ inside X. Gasparis [102] showed a similar result for asymptotically ℓ_1 spaces, and therefore an Asplund separably injective space cannot contain asymptotically ℓ_1 spaces. Rosenthal's conjecture is that any operator $T : C[0, 1] \longrightarrow X$ that fixes an infinite dimensional subspace not containing c_0 would also fix a copy of $C[0, 1]$. If this were true, an Asplund separably injective space would be c_0-saturated. Gasparis [103] solves affirmatively Rosenthal's conjecture under the conditions that the operator is contractive and its restriction to the subspace is an isometry. One therefore has:

Proposition 6.16 *Every infinite dimensional Asplund 1-separably injective space is c_0-saturated.*

It seems very likely that infinite dimensional Asplund 1-separably injective spaces do not exist.

6.8 Grothendieck Spaces

As we have already mentioned, it is an open problem that seems to have been first posed by Lindenstrauss—see [167, 226]—to characterize Grothendieck $C(K)$ spaces in terms of topological properties of K. An obvious necessary condition is that every convergent sequence in K is eventually constant. The condition is insufficient since $C(\beta\mathbb{N} \times \beta\mathbb{N})$ contains complemented copies of c_0. There is another example due to Schlumprecht [227, 5.4] of a $C(K)$-space with the Gelfand-Phillips property (something that a Grothendieck space cannot have) without non-stationary convergent sequences in K. Koszmider remarks in [167] that the class of compact spaces where every convergent sequence is eventually constant does not admit a characterization by means of isomorphic properties of the Banach space $C(K)$. To show this, consider an example of Schachermayer in [226] of the Stone compact associated to the Boolean algebra $\mathcal{A}$ of all subsets A of $\mathbb{N}$ such that $2n \in A$ if and only if $2n + 1 \in A$ for all but finitely many $n \in \mathbb{N}$. The compact space $S(\mathcal{A})$ does not contain non-stationary convergent sequences for almost the same reason as $\beta\mathbb{N}$ does not: in fact, $S(\mathcal{A}) \setminus \mathbb{N}$ is homeomorphic to $\beta\mathbb{N} \setminus \mathbb{N}$. The space $C(S(\mathcal{A}))$ is not

Grothendieck since $T(f) = (f(2n+1)-f(2n))_n$ defines an operator $C(S(\mathcal{A})) \longrightarrow c_0$ which is onto because all the norm one finitely supported sequences are in the image of the closed unit ball. On the other hand $C(S(\mathcal{A})) \sim \ell_\infty \oplus c_0 \sim C(\beta\mathbb{N} \sqcup \alpha\mathbb{N})$, while the compact $\beta\mathbb{N} \sqcup \alpha\mathbb{N}$ has non-stationary convergent sequences.

A Banach space characterization of Lindenstrauss spaces which are Grothendieck spaces is simple: not containing complemented copies of c_0. A Banach space characterization of Grothendieck $\mathscr{L}_\infty$-spaces seems to be unknown. We conjecture

Conjecture 3 Every $\mathscr{L}_\infty$ space that contains no complemented separable subspaces is a Grothendieck space.

The next proposition is implicit in Lindenstrauss [177]:

Proposition 6.17 *An $\mathscr{L}_\infty$-space X is Grothendieck if and only if every operator $T : X \to S$ with S separable can be extended everywhere.*

Proof Let S be a separable space, let $j : X \to E$ be an embedding and let $t : X \to S$ be an operator. Since X is Grothendieck, t must be weakly compact, hence t^{**} : $X^{**} \to S$ is well defined. Since X is an $\mathscr{L}_\infty$-space, X^{**} is injective and thus it is complemented in E^{**}. Therefore t^{**} admits an extension $T : E^{**} \to S$, whose restriction to E is an extension of t. The converse is clear just embed X into some $\ell_\infty(\Gamma)$. □

We have already mentioned that Grothendieck spaces of type $\mathscr{L}_\infty$ do not necessarily contain ℓ_∞: Talagrand [239] constructed, under **CH**, a Grothendieck C-space that does not have ℓ_∞ as a quotient; while Haydon [125] obtained an independent construction, in **ZFC**, of a Grothendieck C-space that does not admit ℓ_∞ as a subspace. The density character of Grothendieck spaces was treated in Brech [48] who constructed by forcing an example of a Grothendieck space $C(\mathscr{B}r)$ of density $\aleph_1$ in a certain model in which $\mathfrak{c} = \aleph_2$. In particular, $C(\mathscr{B}r)$ is a subspace of $\ell : \infty$ and cannot contain ℓ_∞. More examples of Grothendieck spaces without copies of ℓ_∞ and additional properties have appeared after Koszmider's construction of C-spaces with few operators in [166, 167]. The example of Brech shows that the assumption of the existence of a nonreflexive Grothendieck space of density $\aleph_1$ is weaker than **CH**: there are models of **ZFC** where no nonreflexive Grothendieck space of density $\aleph_1$ exists. In particular, if $\mathfrak{s}$ denotes the smallest cardinal κ such that $[0, 1]^\kappa$ is not sequentially compact then one has:

Proposition 6.18 *Every Grothendieck space with density character strictly smaller than $\mathfrak{s}$ must be reflexive.*

Proof Assume that X is a Grothendieck space with density character smaller than $\mathfrak{s}$. Its dual unit ball in the weak* topology will be a compact having weight smaller than $\mathfrak{s}$, hence sequentially compact. By the Grothendieck character, it will also be weakly sequentially compact and X^* should be reflexive, as well as X. □

As it is well-known, $\aleph_0 < \mathfrak{s} \leq \mathfrak{c}$ and $\mathfrak{s} = \mathfrak{c}$ under Martin's axiom, and in particular $\aleph_1 < \mathfrak{s}$ is consistent. If $\mathfrak{p}$ is defined as the least cardinality of a family $\mathcal{F}$

of infinite subsets of $\mathbb{N}$ which is closed under finite intersections and such that for every infinite subset A of $\mathbb{N}$ there exists $B \in \mathcal{F}$ such that $A \setminus B$ is infinite (in other words, the smallest cardinal of a filter base in $\mathbb{P}(\mathbb{N})/$ fin whose filter is not contained in a principal filter) then consistently $\mathfrak{p} < \mathfrak{s}$ (see [86]).

The following point regarding the relation between separable injectivity and Grothendieck character remains unsolved:

Problem 19 Is every λ-separably injective space, with $\lambda < 2$, a Grothendieck space?

This has obvious connections with Problem 1. Another problem connecting Grothendieck spaces and cardinals is the following: Let $(E_i)_{i \in I}$ be a family of Banach spaces containing no complemented copy of c_0. Can $\ell_\infty(I, E_i)$ contain a complemented copy of c_0? Leung and Räbiger show in [174] that if $|I|$ is not real-valued measurable and $(E_i)_{i \in I}$ is a family of Grothendieck spaces that are Lindenstrauss spaces then $\ell_\infty(I, E_i)$ is a Grothendieck space and so it cannot contain a complemented copy of c_0. See Sect. 2.2.6.

Although the existence of real-valued measurable cardinals cannot be proved in ZFC, such cardinals, if they exist, need not to be very large: Ulam proved in [243] that if real-valued measurable cardinals do exist then the continuum is one (cf. [142, Theorem 10.1]).

Appendix A

A.1 Dunford-Pettis and Pełczyński Properties

The following properties are important in the study of the C-spaces, that is, the spaces of type $C(K)$ for some compact Hausdorff space K.

Definition A.1 A Banach space X is said to have the Dunford-Pettis property (in short, DPP) if every weakly compact operator defined on X sends weakly convergent sequences to convergent sequences.

The following result is due to Grothendieck [115]:

Proposition A.2 *C-spaces have the DPP.*

It is clear that a complemented subspace of a space with the DPP also enjoys the DPP. And that an infinite dimensional space with DPP cannot be reflexive: Otherwise, weakly convergent sequences in E must be convergent and E and this makes the unit ball of E compact. A general background about the Dunford-Pettis property can be found in [82] and in [61, Chap. 6].

Definition A.3 A Banach space X is said to have Pełczyński's property (V) if every operator defined on X is either weakly compact or an isomorphism on a copy of c_0.

Proposition A.4 *C-spaces have Pełczyński's property (V).*

Quotients of spaces enjoying Pełczyński's property (V) have the same property. Since Johnson and Zippin proved in [147] that every separable Lindenstrauss space is a quotient of $C[0, 1]$, Lindenstrauss spaces also have Pełczyński's property (V). The combination of property (V) and DPP yields.

Proposition A.5 *Every infinite dimensional complemented subspace of a C-space contains c_0.*

© Springer International Publishing Switzerland 2016
A. Avilés et al., *Separably Injective Banach Spaces*, Lecture Notes
in Mathematics 2132, DOI 10.1007/978-3-319-14741-3

Proof Let P be a projection on a space enjoying both DPP and (V) whose range, denoted by E, is infinite dimensional. Then P cannot be weakly compact unless E is reflexive, which is not since it has DPP. Thus, by property (V), P must be an isomorphism on some copy of c_0.   □

A.2 $\mathscr{L}_\infty$-Spaces

Definition A.6 A Banach space X is said to be an $\mathscr{L}_{\infty,\lambda}$-*space* $(1 \le \lambda < \infty)$ if every finite dimensional subspace of X is contained in another finite dimensional subspace of X whose Banach-Mazur distance to the corresponding space ℓ_∞^m is at most λ. The space X is said to be an $\mathscr{L}_\infty$-space if it is an $\mathscr{L}_{\infty,\lambda}$-space for some λ.

As usual, A Banach space is said to be an $\mathscr{L}_{\infty,\lambda+}$-*space* if it is an $\mathscr{L}_{\infty,\lambda'}$-space for each $\lambda' > \lambda$. The basic theory and examples of $\mathscr{L}_\infty$-spaces can be found in [181, Chap. 5]; to see more exotic examples we refer to [12, 45, 240] or [57]. The $\mathscr{L}_\infty$ spaces can be considered as the local version of C-spaces and for some time questions such as if every $\mathscr{L}_\infty$-space must be isomorphic to some C-space, or can be renormed to be an $\mathscr{L}_{\infty,1+}$ space, were open. The following facts are well known by now:

- $\mathscr{L}_{\infty,1+}$-spaces are exactly the Lindenstrauss spaces, i.e., the isometric preduals of $L_1(\mu)$-spaces [175].
- A Banach space is an $\mathscr{L}_{\infty,1}$-space if and only if it is a polyhedral Lindenstrauss space [181, p.199].
- $\mathscr{L}_\infty$-spaces have the DPP (Proposition A.14).
- There exist $\mathscr{L}_\infty$-spaces without Pełczyński's property (V). These cannot therefore be renormed to be Lindenstrauss spaces [46, 57].
- Every Banach space X can be embedded into some $\mathscr{L}_\infty$-space $\mathscr{L}_\infty(X)$ so that the quotient space $\mathscr{L}_\infty(X)/X$ has the Schur and Radon-Nikodym properties. This was proved by Bourgain and Pisier [47] for separable X and in full generality by López-Abad [183].
- The class of $\mathscr{L}_\infty$ spaces includes Schur spaces [46], ℓ_2-saturated spaces [46] (see also [106]), and hereditarily indecomposable spaces [12, 240].

One important fact for this monograph (Proposition A.13) is that a Banach space is an $\mathscr{L}_\infty$-space if and only if its bidual is an $\mathscr{L}_\infty$ space. The key to prove that is the so-called principle of local reflexivity [179, Theorem 3.1]:

Theorem A.7 (Principle of Local Reflexivity) *Let X be a Banach space, F a finite dimensional subspace of X^{**}, G a finite dimensional subspace of X^* and $\varepsilon > 0$. Then there is an operator $T : F \to X$ so that:*

- $\|T\| \|T^{-1}\| \le 1 + \varepsilon$,
- $Tx = x$ *for every $x \in F \cap X$, and*
- $x^{**}(x^*) = x^*(Tx^{**})$ *for every $x^* \in G$ and every $x^{**} \in F$.*

A consequence of the principle of local reflexivity is:

Proposition A.8 *The second dual X^{**} of a Banach space X embeds isometrically as a 1-complemented subspace of some ultrapower of X.*

Proof Let $\mathfrak{F}(X^{**})$ and $\mathfrak{F}(X^*)$ denote the set of all finite dimensional subspaces of X^{**} and X^* and consider the order in $\mathfrak{F}(X^{**}) \times \mathfrak{F}(X^*) \times (0, \infty)$ given by $(F, G, \varepsilon) \leq (F', G', \varepsilon')$ if $F \subset F', G \subset G'$ and $\varepsilon' \leq \varepsilon$. Let $\mathcal{U}$ be a free ultrafilter refining the order filter on $\mathfrak{F}(X^{**}) \times \mathfrak{F}(X^*) \times (0, \infty)$. Given $F \in \mathfrak{F}(X^{**}), G \in \mathfrak{F}(X^*)$ and $\varepsilon > 0$ we consider the operator $T_{(F,G,\varepsilon)} : F \to X$ provided by the principle of local reflexivity. We define a map $\Delta : X^{**} \to X_{\mathcal{U}}$ by letting $\Delta(x^{**}) = [(x_{(F,G,\varepsilon)})]$, where

$$x_{(F,G,\varepsilon)} = \begin{cases} T_{(F,G,\varepsilon)}(x^{**}) & \text{if } x^{**} \in F \\ 0 & \text{otherwise.} \end{cases} \tag{A.1}$$

Clearly, Δ is a linear isometry of X^{**} into $X_{\mathcal{U}}$. Note that $x_{(F,G,\varepsilon)}$ and $T_{(F,G,\varepsilon)}(x^{**})$ agree "eventually" and so the linearity of Δ is not a problem due to our choice of $\mathcal{U}$.

To complete the proof consider the operator $\Pi : X_{\mathcal{U}} \to X^{**}$ given by

$$\Pi[(x_{(F,G,\varepsilon)})] = \text{weak*} - \lim_{\mathcal{U}} x_{(F,G,\varepsilon)}.$$

Clearly, Π is a well-defined, contractive operator. Besides, $\Pi \circ \Delta = 1_{X^{**}}$ since, if $x_{(F,G,\varepsilon)}$ are as in (A.1), then one has $x_{(F,G,\varepsilon)} \to x^{**}$ weakly* along $\mathcal{U}$ because each $x^* \in X^*$ falls eventually in G. □

Kalton and Fakhouri [93, 154] "generalized" the situation above as follows:

Definition A.9 A closed subspace E of X is said to be *locally complemented* if there is $\lambda > 0$ so that for every finite-dimensional subspace $F \subset X$ there is an operator $T : F \to E$ so that $\|T\| \leq \lambda$ and $T(x) = x$ for every $x \in F \cap E$.

We refer to [111, Sect. 4] for ultrapower characterizations of locally complemented subspaces and related concepts. With such notion one has:

Proposition A.10 *An $\mathscr{L}_\infty$-space is locally complemented in every superspace.*

Proof Let E be an $\mathscr{L}_{\infty,\lambda}$-space embedded into a superspace X. For a given finite dimensional subspace $F \subset X$, consider the inclusion map $F \cap E \to X$. Pick a subspace $G \subset E$ containing $F \cap E$ and λ-isomorphic to some ℓ_∞^m. Then the inclusion $F \cap E \to G$ can be extended to an operator $T : X \to G$, with $\|T\| \leq \lambda$, whose restriction to F is the operator we are looking for. □

The general version of the argument above is:

Lemma A.11 *E is locally complemented subspace of X if and only if $E^{\perp\perp} = E^{**}$ is complemented in X^{**}.*

Proof Suppose E is λ-locally complemented in X. For each $F \in \mathfrak{F}(X)$ we consider an operator $T_F : F \to E$ as in the definition of local complementation, that is, with

$T_F(x) = x$ for $x \in F \cap E$ and $\|T_F\| \le \lambda$. Let $\mathscr{U}$ be a free ultrafilter on $\mathfrak{F}(X)$ refining the order filter. We define an operator $T : X \to E^{**}$ as

$$T(x^{**}) = \text{weak}^*\text{-}\lim_{\mathscr{U}} T_F(x^{**}).$$

Since $T(x) = x$ for all $x \in E$, the canonical embedding $\delta : E \longrightarrow E^{**}$ factorizes as $\delta = Tj$ through X. Observe the diagram

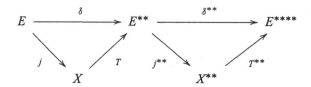

If $\pi : E^{****} \longrightarrow E^{**}$ denotes a projection through δ^{**} then $\pi T^{**} : X^{**} \longrightarrow E^{**}$ is a projection through j^{**}, so E^{**} is complemented in X^{**}.

The converse, thanks to the principle of local reflexivity, is clear: E is locally complemented in E^{**}, which is complemented in X^{**}, so E is locally complemented in X^{**}, hence in X.   □

Lemma A.12 *A locally complemented subspace of an $\mathscr{L}_\infty$-space is an $\mathscr{L}_\infty$-space.*

Proof By Proposition A.10 every $\mathscr{L}_\infty$-space is a locally complemented subspace of a C-space. Since "to be a locally complemented subspace of" is a transitive relation it suffices to prove that each locally complemented subspace of a C-space is an $\mathscr{L}_\infty$-space.

On the other hand the bidual of any C-space is again a C-space since it is isometrically isomorphic to some $L_\infty(\mu)$. Hence, if E is locally complemented in a C-space, then E^{**} is complemented in a C-space. Thus, if we show that a complemented subspace of a C-space is an $\mathscr{L}_\infty$-space, we are done, because an immediate consequence of the principle of local reflexivity is that if E^{**} is an $\mathscr{L}_{\infty,\lambda}$-space then E is an $\mathscr{L}_{\infty,\lambda+}$-space.

To prove that a complemented subspace of a C-space is an $\mathscr{L}_\infty$-space we present a streamlined version of [179, Theorem 3.2]. So, assume E is an infinite dimensional subspace of $C(K)$ and let Q be a projection of $C(K)$ with range E. Let F be a finite dimensional subspace of E and let us fix $\varepsilon \in (0, 1)0$. Then F is contained in a finite dimensional subspace G of $C(K)$ which is $(1 + \varepsilon)$-isomorphic to ℓ_∞^m. Let E_0 be a complement of $Q[G]$ in E, that is, $E = Q[G] + E_0$ and $Q[G] \cap E_0 = 0$. As E_0 is complemented in E it is also complemented in the space $C(K)$.

By Proposition A.5, every (infinite dimensional) complemented subspace of a C-space contains c_0 and every Banach space that contains c_0 contains almost-isometric copies of c_0. Thus, E_0 contains almost isometric copies of c_0. Pick then a normalized $(1+\varepsilon)$-isomorphic embedding $\tau : \ell_\infty^m \to E_0$. Also, let $u : \ell_\infty^m \to G$ be a normalized surjective $(1 + \varepsilon)$-isomorphism and extend $u^{-1} : G \to \ell_\infty^m$ to an operator $P : C(K) \to \ell_\infty^m$ with $\|P\| = \|u^{-1}\| \le 1 + \varepsilon$. Of course P can be seen as a "projection" since $P \circ u = 1_{\ell_\infty^m}$. Finally, consider the operator $T : \ell_\infty^m \to E$ defined by

$$T = Q \circ u + \tau(1_{\ell_\infty^m} - P \circ Q \circ u).$$

We conclude the proof by showing:

- The range of T contains F: Pick $f \in F$ and take $g = u^{-1}(f) \in \ell_\infty^m$. Then $P(Q(ug)) = P(f) = g$ and

$$T(g) = Q(f) + \tau(g - PQ(ug)) = f + \tau(g - P(f)) = f.$$

- T is an isomorphism and $\|T\|\|T^{-1}\|$ depends only on $\|Q\|$. Note that $\|T\| \leq 1 + 3\|Q\|$. To obtain a lower bound for the action of T observe that the factor $\tau[\ell_\infty^m]$ is $(1 + \varepsilon)$-injective and so it is complemented in $Q[\ell_\infty^m] + \tau[\ell_\infty^m]$ by a projection of norm at most $1 + \varepsilon \leq 2$. It follows that

$$\max(\|f\|, \|g\|) \leq 3\|f + g\|$$

for every $f \in Q[\ell_\infty^m]$ and every $g \in \tau[\ell_\infty^m]$. Hence, for $x \in \ell_\infty^m$, we have

$$\|Tx\| = \|Q(u(x)) + \tau(x - PQ(ux))\|$$
$$\geq \tfrac{1}{3} \max \left(\|Q(u(x))\|, \|\tau(x - PQ(ux))\| \right)$$
$$\geq \tfrac{1}{3} \max \left(\|Q(u(x))\|, (1 + \varepsilon)^{-1}\|x - PQ(ux)\| \right).$$

Now, if $\|Q(ux)\| \geq \frac{1}{2(1+\varepsilon)}\|ux\|$, then

$$\|Tx\| \geq \frac{\|u(x)\|}{3 \cdot 2 \cdot (1 + \varepsilon)} \geq \frac{\|x\|}{3 \cdot 2 \cdot (1 + \varepsilon)^2} \geq \frac{\|x\|}{24}.$$

If, however, $\|Q(ux)\| \leq \frac{1}{2(1+\varepsilon)}\|ux\|$ then, choosing the second quantity in the maximum,

$$\|Tx\| \geq \frac{\|x - PQ(ux)\|}{3(1 + \varepsilon)} \geq \frac{\|x\| - \|PQ(ux)\|}{3(1 + \varepsilon)}$$
$$\geq \frac{\|x\| - \|P\|\|Q(ux)\|}{3(1 + \varepsilon)} \geq \frac{\|x\|}{3 \cdot 2 \cdot (1 + \varepsilon)}.$$

Thus, $\|Tx\| \geq \frac{1}{24}\|x\|$ in any case. $\square$

Therefore

Proposition A.13 *The space E is an $\mathscr{L}_\infty$-space if and only if E^{**} is an $\mathscr{L}_\infty$-space.*

Proof One assertion is consequence of the principle of local reflexivity; as for the other, if E is an $\mathscr{L}_\infty$-space, it is locally complemented in some $C(K)$-space, so E^{**} is complemented in $C(K)^{**}$, and thus it is an $\mathscr{L}_\infty$-space.   □

Proposition A.14 *$\mathscr{L}_\infty$-spaces have the DPP. In particular, no infinite dimensional $\mathscr{L}_\infty$-space is reflexive.*

Proof From Lemma A.11—and the well-known fact that weakly compact operators extend to the bidual—it follows that a locally complemented subspace of a space with DPP also enjoys DPP. From Propositions A.10 we get that every $\mathscr{L}_\infty$ space is locally complemented in some $\ell_\infty(I)$-space, and from Proposition A.2 that this last space has the DPP.   □

A.3 $\mathscr{L}_1$-Spaces

The $\mathscr{L}_1$-spaces are the local version of the $L_1(\mu)$-spaces. Precisely

Definition A.15 A Banach space X is said to be an $\mathscr{L}_{1,\lambda}$-*space* $(1 \leq \lambda < \infty)$ if every finite dimensional subspace of X is contained in another finite dimensional subspace of X whose Banach-Mazur distance to the corresponding space ℓ_1^m is at most λ. The space X is said to be an $\mathscr{L}_1$-*space* if it is an $\mathscr{L}_{1,\lambda}$-space for some λ.

The basic theory and examples of $\mathscr{L}_1$-spaces can be found in [181, Chap. 5], and for some exotic examples we refer to [45]. We are interested in $\mathscr{L}_1$-spaces because of their duality relations with $\mathscr{L}_\infty$-spaces.

Proposition A.16 *A Banach space is an $\mathscr{L}_1$-space (resp. $\mathscr{L}_\infty$-space) if and only if its dual is an $\mathscr{L}_\infty$-space (resp. $\mathscr{L}_1$-space).*

A proof can be found in [181, Theorem II.5.7]. Since dual $\mathscr{L}_\infty$-spaces are injective (Proposition 1.9) one has:

Corollary A.17 *A Banach space is an $\mathscr{L}_1$-space if and only if its dual is injective.*

There are essential differences in the behaviour of $\mathscr{L}_1$ and $\mathscr{L}_\infty$ spaces. For instance, every infinite-dimensional $\mathscr{L}_1$-space contains a complemented copy of ℓ_1 [181, Theorem II.5.7], a result for which there is no counter-part for $\mathscr{L}_\infty$-spaces. The most striking one is the following fundamental result of Lindenstrauss [176] (see also [158]) for $\mathscr{L}_1$-spaces that has no counter-part for $\mathscr{L}_\infty$-spaces.

Proposition A.18 (Lindenstrauss Lifting) *Let $\mathscr{L}_1$ denote an arbitrary $\mathscr{L}_1$-space and let Y be a space complemented in its bidual. Then $\mathrm{Ext}(\mathscr{L}_1, Y) = 0$, that is, every exact sequence*

$$0 \longrightarrow Y \stackrel{\imath}{\longrightarrow} X \stackrel{\pi}{\longrightarrow} \mathscr{L}_1 \longrightarrow 0$$

splits.

Proof Since $\mathscr{L}_1^*$ is injective (Corollary A.17) the dual sequence

$$0 \longrightarrow \mathscr{L}_1^* \xrightarrow{\;\pi^*\;} X^* \xrightarrow{\;\imath^*\;} Y^* \longrightarrow 0$$

splits and so there is an operator $\kappa : Y^* \to X^*$ such that $\imath^* \circ \kappa = 1_{Y^*}$. If P is any projection of Y^{**} onto Y, then the restriction of $\kappa^* \circ P : X^{**} \to Y^{**} \to Y$ to X provides a left-inverse for $\imath$.

A.4 Homological Tools

We describe now some basic homological constructions.

A.4.1 The Push-Out Construction

In this section we describe the push-out construction for Banach spaces. This construction appears naturally when one considers a couple of operators defined on the same space, in particular in any extension problem. Let us explain why. Given operators $\alpha : Y \to A$ and $\beta : Y \to B$, the associated push-out diagram is

$$\begin{array}{ccc} Y & \xrightarrow{\;\alpha\;} & A \\ {\scriptstyle\beta}\downarrow & & \downarrow{\scriptstyle\beta'} \\ B & \xrightarrow{\;\alpha'\;} & \mathrm{PO} \end{array} \qquad\qquad (A.2)$$

The push-out space $\mathrm{PO} = \mathrm{PO}(\alpha, \beta)$ is quotient of the direct sum space $A \oplus_1 B$ by the closure of the subspace $\Delta = \{(\alpha y, -\beta y) : y \in Y\}$. The map α' is given by the inclusion of B into $A \oplus_1 B$ followed by the natural quotient map $A \oplus_1 B \to (A \oplus_1 B)/\overline{\Delta}$, so that $\alpha'(b) = (0, b) + \overline{\Delta}$ and, analogously, $\beta'(a) = (a, 0) + \overline{\Delta}$.

The preceding diagram is commutative: $\beta'\alpha = \alpha'\beta$. Moreover, it is "minimal" in the sense of having the following universal property: if $\beta'' : A \to C$ and $\alpha'' : B \to C$ are operators such that $\beta''\alpha = \alpha''\beta$, then there is a unique operator $\gamma : \mathrm{PO} \to C$ such that $\alpha'' = \gamma\alpha'$ and $\beta'' = \gamma\beta'$ as illustrated in the following diagram

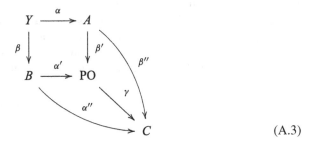

$$(A.3)$$

Clearly, $\gamma((a, b) + \overline{\Delta}) = \beta''(a) + \alpha''(b)$ and one has $\|\gamma\| \le \max\{\|\alpha''\|, \|\beta''\|\}$.

Regarding the behaviour of the maps in Diagram (A.2), one has:

Lemma A.19

1. $\max\{\|\alpha'\|, \|\beta'\|\} \leq 1$.
2. If α is an isomorphic embedding, then Δ is closed.
3. If α is an isometric embedding and $\|\beta\| \leq 1$ then α' is an isometric embedding.
4. If α is an isomorphic embedding then α' is an isomorphic embedding.
5. If $\|\beta\| \leq 1$ and α is an isomorphism then α' is an isomorphism and

$$\|(\alpha')^{-1}\| \leq \max\{1, \|\alpha^{-1}\|\}.$$

Proof (1) and (2) are clear. To prove (3), recall that Δ is closed. Now, if $\|\beta\| \leq 1$,

$$\|\alpha'(b)\| = \|(0, b) + \Delta\| = \inf_{y \in Y}\{\|\alpha y\| + \|b - \beta y\|\} \geq \inf_{y \in Y}\{\|\beta y\| + \|b - \beta y\|\} \geq \|b\|,$$

as required. (4) is clear after (3). To prove the assertion about $(\alpha')^{-1}$ in (5), notice that for all $a \in A$ and $b \in B$ one has $(a, b) + \Delta = (0, b + \beta y) + \Delta$ for any $y \in Y$ such that $\alpha y = a$. Therefore, for all $y' \in Y$ one has

$$\|b + \beta y\| \leq \|b + \beta y + \beta y'\| + \|\beta y'\|$$
$$\leq |b + \beta y + \beta y'\| + \|y'\|$$
$$\leq \|b + \beta y + \beta y'\| + \|\alpha^{-1}\|\|\alpha y'\|$$

from where the assertion follows.   □

Isomorphic vs. Isometric Push-Out Diagram We will say that a commutative diagram

$$
\begin{array}{ccc}
Y & \xrightarrow{\alpha} & A \\
\beta \downarrow & & \downarrow b \\
B & \xrightarrow{a} & X
\end{array}
\tag{A.4}
$$

is a push-out square (or an *isomorphic* push-out square if specification is needed) if it has the universal property reflected in Diagram (A.3), namely that for any couple of operators $\alpha'' : B \to C$ and $\beta'' : A \to C$ satisfying $\beta'' \circ \alpha = \alpha'' \circ \beta$ there exists a unique operator $\gamma : X \to C$ such that $\beta'' = \gamma \circ b$ and $\alpha'' = \gamma \circ a$.

If the maps in (A.4) are isometries and $\|\gamma\| \leq \max\{\|\alpha''\|, \|\beta''\|\}$ for every α'', β'', then we say that (A.4) is an *isometric* push-out square. In this case there is a surjective isometry $\gamma : \mathrm{PO} \to X$ making commutative the diagram

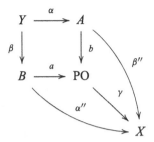

A.4.2 Further Properties of Push-Out Squares

We collect here several properties of push-out squares that are needed in the monograph. The proofs are immediate from the universal property of the push-out construction.

Push-Out Made with Inclusions Let A, B be two subspaces of a Banach space E. The well-known isomorphism between the spaces $\overline{A + B}/A$ and $B/(A \cap B)$ can be reinterpreted by saying that the diagram

$$
\begin{array}{ccc}
A \cap B & \longrightarrow & B \\
\downarrow & & \downarrow \\
A & \longrightarrow & \overline{A + B}
\end{array}
$$

is an isomorphic push-out square. Indeed, assume one has operators $\alpha : A \to X$ and $\beta : B \to X$ agreeing on $A \cap B$. Then the operator $\gamma : A + B \to X$ given by $\gamma(a + b) = \alpha(a) + \beta(b)$ is well defined, agrees with α on A and agrees with β on B.

Concatenation of Squares Consider a commutative diagram

$$
\begin{array}{ccccc}
A & \xrightarrow{\;a\;} & B & \xrightarrow{\;b\;} & C \\
\downarrow & & \downarrow & & \downarrow \\
D & \xrightarrow{\;d\;} & E & \xrightarrow{\;e\;} & F
\end{array}
\tag{A.5}
$$

If the left and right squares are isomorphic (resp. isometric) push-out squares then so is their "concatenation"

$$
\begin{array}{ccc}
A & \xrightarrow{\;ba\;} & E \\
\downarrow & & \downarrow \\
C & \xrightarrow{\;ed\;} & F
\end{array}
$$

As a partial converse, if the preceding diagram is a push-out square, then so is the right square in Diagram (A.5).

Limits Given a commutative diagram made with push-out squares

$$
\begin{array}{ccccccc}
A & \longrightarrow & A_1 & \longrightarrow & A_2 & \longrightarrow & \cdots \\
\downarrow & & \downarrow & & \downarrow & & \\
B & \longrightarrow & \mathrm{PO}_1 & \longrightarrow & \mathrm{PO}_2 & \longrightarrow & \cdots
\end{array}
$$

in which the inductive limits $\varinjlim A_n$ and $\varinjlim \mathrm{PO}_n$ exist (which is always the case if one considers isometric push-out squares) then also

$$
\begin{array}{ccc}
A & \longrightarrow & \varinjlim A_n \\
\downarrow & & \downarrow \\
B & \longrightarrow & \varinjlim \mathrm{PO}_n
\end{array}
$$

is a push-out square.

A.4.3 The Pull-Back Construction

The pull-back construction is the dual of that of push-out in the sense of categories, that is, "reversing arrows". Indeed, let $\alpha : A \to Z$ and $\beta : B \to Z$ be operators acting between Banach spaces. The associated pull-back diagram is

$$
\begin{array}{ccc}
\mathrm{PB} & \xrightarrow{\;'\beta\;} & A \\
{\scriptstyle '\alpha}\downarrow & & \downarrow{\scriptstyle \alpha} \\
B & \xrightarrow{\;\beta\;} & Z
\end{array}
\qquad\qquad (A.6)
$$

The pull-back space is $\mathrm{PB} = \mathrm{PB}(\alpha, \beta) = \{(a, b) \in A \oplus_\infty B : \alpha(a) = \beta(b)\}$. The arrows after primes are the restriction of the projections onto the corresponding factor. Needless to say (A.6) is minimally commutative in the sense that if the operators $''\beta : C \to A$ and $''\alpha : C \to B$ satisfy $\alpha \circ ''\beta = \beta \circ ''\alpha$, then there is a unique operator $\gamma : C \to \mathrm{PB}$ such that $''\beta = '\beta\gamma$ and $''\beta = '\beta\gamma$, that is, the following diagram is commutative:

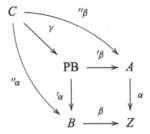

Clearly, $\gamma(c) = ({''\beta}(c), {''\alpha}(c))$ and $\|\gamma\| \leq \max\{\|{''\alpha}\|, \|{''\beta}\|\}$. Also, the map ${'\alpha}$ is onto if α is.

A.4.4 Short Exact Sequences

A short exact sequence of Banach spaces is a diagram

$$0 \xrightarrow{} Y \xrightarrow{j} X \xrightarrow{q} Z \xrightarrow{} 0 \tag{A.7}$$

where Y, X and Z are Banach spaces and the arrows are operators in such a way that the kernel of each arrow coincides with the image of the preceding one. By the open mapping theorem j embeds Y as a closed subspace of X and Z is isomorphic to the quotient $X/j[Y]$. The middle space X in an exact sequence (A.7) is called a twisted sum of Y an Z.

We say that an exact sequence $0 \to Y \to X_1 \to Z \to 0$ is equivalent to (A.7) if there exists an operator $T : X \to X_1$ making commutative the diagram

$$
\begin{array}{ccccccccc}
0 & \longrightarrow & Y & \longrightarrow & X & \longrightarrow & Z & \longrightarrow & 0 \\
  &   & \| &   & \downarrow{\scriptstyle T} & & \| & & \\
0 & \longrightarrow & Y & \longrightarrow & X_1 & \longrightarrow & Z & \longrightarrow & 0
\end{array}
\tag{A.8}
$$

The operator T above must actually be an isomorphism in view of the classical "three-lemma" asserting that in a commutative diagram of vector spaces and linear maps

$$
\begin{array}{ccccccccc}
0 & \longrightarrow & A & \longrightarrow & B & \longrightarrow & C & \longrightarrow & 0 \\
  &   & \downarrow{\scriptstyle u} & & \downarrow{\scriptstyle v} & & \downarrow{\scriptstyle w} & & \\
0 & \longrightarrow & A_1 & \longrightarrow & B_1 & \longrightarrow & C_1 & \longrightarrow & 0
\end{array}
$$

having exact rows, if u and w are surjective (resp. injective) then so is v. Hence, the operator T in Diagram (A.8) is a bijection and therefore it is a linear homeomorphism, according to the open mapping theorem.

The exact sequence (A.7) is said to be *trivial*, or that *splits* if there is an operator $p : X \to Y$ such that $pj = 1_Y$ (i.e., $j(Y)$ is complemented in X); equivalently, there is an operator $s : Z \to X$ such that $qs = 1_Z$. Of course that an exact sequence (A.7) splits if and only if it is equivalent to the direct sum sequence $0 \to Y \to Y \oplus Z \to Z \to 0$.

For every pair of Banach spaces Z and Y, we denote by $\mathrm{Ext}(Z, Y)$ the space of all exact sequences $0 \to Y \to X \to Z \to 0$, modulo equivalence. We write $\mathrm{Ext}(Z, Y) = 0$ to mean that every sequence of this form is trivial. The reason for this notation is that $\mathrm{Ext}(Z, Y)$ has a natural linear structure [55, 61] for which the (class of the) trivial exact sequence is the zero element.

Two exact sequences $0 \to Y \to X \to Z \to 0$ and $0 \to Y_1 \to X_1 \to Z_1 \to 0$ are said to be *isomorphically equivalent* if there exist isomorphisms α, β, γ making a commutative diagram

$$
\begin{array}{ccccccccc}
0 & \longrightarrow & Y & \longrightarrow & X & \longrightarrow & Z & \longrightarrow & 0 \\
 & & \downarrow{\scriptstyle \alpha} & & \downarrow{\scriptstyle \beta} & & \downarrow{\scriptstyle \gamma} & & \\
0 & \longrightarrow & Y_1 & \longrightarrow & X_1 & \longrightarrow & Z_1 & \longrightarrow & 0
\end{array}
$$

A.4.5   Push-Out and Pull-Back Exact Sequences

Suppose we are given an exact sequence (A.7) and an operator $t : Y \to B$. Consider the push-out of the couple (t, j) and draw the corresponding arrows:

$$
\begin{array}{ccccccc}
0 & \longrightarrow & Y & \overset{j}{\longrightarrow} & X & \longrightarrow & Z & \longrightarrow & 0 \\
 & & {\scriptstyle t}\downarrow & & \downarrow{\scriptstyle t'} & & & \\
 & & B & \overset{j'}{\longrightarrow} & \mathrm{PO} & & &
\end{array}
$$

By Lemma A.19(4), j' is an isomorphic embedding. Now, the operator $q : X \to Z$ and the null operator $0 : B \to Z$ satisfy the identity $qj = 0t = 0$; thus, the universal property of the push-out gives a unique operator $\varpi : \mathrm{PO} \to Z$ making the following diagram commutative:

$$
\begin{array}{ccccccccc}
0 & \longrightarrow & Y & \overset{j}{\longrightarrow} & X & \overset{q}{\longrightarrow} & Z & \longrightarrow & 0 \\
 & & {\scriptstyle t}\downarrow & & \downarrow{\scriptstyle t'} & & \| & & \\
0 & \longrightarrow & B & \overset{j'}{\longrightarrow} & \mathrm{PO} & \overset{\varpi}{\longrightarrow} & Z & \longrightarrow & 0
\end{array}
\qquad (A.9)
$$

An explicit definition for this operator is: $\varpi((x, b) + \Delta) = q(x)$. Elementary considerations show that the lower sequence in the preceding Diagram (A.9) is exact, and thus it will be referred to as "the push-out sequence". In fact, the universal property of the push-out makes this diagram unique, in the sense that for any other commutative diagram of exact sequences

$$
\begin{array}{ccccccccc}
0 & \longrightarrow & Y & \xrightarrow{\;j\;} & X & \xrightarrow{\;q\;} & Z & \longrightarrow & 0 \\
 & & \downarrow{\scriptstyle t} & & \downarrow & & \| & & \\
0 & \longrightarrow & B & \longrightarrow & X' & \longrightarrow & Z & \longrightarrow & 0
\end{array}
\qquad \text{(A.10)}
$$

the lower exact sequence turns out to be equivalent to the push-out sequence in (A.9). For this reason we usually refer to a diagram like (A.10) as a push-out diagram. One has:

Lemma A.20 *In a push-out diagram (A.10) the push-out sequence splits if and only if t extends to X in the sense that there is an operator $T : X \to B$ such that $t = T \circ j$.*

Proof The universal property of the push-out construction. □

Proceeding dually one obtains the pull-back sequence. Consider again (A.7) and an operator $u : A \to Z$. Let us form the pull-back diagram of the couple (q, u) to get a diagram

$$
\begin{array}{ccccccccc}
0 & \longrightarrow & Y & \xrightarrow{\;j\;} & X & \xrightarrow{\;q\;} & Z & \longrightarrow & 0 \\
 & & & & \uparrow{\scriptstyle 'u} & & \uparrow{\scriptstyle u} & & \\
 & & & & \text{PB} & \xrightarrow{\;'q\;} & A & &
\end{array}
$$

Recalling that $'q$ is onto and taking $i(y) = (0, j(y))$, it is easily seen that the following diagram is commutative:

$$
\begin{array}{ccccccccc}
0 & \longrightarrow & Y & \xrightarrow{\;j\;} & X & \xrightarrow{\;q\;} & Z & \longrightarrow & 0 \\
 & & \| & & \uparrow{\scriptstyle 'u} & & \uparrow{\scriptstyle u} & & \\
0 & \longrightarrow & Y & \xrightarrow{\;i\;} & \text{PB} & \xrightarrow{\;'q\;} & A & \longrightarrow & 0
\end{array}
\qquad \text{(A.11)}
$$

The lower sequence is exact, and we shall refer to it as the pull-back sequence. Given a commutative diagram

$$
\begin{array}{ccccccccc}
0 & \longrightarrow & Y & \xrightarrow{\;j\;} & X & \xrightarrow{\;q\;} & Z & \longrightarrow & 0 \\
 & & \| & & \uparrow & & \uparrow{\scriptstyle u} & & \\
0 & \longrightarrow & Y & \longrightarrow & X' & \longrightarrow & A & \longrightarrow & 0
\end{array}
\qquad \text{(A.12)}
$$

the lower exact sequence is equivalent to the pull-back sequence in (A.11). For this reason any diagram like (A.12) is called a pull-back diagram. The splitting criterion is now as follows.

Lemma A.21 *In a pull-back diagram (A.12) the pull-back sequence splits if and only if u lifts to X in the sense that there is an operator $L : A \to X$ such that $u = q \circ L$.*

Proof The universal property of the pull-back construction.   □

A.4.6 Commutativity of Pull-Back and Push-Out Operations

Let us show that the pull-back and push-out processes are commutative. Making pull-back first and then push-out yields

$$
\begin{array}{ccccccccc}
0 & \longrightarrow & Y & \xrightarrow{\ j\ } & X & \xrightarrow{\ q\ } & Z & \longrightarrow & 0 \\
 & & \| & & \uparrow{\bar{\gamma}} & & \uparrow{\gamma} & & \\
0 & \longrightarrow & Y & \xrightarrow{\ j_B\ } & PB & \xrightarrow{\ q_B\ } & Z_1 & \longrightarrow & 0 \\
 & & {\scriptstyle\alpha}\downarrow & & \downarrow{\scriptstyle\alpha} & & \| & & \\
0 & \longrightarrow & Y_1 & \xrightarrow{\ j_{OB}\ } & PO(PB) & \xrightarrow{\ q_{OB}\ } & Z_1 & \longrightarrow & 0
\end{array}
$$

While making push-out first and then pull-back yields the commutative diagram

$$
\begin{array}{ccccccccc}
0 & \longrightarrow & Y & \xrightarrow{\ j\ } & X & \xrightarrow{\ q\ } & Z & \longrightarrow & 0 \\
 & & {\scriptstyle\alpha}\downarrow & & {\scriptstyle\bar{\alpha}}\downarrow & & \| & & \\
0 & \longrightarrow & Y_1 & \xrightarrow{\ j_O\ } & PO & \xrightarrow{\ q_O\ } & Z & \longrightarrow & 0 \\
 & & \| & & \uparrow{\underline{\gamma}} & & \uparrow{\gamma} & & \\
0 & \longrightarrow & Y_1 & \xrightarrow{\ j_{BO}\ } & PB(PO) & \xrightarrow{\ q_{BO}\ } & Z_1 & \longrightarrow & 0
\end{array}
$$

Let us show that the final resulting sequences are equivalent; i.e., that there is an operator $T : PO(PB) \longrightarrow PB(PO)$ making commutative the diagram

$$
\begin{array}{ccccccccc}
0 & \longrightarrow & Y_1 & \xrightarrow{\ j_{OB}\ } & PO(PB) & \xrightarrow{\ q_{OB}\ } & Z_1 & \longrightarrow & 0 \\
 & & \| & & \downarrow{T} & & \| & & \\
0 & \longrightarrow & Y_1 & \xrightarrow{\ j_{BO}\ } & PB(PO) & \xrightarrow{\ q_{BO}\ } & Z_1 & \longrightarrow & 0
\end{array}
\tag{A.13}
$$

There are two ways to get such map: one is relying on the fact that PO(PB) is a push-out and the other relying on the fact that PB(PO) is a pull-back. Let us make this last case. Thus, consider the pull-back square

$$
\begin{array}{ccc}
PO & \xrightarrow{\ q_0\ } & Z \\[4pt]
\underline{\gamma}\big\uparrow & & \big\uparrow\gamma \\[4pt]
PB(PO) & \xrightarrow[\ q_{BO}\ ]{} & Z_1
\end{array}
\qquad\qquad (A.14)
$$

Let us form another commutative square:

$$
\begin{array}{ccc}
PO & \xrightarrow{\ q_0\ } & Z \\[4pt]
\delta\big\uparrow & & \big\uparrow\gamma \\[4pt]
PO(PB) & \xrightarrow[\ q_{OB}\ ]{} & Z_1
\end{array}
\qquad\qquad (A.15)
$$

in which the arrow δ is obtained from the push-out square

$$
\begin{array}{ccc}
Y & \xrightarrow{\ j_B\ } & PB \\[4pt]
\alpha\big\downarrow & & \big\downarrow\alpha \\[4pt]
Y_1 & \xrightarrow[\ j_{OB}\ ]{} & PO(PB)
\end{array}
$$

in combination with the fact that the square

$$
\begin{array}{ccc}
Y & \xrightarrow{\ j_B\ } & PB \\[4pt]
\alpha\big\downarrow & & \big\downarrow\overline{\alpha\gamma} \\[4pt]
Y_1 & \xrightarrow[\ j_0\ ]{} & PO
\end{array}
$$

is also commutative: $\overline{\alpha\gamma}j_B = \overline{\alpha}j = j_0\alpha$. Thus, there is a unique operator δ : PO(PB) $\to$ PO such that $\delta\alpha = \overline{\alpha\gamma}$ and $j_{OB} = j_0$.

A combination of diagrams (A.14) and (A.15) immediately yields the existence of an operator T : PO(PB) $\to$ PB(PO) such that $q_{BO}T = q_{OB}$ and $\gamma T = \delta$. The first of those equalities, $q_{BO}T = q_{OB}$ is the commutativity of the right square in diagram (A.13). Let us prove the commutativity of the left square; i.e., $Tj_{OB} = j_{BO}$: Since $q_{BO}T = q_{OB}$ it is clear that $q_{BO}Tj_{OB} = q_{OB}j_{OB} = 0$ and therefore some operator u : $Y_1 \to Y_1$ must exist so that $Tj_{OB} = j_{BO}u$. But since $j_0 u = \gamma j_{BO}u = \gamma Tj_{OB} = \delta j_{OB} = j_0$ it follows that $u = 1_{Y_1}$, and this concludes the proof.

A.4.7   The Diagonal Principle

Diagonal principles were introduced and studied in [63]. They are useful tools when dealing with problems involving the extension/lifting of embeddings. A homological proofs for these principles can be seen in [63], while an elementary proof for the diagonal principle, extension case, can be traced back to Klee [165]; see also [37]. We present here a straightforward proof for the extension case.

Proposition A.22 *Let $\imath : Y \to X$ and $j : Y \to X'$ be into isomorphisms between Banach spaces. Then there exist operators $I : X' \to X$ and $J : X \to X'$ such that $Ij = \imath$ and $J\imath = j$ if and only if there exists an automorphism $\tau : X \oplus X' \to X \oplus X'$ such that $\tau(\imath y, 0) = (0, jy)$.*

Proof The maps $\tau_j : X \oplus X' \to X \oplus X'$ given by $\tau_j(x, x') = (x, x' + J(x))$ and $\tau_\imath : X \oplus X' \to X \oplus X'$ given by $\tau_\imath(x, x') = (x - I(x'), x')$, are both isomorphisms. So, $\tau = \tau_\imath \tau_j$ turns out to be the isomorphism we are looking for:

$$\tau(x, x') = \tau_\imath \tau_j(x, x') = \tau_\imath(x, x' + J(x))(x - I(x' + J(x)), x' + J(x))$$

and thus

$$\tau(\imath y, 0) = (\imath y - I(0 + J(\imath y)), 0 + J(\imath y)) = (0, jy).$$

The converse is also true. Indeed, if $\tau : X \oplus X' \to X \oplus X'$ is an automorphism such that $\tau(\imath y, 0) = (0, jy)$, the restriction of the projection $\pi : X \oplus X' \to X'$ to X is obviously an extension of j through $\imath$.   □

A.4.8   The Homology Sequence

Given an exact sequence $0 \longrightarrow Y \overset{j}{\longrightarrow} X \overset{q}{\longrightarrow} Z \longrightarrow 0$ and another Banach space B, taking operators with values in B one gets the exact sequence

$$0 \longrightarrow L(Z, B) \overset{q^*}{\longrightarrow} L(X, B) \overset{j^*}{\longrightarrow} L(Y, B)$$

where q^* (resp. j^*) means composition with q (resp. j) on the right. The sequence, so homological algebra says, can be continued to form a "long exact sequence"

$$0 \longrightarrow L(Z, B) \overset{q^*}{\longrightarrow} L(X, B) \overset{j^*}{\longrightarrow} L(Y, B) \overset{\beta}{\longrightarrow} \mathrm{Ext}(Z, B) \longrightarrow \mathrm{Ext}(X, B) \longrightarrow \mathrm{Ext}(Y, B)$$

The dual thing happens taking operators from a given Banach space A: one gets the exact sequence

$$0 \longrightarrow L(A, Y) \overset{j_*}{\longrightarrow} L(A, X) \overset{q_*}{\longrightarrow} L(A, Z)$$

where q_* (resp. j_*) means composition with q (resp. j) on the left. The sequence, so homological algebra says, can be continued to form a "long exact sequence"

$$0 \longrightarrow L(A, Y) \xrightarrow{j_*} L(A, X) \xrightarrow{q_*} L(A, Z) \xrightarrow{\alpha} \text{Ext}(A, Y) \longrightarrow \text{Ext}(A, X) \longrightarrow \text{Ext}(A, Z)$$

A detailed description of such sequences can be seen in [55]. Here we only indicate the action of the arrows α and β, which suffices for every (but one) purpose in this monograph.

- Given $u \in L(A, Z)$, $\alpha(u)$ is (the class in $\text{Ext}(A, Y)$) of the lower row in the pull-back diagram (A.11).
- Given $t \in L(Y, B)$, $\beta(t)$ is (the class in $\text{Ext}(Z, B)$ of) the lower extension of the push-out diagram (A.9).

A.5 Ordinals and Cardinals

Ordinals are a generalization of natural numbers that allows to use induction arguments even on uncountable sets. Ordinals are constructed *inductively* starting from the ordinal 0 by two procedures: given an ordinal α we can construct its successor $\alpha + 1$, and (this is how one goes beyond natural numbers) given any set of ordinals, we can construct the least ordinal which is above all of them. An ordinal can be formally defined as a set γ where the inclusion and the order relations coincide, in the sense that for every $\alpha, \beta \in \gamma \cup \{\gamma\}$ we have that $\alpha \in \beta$ if and only if $\alpha \subsetneq \beta$. Such a definition is good to formalize the theory, but basic intuition and working knowledge about ordinals are better summarized in the following list of properties:

1. Each ordinal is a set.
2. Given two ordinals α, β, we have that $\alpha \in \beta$ if and only if $\alpha \subsetneq \beta$. In that case, we write $\alpha < \beta$.
3. Each ordinal β coincides with the set of all ordinals below β, that is, $\beta = \{\alpha \text{ ordinal} : \alpha < \beta\}$.
4. The first ordinals are the finite ordinals: the natural numbers, $0 = \varnothing$, $1 = \{0\}$, $2 = \{0, 1\}$, $3 = \{0, 1, 2\}$, etc. But ordinals further continue, the first infinite ordinal is $\omega = \{0, 1, 2, \ldots\}$, then $\omega + 1 = \{0, 1, 2, \ldots, \omega\}$, etc.
5. Ordinals are well ordered. This means that every nonempty class of ordinals A has a minimum element, an ordinal α such that $\alpha \leq \beta$ for all $\beta \in A$. In particular, $<$ is also a linear order: for any ordinals α, β, either $\alpha < \beta$ or $\beta < \alpha$ of $\alpha = \beta$.
6. For every ordinal α there exists an immediate successor ordinal $\alpha + 1$: just apply the principle of well order to the class ordinals larger than α. Ordinals of the form $\alpha + 1$ are called successor ordinals, and the rest are called limit ordinals. A limit ordinal β is the supremum (also the union) of all ordinals below β.

7. The well ordering of ordinals implies the induction principle: If we want to prove a statement $P(\alpha)$ for all ordinals α it is enough to prove that, for every ordinal β, if $P(\alpha)$ holds for all $\alpha < \beta$, then $P(\beta)$ must hold as well.

8. The induction principle can be used also to make definitions. If we want to define a function F on all ordinals, it is enough to describe, for every ordinal β how $F(\beta)$ is constructed if we suppose that $F(\alpha)$ is already given for all $\alpha < \beta$.

9. Inductive arguments as in the items above are often splitted into two cases, depending on whether β is a successor or a limit ordinal. Although formally not needed, the initial case of $\beta = 0$ is sometimes considered for clarity.

A cardinal is an ordinal κ that cannot be bijected with any ordinal $\alpha < \kappa$.

1. Every set X can be bijected with a unique cardinal κ, that is called the cardinality of X, and is denoted as $\kappa = |X|$.

2. By the well ordering principle of ordinals, given a cardinal κ, there exists the minimum cardinal above κ, that we denote by κ^+.

3. All natural numbers are cardinals. The first infinite cardinal is ω, that is also called ω_0 or $\aleph_0$ when viewed as a cardinal. The cardinal ω^+ is denoted as ω_1 or $\aleph_1$. This notation extends by declaring $\aleph_{n+1} = \aleph_n^+$, and so we have $\aleph_0, \aleph_1, \aleph_2, \aleph_3, \ldots$.

4. Another cardinal is $\aleph_\omega$, the supremum of the $\aleph_n$ for $n < \omega$.

5. Given a cardinal κ, the cardinal 2^κ is the cardinality of the cartesian product $\{0, 1\}^\kappa$, or equivalently of the power set of κ. It always happens that $\kappa < 2^\kappa$.

6. The cardinal $\mathfrak{c} = 2^{\aleph_0}$ is the cardinality of the continuum.

7. The cofinality of a limit ordinal α is the least cardinal λ for which there is a subset of α of cardinality λ whose supremum is α. Thus for example, $\aleph_\omega$ has cofinality $\aleph_0$. On the other hand, the cofinality of $\mathfrak{c}$ is strictly greater than $\aleph_0$.

A.6   Direct and Inverse Limits

In Banach space theory it is natural to construct a space as the "union" of a family of larger and larger spaces as follows:

Let (X_α) be a family of Banach spaces indexed by a directed set Γ whose preorder is denoted by $\leq$. Suppose that, for each $\alpha, \beta \in \Gamma$ with $\alpha \leq \beta$ we have an isometry $f_\alpha^\beta : X_\alpha \to X_\beta$ in such a way that f_α^α is the identity on X_α for every $\alpha \in \Gamma$ and $f_\beta^\gamma \circ f_\alpha^\beta = f_\alpha^\gamma$ provided $\alpha \leq \beta \leq \gamma$. Then $(X_\alpha, f_\alpha^\beta)$ is said to be a directed system of Banach spaces.

The *direct limit* of the system is constructed as follows. First we take the disjoint union $\bigsqcup_\alpha X_\alpha$ and we define an equivalence relation $\asymp$ by identifying $x_\alpha \in X_\alpha$ and $x_\beta \in X_\beta$ if there is $\gamma \in \Gamma$ such that $f_\alpha^\gamma(x_\alpha) = f_\beta^\gamma(x_\beta)$. Then we may use the natural inclusion maps $\iota_\gamma : X_\gamma \to \bigsqcup_\alpha X_\alpha$ to transfer the linear structure and the norm from the spaces X_α to $\bigsqcup_\alpha X_\alpha / \asymp$ thus obtaining a normed space whose completion is called the direct limit of the system and is denoted by $\varinjlim X_\gamma$. The universal property behind this construction is the following: if we are given a system of contractive operators $u_\gamma : X_\gamma \to Y$, where Y is a Banach space, which are compatible with the f_α^β in the sense that $u_\alpha = u_\beta \circ f_\alpha^\beta$ for $\alpha \le \beta$, then there is a unique contractive operator $u : \varinjlim X_\gamma \to Y$ such that $u \circ \iota_\alpha = u_\alpha$ for every $\alpha \in \Gamma$. That operator is often called the direct limit of the family (u_α).

A closely related notion is that of an ordinal-indexed sequence of Banach spaces. Let ξ be an ordinal. Then a ξ-sequence of Banach spaces is a family of Banach spaces $(X_\alpha)_{\alpha \le \xi}$ such that $X_\alpha \subset X_\beta$ for $\alpha \le \beta \le \xi$. If, besides, one has

$$X_\beta = \overline{\bigcup_{\alpha < \beta} X_\alpha}$$

for each limit ordinal $\beta \le \xi$, then $(X_\alpha)_{\alpha \le \xi}$ is said to be continuous.

The following construction is typical. Let X be a Banach space and let ξ be the least ordinal such that $|\xi| = \mathrm{dens}(X)$. By the very definition there is a subset $\{x_\alpha : \alpha < \xi\}$ spanning a dense subspace of X. Then

$$X_\alpha = \overline{\mathrm{span}}\{x_\beta : \beta < \alpha\} \qquad (\alpha \le \xi)$$

is a continuous ξ-sequence, and $X_\xi = X$.

The simplest nontrivial example is the writing of a separable Banach space as the closure of the union of an increasing sequence of finite dimensional subspaces; that is, a continuous ω-sequence. The next interesting case, which already involves Zorn lemma, occurs when $\mathrm{dens}(X) = \aleph_1$, so that the nonseparable space $X = X_{\omega_1}$ can be seen as a continuous ω_1-sequence of separable Banach spaces.

The dual construction in topology is that of *inverse limit* of compact spaces. An inverse system is a family $(K_\alpha)_\alpha$ of compact spaces indexed by directed set Γ together with continuous surjections $\pi_\alpha^\beta : K_\beta \longrightarrow K_\alpha$ for $\alpha \le \beta$, with the property that $\pi_\alpha^\beta \circ \pi_\beta^\gamma = \pi_\alpha^\gamma$ when $\alpha \le \beta \le \gamma$ and $\pi_\alpha^\alpha = 1_{K_\alpha}$ for every $\alpha \in \Gamma$. The inverse limit of the system is

$$\varprojlim K_\alpha = \left\{ (x_\alpha) \in \prod_{\alpha \in \Gamma} K_\alpha : \pi_\beta^\gamma(x_\gamma) = x_\beta \text{ for } \beta \le \gamma \right\}$$

which is a compact space equipped with the relative product topology.

References

1. F. Albiac, N.J. Kalton, *Topics in Banach Space Theory*. Graduate Texts in Mathematics, vol. 233 (Springer, New York, 2006)
2. F. Albiac, N.J. Kalton, A characterization of real C(K) spaces. Am. Math. Mon. **114**, 737–743 (2007)
3. A. Alexiewicz, W. Orlicz, Analytic operations in real Banach spaces. Stud. Math. **14**(1953), 57–78 (1954)
4. D. Amir, Continuous function spaces with the bounded extension property. Bull. Res. Counc. Isr. Sect. F **10**, 133–138 (1962)
5. D. Amir, Projections onto continuous function spaces. Proc. Am. Math. Soc. **15**, 396–402 (1964)
6. D. Amir, J. Lindenstrauss, The structure of weakly compact sets in Banach spaces. Ann. Math. **88**, 35–46 (1968)
7. T. Ando, Closed range theorems for convex sets and linear liftings. Pac. J. Math. **44**, 393–410 (1973)
8. S.A. Argyros, Weak compactness in $L^1(\lambda)$ and injective Banach spaces. Isr. J. Math. **37**, 21–33 (1980)
9. S.A. Argyros, On the dimension of injective Banach spaces. Proc. Am. Math. Soc. **78**, 267–268 (1980)
10. S.A. Argyros, On nonseparable Banach spaces. Trans. Am. Math. Soc. **270**, 193–216 (1982)
11. S.A. Argyros, On the space of bounded measurable functions. Q. J. Math. Oxf. Ser. (2) **34**, 129–132 (1983)
12. S.A. Argyros, R. Haydon, A hereditarily indecomposable $\mathscr{L}_\infty$-space that solves the scalar-plus-compact problem. Acta Math. **206**, 1–54 (2011)
13. S.A. Argyros, J.M.F. Castillo, A.S. Granero, M. Jiménez, J.P. Moreno, Complementation and embeddings of $c_0(I)$ in Banach spaces. Proc. Lond. Math. Soc. **85**, 742–772 (2002)
14. S.A. Argyros, I. Gasparis, P. Motakis. On the structure of separable $\mathscr{L}_\infty$-spaces. arXiv: 1504.08223v1 (2015)
15. N. Aronszajn, P. Panitchpakdi, Extension of uniformly continuous transformations and hyperconvex metric spaces. Pac. J. Math. **6**, 405–439 (1956); Correction. Ibid. **7**, 1729 (1957)
16. A. Avilés, C. Brech, A Boolean algebra and a Banach space obtained by push-out iteration. Top. Appl. **158**, 1534–1550 (2011)
17. A. Avilés, Y. Moreno, Automorphisms in spaces of continuous functions on Valdivia compacta. Top. Appl. **155**, 2027–2030 (2008)
18. A. Avilés, S. Todorcevic, Multiple gaps. Fundam. Math. **213**, 15–42 (2011)

© Springer International Publishing Switzerland 2016
A. Avilés et al., *Separably Injective Banach Spaces*, Lecture Notes in Mathematics 2132, DOI 10.1007/978-3-319-14741-3

19. A. Avilés, F. Cabello Sánchez, J.M.F. Castillo, M. González, Y. Moreno, Banach spaces of universal disposition. J. Funct. Anal. **261**, 2347–2361 (2011)

20. A. Avilés, F. Cabello Sánchez, J.M.F. Castillo, M. González, Y. Moreno. On separably injective Banach spaces. Adv. Math. **234**, 192–216 (2013)

21. A. Avilés, F. Cabello Sánchez, J.M.F. Castillo, M. González, Y. Moreno, On ultraproducts of Banach space of type $\mathscr{L}_\infty$. Fundam. Math. **222**, 195–212 (2013)

22. A. Avilés, F. Cabello Sánchez, J.M.F. Castillo, M. González, Y. Moreno, ℵ-injective Banach spaces and ℵ-projective compacta. Rev. Mat. Iberoramericana **31**(2), 575–600 (2015)

23. W.G. Bade, Complementation problems for the Baire classes. Pac. J. Math. **45**, 1–11 (1973)

24. W. Bade, P. Curtis, The Wedderburn decomposition of commutative Banach algebras. Am. J. Math. **82**, 851–866 (1960)

25. J.W. Baker, Projection constants for $C(S)$ spaces with the separable projection property. Proc. Am. Math. Soc. **41**, 201–204 (1973)

26. P. Bankston, Reduced coproducts of compact Hausdorff spaces. J. Symb. Log. **52**, 404–424 (1987)

27. P. Bankston, A survey of ultraproduct constructions in general topology. Top. Atlas **8**, 1–32 (1993)

28. S. Baratella, S.-A. Ng, Isometry games in Banach spaces. Bull. Belg. Math. Soc. Simon Stevin **15**, 509–521 (2008)

29. M. Bell, W. Marciszewski, On scattered Eberlein compact spaces. Isr. J. Math. **158**, 217–224 (2007)

30. I. Ben Yaacov, C.W. Henson, Generic orbits and type isolation in the Gurarij space. arXiv:1211.4814v1 (2015)

31. Y. Benyamini, Separable G spaces are isomorphic to $C(K)$ spaces. Isr. J. Math. **14**, 287–293 (1973)

32. Y. Benyamini, An M-space which is not isomorphic to a $C(K)$-space. Isr. J. Math. **28**, 98–104 (1977)

33. Y. Benyamini, An extension theorem for separable Banach spaces. Isr. J. Math **29**, 24–30 (1978)

34. Y. Benyamini, J. Lindenstrauss, A predual of ℓ_1 which is not isomorphic to a $C(K)$ space. Isr. J. Math. **13**, 246–254 (1972)

35. Y. Benyamini, M.E. Rudin, M. Wage, Continuous images of weakly compact subsets of Banach spaces. Pac. J. Math. **70**(2), 309–324 (1977)

36. C. Bessaga, A. Pełczyński, Spaces of continuous functions (IV) (On isomorphically classification of spaces of continuous functions). Stud. Math. **19**, 53–62 (1960)

37. C. Bessaga, A. Pełczyński, Selected topics in inifinite-dimensional convexity. Monografie Matematyczne, vol. 58 (Polish Scientific Publishers, Warszawa, 1975)

38. J.L. Blasco, C. Ivorra, Injective spaces of real-valued functions with the Baire property. Isr. J. Math. **91**, 341–348 (1995)

39. J.L. Blasco, C. Ivorra, On constructing injective spaces of type $C(K)$. Indag. Math. (N.S.) **9**(2), 161–172 (1998)

40. A. Blaszczyk, A.R. Szymański, Concerning Parovičenko's theorem. Bull. Acad. Polon. Sci. Math. **28**, 311–314 (1980)

41. J. Bochnak, Analytic functions in Banach spaces. Stud. Math. **35**, 273–292 (1970)

42. K. Borsuk, Über Isomorphie der Funktionalräume. Bull. Acad. Polon. Sci. Math. 1–10 (1933)

43. J. Bourgain, A result on operators on $C[0,1]$. J. Oper. Theory **3**, 275–289 (1980)

44. J. Bourgain, A counterexample to a complementation problem. Compos. Math. **43**, 133–144 (1981)

45. J. Bourgain, *New Classes of $\mathscr{L}^p$-Spaces*. Lecture Notes in Mathematics, vol. 889 (Springer, Berlin, 1981)

46. J. Bourgain, F. Delbaen, A class of special $\mathscr{L}_\infty$ spaces. Acta Math. **145**, 155–176 (1980)

47. J. Bourgain, G. Pisier, A construction of $\mathscr{L}_\infty$-spaces and related Banach spaces. Bol. Soc. Bras. Mat. **14**, 109–123 (1983)

48. C. Brech, On the density of Banach spaces $C(K)$ with the Grothendieck property. Proc. Am. Math. Soc. **134**, 3653–3663 (2006)
49. C. Brech, P. Koszmider, On universal Banach spaces of density continuum. Isr. J. Math. **190**, 93–110 (2012)
50. F. Cabello Sánchez, Regards sur le problème des rotations de Mazur. Extracta Math. **12**, 97–116 (1997)
51. F. Cabello Sánchez, Transitivity of M-spaces and Wood's conjecture. Math. Proc. Camb. Philos. Soc. **124**, 513–520 (1998)
52. F. Cabello Sánchez, Yet another proof of Sobczyk's theorem, in *Methods in Banach Space Theory*. London Mathematical Society Lecture Notes, vol. 337 (Cambridge University Press, Cambridge, 2006), pp. 133–138
53. F. Cabello Sánchez, J.M.F. Castillo, Duality and twisted sums of Banach spaces. J. Funct. Anal. **175**, 1–16 (2000)
54. F. Cabello Sánchez, J.M.F. Castillo, Uniform boundedness and twisted sums of Banach spaces. Houst. J. Math. **30**, 523–536 (2004)
55. F. Cabello Sánchez, J.M.F. Castillo, The long homology sequence for quasi-Banach spaces, with applications. Positivity **8**, 379–394 (2004)
56. F. Cabello Sánchez, J.M.F. Castillo, D.T. Yost, Sobczyk's theorems from A to B. Extracta Math. **15**, 391–420 (2000)
57. F. Cabello Sánchez, J.M.F. Castillo, N.J. Kalton, D.T. Yost, Twisted sums with $C(K)$-spaces. Trans. Am. Math. Soc. **355**, 4523–4541 (2003)
58. F. Cabello Sánchez, J. Garbulińska-Węgrzyn, W. Kubiś, Quasi-Banach spaces of almost universal disposition. J. Funct. Anal. **267**, 744–771 (2014)
59. L. Candido, P. Koszmider, On complemented copies of $c_0(\omega_1)$ in $C(K^n)$ spaces. arXiv: 1501.01785v2 (2015)
60. P.G. Casazza, Approximation properties, in *Handbook of the Geometry of Banach Spaces*, vol. 1, ed. by W.B. Johnson, J. Lindenstrauss (North-Holland, Amsterdam, 2001), pp. 271–316
61. J.M.F. Castillo, M. González, *Three-Space Problems in Banach Space Theory*. Lecture Notes in Mathematical, vol. 1667 (Springer, Berlin, 1997)
62. J.M.F. Castillo, M. González, Continuity of linear maps on $\mathscr{L}_1$-spaces. J. Math. Anal. Appl. **385**, 12–15 (2012)
63. J.M.F. Castillo, Y. Moreno, On the Lindenstrauss-Rosenthal theorem. Isr. J. Math. **140**, 253–270 (2004)
64. J.M.F. Castillo, Y. Moreno, Twisted dualites for Banach spaces, in *Banach Spaces and Their Applications in Analysis*. Proceedings of the Meeting in Honor of Nigel Kalton's 60th Birthday (Walter de Gruyter, Berlin, 2007), pp. 59–76
65. J.M.F. Castillo, Y. Moreno, Sobczyk's theorem and the bounded approximation property. Stud. Math. **201**, 1–19 (2010)
66. J.M.F. Castillo, Y. Moreno, On the bounded approximation property in Banach spaces. Isr. J. Math. **198**, 243–259 (2013)
67. J.M.F. Castillo, A. Plichko, Banach spaces in various positions. J. Funct. Anal. **259**, 2098–2138 (2010)
68. J.M.F. Castillo, M. Simões, Property (V) still fails the 3-space property. Extracta Math. **27**, 5–11 (2012)
69. J.M.F. Castillo, J. Suárez, On $\mathscr{L}_\infty$-envelopes of Banach spaces. J. Math. Anal. Appl. **394**, 152–158 (2012)
70. J.M.F. Castillo, Y. Moreno, J. Suárez, On Lindenstrauss-Pełczyński spaces. Stud. Math. **174**, 213–231 (2006)
71. J.M.F. Castillo, Y. Moreno, J. Suárez, On the structure of Lindenstrauss-Pełczyński spaces. Stud. Math. **194**, 105–115 (2009)
72. P. Cembranos, $C(K, X)$ contains a complemented copy of c_0. Proc. Am. Math. Soc. **91**, 556–558 (1984)

73. P. Cembranos, J. Mendoza, *Banach Spaces of Vector-Valued Functions*. Lecture Notes in Mathematics, vol. 1676 (Springer, Berlin, 1997)

74. P. Cembranos, J. Mendoza, The Banach spaces $\ell_\infty(c_0)$ and $c_0(\ell_\infty)$ are not isomorphic. J. Math. Anal. Appl. **367**, 461–463 (2010)

75. C.C. Chang, H.J. Keisler, *Model Theory*. Studies in Logic and the Foundations of Mathematics, vol. 73, 3rd edn. (North-Holland, Amsterdam, 1990)

76. M.-D. Choi, E.G. Effros, Lifting problems and the cohomology of C*-algebras. Can. J. Math. **29**, 1092–1111 (1977)

77. W.W. Comfort, S. Negrepontis, *The Theory of Ultrafilters* (Springer, Berlin, 1974)

78. F.K. Dashiell Jr., Isomorphism problems for the Baire classes. Pac. J. Math. **52**, 29–43 (1974)

79. F.K. Dashiell Jr., Nonweakly compact operators from order-Cauchy complete $C(S)$ lattices, with application to Baire classes. Trans. Am. Math. Soc. **266**, 397–413 (1981)

80. F.K. Dashiell Jr., J. Lindenstrauss, Some examples concerning strictly convex norms on $C(K)$ spaces. Isr. J. Math. **16**, 329–342 (1973)

81. R. Deville, G. Godefroy, V. Zizler, *Smoothness and Renormings in Banach Spaces*. Pitman Monographs and Surveys in Pure and Applied Mathematics, vol. 64 (Wiley, New York, 1993)

82. J. Diestel, A survey of results related to the Dunford-Pettis property. Contemp. Math. **2**, 15–60 (1980)

83. J. Diestel, J.H. Fourie, J. Swart, *The Metric Theory of Tensor Products: Grothendieck's Résumé Revisited* (American Mathematical Society, Providence, 2008)

84. S.J. Dilworth, B. Randrianantoanina, On an isomorphic Banach-Mazur rotation problem and maximal norms in Banach spaces. J. Funct. Anal. **268**, 1587–1611 (2015)

85. J. Dixmier, Sur certains espaces considérés par M.H. Stone. Summa Bras. Math. **2**, 151–182 (1951)

86. E.K. van Douwen, The integers and topology, in *Handbook of Set-Theoretic Topology* (Elsevier, New York, 1984)

87. A. Dow, K.P. Hart, Applications of another characterization of $\beta\mathbb{N} \setminus \mathbb{N}$. Top. Appl. **122**, 105–133 (2002)

88. A. Dow, J. Vermeer, Extremally disconnected spaces, subspaces and retracts. Top. Appl. **50**, 263–282 (1993)

89. A. Dow, H. Jumila, J. Pelant, Chain condition and weak topologies. Top. Appl. **156**, 1327–1344 (2009)

90. P.N. Dowling, On ℓ_∞-subspaces of Banach spaces. Collect. Math. **51**, 255–260 (2000)

91. J. Dugundji, An extension of Tietze's theorem. Pac. J. Math. **1**, 353–367 (1951)

92. M. Fabian, *Gâteaux Differentiability of Convex Functions and Topology. Weak Asplund Spaces*. Canadian Mathematical Society Series of Monographs and Advanced Texts. A Wiley-Interscience Publication (Wiley, New York, 1997)

93. H. Fakhoury, Sélections linéaires associées au théorème de Hahn-Banach. J. Funct. Anal. **11**, 436–452 (1972)

94. V. Ferenczi, C. Rosendal, On isometry groups and maximal symmetry. Duke Math. J. **162**(10), 1771–1831 (2013)

95. T. Figiel, W.B. Johnson, A. Pełczyński, Some approximation properties of Banach spaces and Banach lattices. Isr. J. Math **183**, 199–232 (2011)

96. V.P. Fonf, P. Wojtaszczyk, Characteristic properties of the Gurariy space. Isr. J. Math. **203**, 109–140 (2014)

97. V.P. Fonf, J. Lindenstrauss, R.R. Phelps, Infinite dimensional convexity, in *Handbook of the Geometry of Banach Spaces*, vol. 1, ed. by W.B. Johnson, J. Lindenstrauss (North-Holland, Amsterdam, 2001), pp. 599–670

98. T.E. Frayne, A.C. Morel, D.S. Scott, Reduced direct products. Fund. Math. **51**, 195–228 (1962)

99. H. Gaifman, Concerning measures on Boolean algebras. Pac. J. Math. **7**, 552–561 (1955)

100. E. Galego, J. Hagler, Copies of $c_0(\Gamma)$ in $C(K, X)$ spaces. Proc. Am. Math. Soc. **140**, 3843–3852 (2012)

101. J. Garbulińska-Węgrzyn, W. Kubiś, Remarks on Gurariĭ spaces. Extracta Math. **26**, 235–269 (2011)
102. I. Gasparis, Operators on $C[0, 1]$ preserving copies of asymptotic ℓ_1 spaces. Math. Ann. **333**, 831–858 (2005)
103. I. Gasparis, On a problem of H.P. Rosenthal concerning operators on $C[0, 1]$. Adv. Math. **218**, 1512–1525 (2008)
104. I. Gasparis, A new isomorphic ℓ_1 predual not isomorphic to a complemented subspace of a $C(K)$ space. Bull. Lond. Math. Soc. **45**, 789–799 (2013)
105. I. Gasparis, A note on operators fixing cotype subspaces of $C[0, 1]$. Proc. Am. Math. Soc. **142**, 1633–1639 (2014)
106. I. Gasparis, M.K. Papadiamantis, D.Z. Zisimopoulou, More ℓ_r saturated $\mathscr{L}^\infty$-spaces. Serdica Math. J. **36**, 149–170 (2010)
107. L. Gillman, M. Jerison, *Rings of Continuous Functions* (Reprint of the 1960 ed.). Graduate Texts in Mathematics, vol. 43 (Springer, New York, 1976)
108. A.M. Gleason, Projective topological spaces. Ill. J. Math. **2**, 482–489 (1958)
109. G. Godefroy, N.J. Kalton, G. Lancien, Subspaces of $c_0(\mathbb{N})$ and Lipschitz isomorphisms. Geom. Funct. Anal. **10**, 798–820 (2000)
110. M. González, On essentially incomparable Banach spaces. Math. Z. **215**, 621–629 (1994)
111. M. González, A. Martínez-Abejón, Local duality for Banach spaces. Expo. Math. **33**, 135–183 (2015)
112. D.B. Goodner, Projections in normed linear spaces. Trans. Am. Math. Soc. **69**, 89–108 (1950)
113. A.S. Granero, M. Jiménez Sevilla, J.P. Moreno, On ω-independence and the Kunen-Shelah property. Proc. Edinb. Math. Soc. **45**, 391–395 (2002)
114. A.S. Granero, M. Jiménez Sevilla, A. Montesinos, J.P. Moreno, A. Plichko, On the Kunen-Shelah properties in Banach spaces. Stud. Math. **157**, 97–120 (2003)
115. A. Grothendieck, Sur les applications linéaires faiblement compactes d'espaces du type $C(K)$. Can. J. Math. **5**, 129–173 (1953)
116. B. Grünbaum, Some applications of expansion constants. Pac. J. Math. **10**, 193–201 (1960)
117. J.B. Guerrero, A. Rodríguez-Palacios, Transitivity of the norm on Banach spaces. Extracta Math. **17**, 1–58 (2002)
118. V.I. Gurariĭ, Spaces of universal placement, isotropic spaces and a problem of Mazur on rotations of Banach spaces. (Russian) Sibirsk. Mat. Ž. **7**, 1002–1013 (1966)
119. R. Gurevič, On ultracoproducts of compact Hausdorff spaces. J. Symb. Log. **53**, 294–300 (1983)
120. P. Hájek, V. Montesinos Santalucía, J. Vanderwerff, V. Zizler, *Biorthogonal Systems in Banach Spaces*. CMS Books in Mathematics (Springer, New York, 2008)
121. P. Harmand, D. Werner, W. Werner, *M-Ideals in Banach Spaces and Banach Algebras*. Lecture Notes in Mathematics, vol. 1547 (Springer, New York, 1993)
122. V.S. Hasanov, Some universally complemented subspaces of $m(\Gamma)$. Mat. Zametki **27**, 105–108 (1980)
123. F. Haussdorff, Summen von $\aleph_1$ Mengen. Fund. Math. **26**, 241–255 (1936)
124. R.G. Haydon, On dual L_1-spaces and injective bidual Banach spaces. Isr. J. Math. **31**, 142–152 (1978)
125. R.G. Haydon, A nonreflexive Grothendieck space that does not contain ℓ_∞. Isr. J. Math. **40**, 65–73 (1981)
126. S. Heinrich, Ultraproducts in Banach space theory. J. Reine Angew. Math. **313**, 72–104 (1980)
127. S. Heinrich, Ultraproducts of L_1-predual spaces. Fund. Math. **113**, 221–234 (1981)
128. S. Heinrich, C.W. Henson, Banach space model theory. II. Isomorphic equivalence. Math. Nachr. **125**, 301–317 (1986)
129. M. Henriksen, Some remarks on a paper of Aronszajn and Panitchpakdi. Pac. J. Math. **7**, 1619–1621 (1957)
130. C.W. Henson, Nonstandard hulls of Banach spaces. Isr. J. Math. **25**, 108–144 (1976)
131. C.W. Henson, Background for Three Lectures on: Nonstandard analysis and ultraproducts in Banach spaces and functional analysis. Manuscript, 16 p.

132. C.W. Henson, J. Iovino, *Ultraproducts in Analysis*. London Mathematical Society Lecture Notes Series, vol. 262 (Cambridge University Press, Cambridge, 2002), pp. 1–113
133. C.W. Henson, L.C. Moore, Nonstandard hulls of the classical Banach spaces. Duke Math. J. **41**, 277–284 (1974)
134. C.W. Henson, L.C. Moore, Nonstandard analysis and the theory Banach spaces, in *Nonstandard Analysis–Recent Developments*. Lecture Notes in Mathematics, vol. 983 (Springer, Berlin, 1983), pp. 27–112
135. C.W. Henson, L.C. Moore, The Banach spaces $\ell_p(n)$ for large p and n. Manuscripta Math. **44**, 1–33 (1983)
136. W. Hodges, Model theory, in *Encyclopedia of Mathematics and Its Applications*, vol. 42 (Cambridge University Press, Cambridge, 1993)
137. O. Hustad, A note on complex spaces. Isr. J. Math. **16**, 117–119 (1973)
138. J. Isbell, Z. Semadeni, Projections constants and spaces of continuous functions. Trans. Am. Math. Soc. **107**, 38–48 (1963)
139. R.C. James, Uniformly nonsquare Banach spaces. Ann. Math. **80**, 542–550 (1964)
140. H. Jarchow, The three space problem and ideals of operators. Math. Nachr. **119**, 121–128 (1984)
141. H.M. Jebreen, F.B. Jamjoom, D.T. Yost, Colocality and twisted sums of Banach spaces. J. Math. Anal. Appl. **323**, 864–875 (2006)
142. T. Jech, *Set Theory*. The third millennium edition, revised and expanded. Springer Monographs in Mathematics (Springer, Berlin, 2003)
143. M. Jiménez Sevilla, J.P. Moreno, Renorming Banach spaces with the Mazur intersection property. J. Funct. Anal. **144**, 486–504 (1997)
144. W.B. Johnson, J. Lindenstrauss, Some remarks on weakly compactly generated Banach spaces. Isr. J. Math. **17**, 219–230 (1974)
145. W.B. Johnson, J. Lindenstrauss, Basic concepts in the theory of Banach spaces, in *Handbook of the Geometry of Banach Spaces*, vol. 1, ed. by W.B. Johnson, J. Lindenstrauss (North-Holland, Amsterdam 2001), pp. 1–84
146. W.B. Johnson, T. Oikhberg, Separable lifting property and extensions of local reflexivity. Ill. J. Math. **45**, 123–137 (2001)
147. W.B. Johnson, M. Zippin, Separable L_1 preduals are quotients of $C(\Delta)$. Isr. J. Math. **16**, 198–202 (1973)
148. W.B. Johnson, H.P. Rosenthal, M. Zippin, On bases, finite dimensional decompositions and weaker structures in Banach spaces. Isr. J. Math. **9**, 488–506 (1971)
149. W.B. Johnson, T. Kania, G. Schechtman, Closed ideals of operators on and complemented subspaces of Banach spaces of functions with countable support (2015). arXiv:1502.03026
150. M.I. Kadec, On complementably universal Banach spaces. Stud. Math. **40**, 85–89 (1971)
151. R.V. Kadison, The von Neumann algebra characterization theorems. Expo. Math. **3**, 193–227 (1985)
152. N.J. Kalton, Universal spaces and universal bases in metric linear spaces. Stud. Math. **61**, 161–191 (1977)
153. N.J. Kalton, Transitivity and quotients of Orlicz spaces. Comment. Math. (Special issue in honor of the 75th birthday of W. Orlicz) 159–172 (1978)
154. N.J. Kalton, Locally complemented subspaces and $\mathscr{L}_p$-spaces for $0 < p < 1$. Math. Nachr. **115**, 71–97 (1984)
155. N.J. Kalton, Extension of linear operators and Lipschitz maps into $C(K)$-spaces. N. Y. J. Math. **13**, 317–381 (2007)
156. N.J. Kalton, Automorphism of $C(K)$-spaces and extension of linear operators. Ill. J. Math. **52**, 279–317 (2008)
157. N.J. Kalton, Lipschitz and uniform embeddings into ℓ_∞. Fund. Math. **212**, 53–69 (2011)
158. N.J. Kalton, A. Pełczyński, Kernels of surjections from $\mathscr{L}_1$-spaces with an application to Sidon sets. Math. Ann. **309**, 135–158 (1997)
159. A. Kanamori, *The Higher Infinite. Large Cardinals in Set Theory From Their Beginnings*, 2nd edn. (Springer, Berlin, 2009)

160. L.V. Kantorovič, On semi-ordered linear spaces and their applications to the theory of linear operations. Doklady' AN SSSR **4**, 11–14 (1935) (in Russian)
161. K. Kawamura, On a conjecture of wood. Glasg. Math. J. **47**, 1–5 (2005)
162. H.J. Keisler, Good ideals in fields of sets. Ann. Math. (2) **79**, 338–359 (1964)
163. J.L. Kelley, Banach spaces with the extension property. Trans. Am. Math. Soc. **72**, 323–326 (1952)
164. J.L. Kelley, *General Topology*. Graduate Texts in Mathematics, vol. 27 (Springer, New York, 1975)
165. V. Klee, Some topological properties of convex sets. Trans. Am. Math. Soc. **78**, 30–45 (1955)
166. P. Koszmider, Banach spaces of continuous functions with few operators. Math. Ann. **330**, 151–184 (2004)
167. P. Koszmider, The interplay between compact spaces and the Banach spaces of their continuous functions. Section 52 in *Open Problems in Topology II* (Elsevier, Amsterdam, 2007)
168. W. Kubiś, Metric-enriched categories and approximate Fraïssé limits. Preprint (2013). arXiv:1210.650v3
169. W. Kubiś, Fraïssé sequences: category-theoretic approach to universal homogeneous structures. Ann. Pure Appl. Log. **165**, 1755–1811 (2014)
170. W. Kubiś, S. Solecki, A proof of uniqueness of the Gurarii space. Isr. J. Math. **195**, 449–456 (2013)
171. J. Kupka, A short proof and generalization of a measure theoretic disjointification lemma. Proc. Am. Math. Soc. **45**, 70–72 (1974)
172. K. Kuratowski, *Topology*, vol. I (Academic Press, New York, 1966)
173. A.J. Lazar, J. Lindenstrauss, Banach spaces whose duals are L_1-spaces and their representing matrices. Acta Math. **126**, 165–193 (1971)
174. D.H. Leung, F. Räbiger, Complemented copies of c_0 in l^∞-sums of Banach spaces. Ill. J. Math. **34** (1990) 52–58.
175. J. Lindenstrauss, On the extension of compact operators. Mem. Am. Math. Soc. **48**, (1964)
176. J. Lindenstrauss, On a certain subspace of ℓ_1. Bull. Acad. Polon. Sci. Math. **12**, 539–542 (1964)
177. J. Lindenstrauss, On the extension of operators with range in a $C(K)$ space. Proc. Am. Math. Soc. **15**, 218–225 (1964)
178. J. Lindenstrauss, A. Pełczyński, Contributions to the theory of the classical Banach spaces. J. Funct. Anal. **8**, 225–249 (1971)
179. J. Lindenstrauss, H.P. Rosenthal, The $\mathscr{L}_p$-spaces. Isr. J. Math. **7**, 325–349 (1969)
180. J. Lindenstrauss, H.P. Rosenthal, Automorphisms in c_0, ℓ_1 and m. Isr. J. Math. **9**, 227–239 (1969)
181. J. Lindenstrauss, L. Tzafriri, *Classical Banach Spaces*. Lecture Notes in Mathematics, vol. 338 (Springer, Berlin, 1973)
182. J. Lindenstrauss, L. Tzafriri, *Classical Banach Spaces I* (Springer, Berlin, 1977)
183. J. López-Abad, A Bourgain-Pisier construction for general Banach spaces. J. Funct. Anal. **265**, 1423–1441 (2013)
184. J. López-Abad, S. Todorcevic, Generic Banach spaces and generic simplexes. J. Funct. Anal. **261**, 300–386 (2011)
185. W. Lusky, The Gurariĭ spaces are unique. Arch. Math. **27**, 627–635 (1976)
186. W. Lusky, Separable Lindenstrauss spaces, *Functional Analysis: Surveys and Recent Results* (North-Holland, Amsterdam, 1977), pp. 15–28
187. W. Lusky, A note on Banach spaces containing c_0 or C_∞. J. Funct. Anal. **62**, 1–7 (1985)
188. W. Lusky, On nonseparable simplex spaces. Math. Scand. **61**, 276–285 (1987)
189. W. Lusky, Three-space problems and basis extensions. Isr. J. Math. **107**, 17–27 (1988)
190. W. Lusky, Three-space problems and bounded approximation property. Stud. Math. **159**, 417–434 (2003)
191. W. Lusky, Every separable L_1-predual is complemented in a C^*-algebra. Stud. Math. **160**, 103–116 (2004)

192. W. Marciszewski, On Banach spaces $C(K)$ isomorphic to $c_0(\Gamma)$. Stud. Math. **156**, 295–302 (2003)
193. W. Marciszewski, R. Pol, On Banach spaces whose norm-open sets are F_σ sets in the weak topology. J. Math. Anal. Appl. **350**, 708–722 (2009)
194. V. Mascioni, Topics in the theory of complemented subspaces in Banach spaces. Expo. Math. **7**, 3–47 (1989)
195. S. Mazur, Über convexe Mengen in linearen normierten Räumen. Stud. Math. **4**, 70–84 (1933)
196. G. Metafune, On the space ℓ_∞/c_0. Rocky Mt. J. Math. **17**, 583–586 (1987)
197. P. Meyer-Nieberg, *Banach Lattices* (Springer, Berlin, 1991)
198. E. Michael, A. Pełczyński, Separable Banach spaces which admit l_∞^n approximations. Isr. J. Math. **4**, 189–198 (1966)
199. Y. Moreno, A. Plichko, On automorphic Banach spaces. Isr. J. Math. **169**, 29–45 (2009)
200. G.A. Muñoz, Y.A. Sarantopoulos, A. Tonge, Complexifications of real Banach spaces, polynomials and multilinear maps. Stud. Math. **134**, 1–33 (1999)
201. L. Nachbin, A theorem of the Hahn-Banach type for linear transformations. Trans. Am. Math. Soc. **68**, 28–46 (1950)
202. S. Negrepontis, Banach spaces and topology, in *Handbook of Set Theoretic Topology*, ed. by K. Kunen, J.E. Vaughan (North-Holland, Amsterdam, 1984), pp. 1045–1142
203. C.W. Neville, Banach spaces with a restricted Hahn-Banach extension property. Pac. J. Math. **63**, 201–212 (1976)
204. C.W. Neville, S.P. LLoyd, $\aleph$-projective spaces. Ill. J. Math. **25**, 159–168 (1981)
205. P. Niemiec, Universal valued Abelian groups. Adv. Math. **235**, 398–449 (2013)
206. T. Oikhberg, The non-commutative Gurarii space. Arch. Math. (Basel) **86**(4), 356–364 (2006)
207. V.M. Onieva, Notes on Banach space ideals. Math. Nachr. **126**, 27–33 (1986)
208. M.I. Ostrovskii, Separably injective Banach spaces. Funct. Anal. i Priložen. **20**, 80–81 (1986); English transl.: Funct. Anal. Appl. **20**, 154–155 (1986)
209. I.I. Parovičenko, On a universal bicompactum of weight $\aleph$. Dokl. Akad. Nauk SSSR **150**, 36–39 (1963)
210. J.R. Partington, Subspaces of certain Banach sequence spaces. Bull. Lond. Math. Soc. **13**, 162–166 (1981)
211. A. Pełczyński, Projections in certain Banach spaces. Stud. Math. **19**, 209–228 (1960)
212. A. Pełczyński, On $C(S)$-subspaces of separable Banach spaces. Stud. Math. **31**, 513–522 (1968)
213. A. Pełczyński, Universal bases. Stud. Math. **32**, 247–268 (1969)
214. A. Pełczyński, V.N. Sudakov, Remarks on non-complemented subspaces of the space $m(S)$. Colloq. Math. **9**, 85–88 (1962)
215. A. Pełczyński, P. Wojtaszczyk, Banach spaces with finite dimensional expansions of identity and universal bases of finite dimensional spaces. Stud. Math. **40**, 91–108 (1971)
216. R.S. Phillips, On linear transformations. Trans. Am. Math. Soc. **48**, 516–541 (1940)
217. M.M. Popov, Codimension of subspaces of L_p for $0 < p < 1$. Funktional. Anal. i Priložen. **18**(2), 94–95 (1984)
218. T.C. Przymusiński, Perfectly normal compact spaces are continuous images of $\beta\mathbb{N} \setminus \mathbb{N}$. Proc. Am. Math. Soc. **86**, 541–544 (1982)
219. F. Rambla, A counter-example to Wood's conjecture. J. Math. Anal. Appl. **317**, 659–667 (2006)
220. T.S.S.R.K. Rao, Isometries of $A_C(K)$. Proc. Am. Math. Soc. **85**, 544–546 (1982)
221. S. Rolewicz, Metric linear spaces, in *Mathematics and Its Applications* (East European Series), 2nd edn., vol. 20 (D Reidel Publishing Co, Dordrecht; PWN-Polish Scientific Publishers, Warsaw, 1985)
222. H.P. Rosenthal, On injective Banach spaces and the spaces $L^\infty(\mu)$ for finite measures μ. Acta Math. **124**, 205–248 (1970)
223. H.P. Rosenthal, On relatively disjoint families of measures, with some applications to Banach space theory. Stud. Math. **37**, 13–36 (1970)
224. H.P. Rosenthal, On factors of $C[0, 1]$ with non-separable dual. Isr. J. Math. **13**, 361–378 (1972)

225. H.P. Rosenthal, The complete separable extension property. J. Oper. Theory **43**, 329–374 (2000)

226. W. Schachermayer, On some classical measure-theoretic theorems for non-sigma-complete Boolean algebras. Dissertationes Math. **214**, 33 pp. (1982)

227. T. Schlumprecht, Limited sets in Banach spaces, Dissertation, Munich, 1987

228. G.L. Seever, Measures on F-spaces. Trans. Am. Math. Soc. **133**, 267–280 (1968)

229. Z. Semadeni, Banach spaces non-isomorphic to their Cartesian squares. II. Bull. Acad. Polon. Sci. Sér. Sci. Math. Astr. Phys. **8**, 81–84 (1960)

230. Z. Semadeni, *Banach Spaces of Continuous Functions* (PWN, Warszawa, 1971)

231. S. Shelah, Uncountable constructions for B.A.e.c. groups and Banach spaces. Isr. J. Math. **51**, 273–297 (1985)

232. W. Sierpiński, *Cardinal and Ordinal Numbers* (PWN, Warszawa, 1965)

233. R. Sikorski, *Boolean Algebras*, 3rd edn. (Springer, New York, 1969)

234. B. Sims, "Ultra"-techniques in Banach space theory, in *Queen's Papers in Pure and Applied Mathematics*, vol. 60 (Kingston, Ontario, 1982)

235. A. Sobczyk, Projection of the space m on its subspace c_0. Bull. Am. Math. Soc. **47**, 938–947 (1941)

236. A. Sobczyk, On the extension of linear transformations. Trans. Am. Math. Soc. **55**, 153–169 (1944)

237. J. Stern, Ultrapowers and local properties of Banach spaces. Trans. Am. Math. Soc. **240**, 231–252 (1978)

238. M.A. Swardson, A generalization of F-spaces and some topological characterizations of GCH. Trans. Am. Math. Soc. **279**, 661–675 (1983)

239. M. Talagrand, Un nouveau $C(K)$ qui possède la propriété de Grothendieck. Isr. J. Math. **37**, 181–191 (1980)

240. M. Tarbard, Hereditarily indecomposable, separable $\mathscr{L}_\infty$ Banach spaces with ℓ_1 dual having few but not very few operators. J. Lond. Math. Soc. **85**, 737–764 (2012)

241. A.E. Taylor, Addition to the theory of polynomials in normed linear spaces. Tohoku Math. J. **44**, 302–318 (1938)

242. A.E. Taylor, D.C. Lay, *Introduction to Functional Analysis*, 2nd. edn. (Wiley, New York, 1980)

243. S. Ulam, Zur Masstheorie in der allgemeinen Mengenlehre. Fund. Math. **16**, 140–150 (1930)

244. W. Veech, A short proof of Sobczyk theorem. Proc. Am. Math. Soc. **28**, 627–628 (1971)

245. R.C. Walker, The Stone-Čech compactification, in *Ergebnisse der Mathematik und ihrer Grenzgebiete*, Band 83 (Springer, New York/Berlin, 1974)

246. S. Willard, *General Topology* (Addison-Wesley, Reading, 1970)

247. E.L. Wimmers, The Shelah P-point independence theorem. Isr. J. Math. **43**, 28–48 (1982)

248. P. Wojtaszczyk, Some remarks on the Gurarij space. Stud. Math. **41**, 207–210 (1972)

249. P. Wojtaszczyk, *Banach Spaces for Analysts*. Cambridge Studies in Advanced Mathematics, vol. 25 (Cambridge University Press, Cambridge, 1991)

250. J. Wolfe, Injective Banach spaces of continuous functions. Trans. Am. Math. Soc. **235**, 115–139 (1978)

251. D. Yost, A different Johnson-Lindenstrauss space. N. Z. J. Math. **36**, 1–3 (2007)

252. M. Zippin, The separable extension problem. Isr. J. Math. **26**, 372–387 (1977)

253. M. Zippin, Extension of bounded linear operators, in *Handbook of the Geometry of Banach Spaces*, vol. 2, ed. by W.B. Johnson, J. Lindenstrauss (North-Holland, Amsterdam 2003), pp. 1703–1742

Index

© Springer International Publishing Switzerland 2016
A. Avilés et al., *Separably Injective Banach Spaces*, Lecture Notes
in Mathematics 2132, DOI 10.1007/978-3-319-14741-3

LECTURE NOTES IN MATHEMATICS   🐴 Springer

Editors in Chief: J.-M. Morel, B. Teissier;

Editorial Policy

1. Lecture Notes aim to report new developments in all areas of mathematics and their applications – quickly, informally and at a high level. Mathematical texts analysing new developments in modelling and numerical simulation are welcome.

   Manuscripts should be reasonably self-contained and rounded off. Thus they may, and often will, present not only results of the author but also related work by other people. They may be based on specialised lecture courses. Furthermore, the manuscripts should provide sufficient motivation, examples and applications. This clearly distinguishes Lecture Notes from journal articles or technical reports which normally are very concise. Articles intended for a journal but too long to be accepted by most journals, usually do not have this "lecture notes" character. For similar reasons it is unusual for doctoral theses to be accepted for the Lecture Notes series, though habilitation theses may be appropriate.

2. Besides monographs, multi-author manuscripts resulting from SUMMER SCHOOLS or similar INTENSIVE COURSES are welcome, provided their objective was held to present an active mathematical topic to an audience at the beginning or intermediate graduate level (a list of participants should be provided).

   The resulting manuscript should not be just a collection of course notes, but should require advance planning and coordination among the main lecturers. The subject matter should dictate the structure of the book. This structure should be motivated and explained in a scientific introduction, and the notation, references, index and formulation of results should be, if possible, unified by the editors. Each contribution should have an abstract and an introduction referring to the other contributions. In other words, more preparatory work must go into a multi-authored volume than simply assembling a disparate collection of papers, communicated at the event.

3. Manuscripts should be submitted either online at www.editorialmanager.com/lnm to Springer's mathematics editorial in Heidelberg, or electronically to one of the series editors. Authors should be aware that incomplete or insufficiently close-to-final manuscripts almost always result in longer refereeing times and nevertheless unclear referees' recommendations, making further refereeing of a final draft necessary. The strict minimum amount of material that will be considered should include a detailed outline describing the planned contents of each chapter, a bibliography and several sample chapters. Parallel submission of a manuscript to another publisher while under consideration for LNM is not acceptable and can lead to rejection.

4. In general, **monographs** will be sent out to at least 2 external referees for evaluation.

   A final decision to publish can be made only on the basis of the complete manuscript, however a refereeing process leading to a preliminary decision can be based on a pre-final or incomplete manuscript.

   Volume Editors of **multi-author works** are expected to arrange for the refereeing, to the usual scientific standards, of the individual contributions. If the resulting reports can be forwarded to the LNM Editorial Board, this is very helpful. If no reports are forwarded or if other questions remain unclear in respect of homogeneity etc, the series editors may wish to consult external referees for an overall evaluation of the volume.

5. Manuscripts should in general be submitted in English. Final manuscripts should contain at least 100 pages of mathematical text and should always include

   – a table of contents;
   – an informative introduction, with adequate motivation and perhaps some historical remarks: it should be accessible to a reader not intimately familiar with the topic treated;
   – a subject index: as a rule this is genuinely helpful for the reader.
   – For evaluation purposes, manuscripts should be submitted as pdf files.

6. Careful preparation of the manuscripts will help keep production time short besides ensuring satisfactory appearance of the finished book in print and online. After acceptance of the manuscript authors will be asked to prepare the final LaTeX source files (see LaTeX templates online: https://www.springer.com/gb/authors-editors/book-authors-editors/manuscriptpreparation/5636) plus the corresponding pdf- or zipped ps-file. The LaTeX source files are essential for producing the full-text online version of the book, see http://link.springer.com/bookseries/304 for the existing online volumes of LNM). The technical production of a Lecture Notes volume takes approximately 12 weeks. Additional instructions, if necessary, are available on request from lnm@springer.com.

7. Authors receive a total of 30 free copies of their volume and free access to their book on SpringerLink, but no royalties. They are entitled to a discount of 33.3 % on the price of Springer books purchased for their personal use, if ordering directly from Springer.

8. Commitment to publish is made by a *Publishing Agreement*; contributing authors of multiauthor books are requested to sign a *Consent to Publish form*. Springer-Verlag registers the copyright for each volume. Authors are free to reuse material contained in their LNM volumes in later publications: a brief written (or e-mail) request for formal permission is sufficient.

Addresses:
Professor Jean-Michel Morel, CMLA, École Normale Supérieure de Cachan, France
E-mail: moreljeanmichel@gmail.com

Professor Bernard Teissier, Equipe Géométrie et Dynamique,
Institut de Mathématiques de Jussieu – Paris Rive Gauche, Paris, France
E-mail: bernard.teissier@imj-prg.fr

Springer: Ute McCrory, Mathematics, Heidelberg, Germany,
E-mail: lnm@springer.com

Printed in the United States
By Bookmasters